Physics for Engineers

Physics for Engineers

Alok Singh
Savita Singh
Sudhir Kumar Sharma

Shaftesbury Road, Cambridge CB2 8EA, United Kingdom

One Liberty Plaza, 20th Floor, New York, NY 10006, USA

477 Williamstown Road, Port Melbourne, VIC 3207, Australia

314–321, 3rd Floor, Plot No. 3, Splendor Forum, Jasola District Centre, New Delhi – 110025, India

103 Penang Road, #05–06/07, Visioncrest Commercial, Singapore 238467

Cambridge University Press is part of Cambridge University Press & Assessment, a department of the University of Cambridge.

We share the University's mission to contribute to society through the pursuit of education, learning and research at the highest international levels of excellence.

www.cambridge.org
Information on this title: www.cambridge.org/9781009424059

© Alok Singh, Savita Singh, and Sudhir Kumar Sharma 2025

This publication is in copyright. Subject to statutory exception and to the provisions of relevant collective licensing agreements, no reproduction of any part may take place without the written permission of Cambridge University Press & Assessment.

First published 2025

Printed in India by Magic International Pvt. Ltd., Greater Noida

A catalogue record for this publication is available from the British Library

ISBN 978-1-009-42405-9 Paperback

Cambridge University Press & Assessment has no responsibility for the persistence or accuracy of URLs for external or third-party internet websites referred to in this publication and does not guarantee that any content on such websites is, or will remain, accurate or appropriate.

For EU product safety concerns, contact us at Calle de José Abascal, 56, 1°, 28003 Madrid, Spain, or email eugpsr@cambridge.org.

Contents

Preface		xv
Chapter 1	**Relativistic Mechanics**	**1**
1.1	Introduction	1
1.2	Frame of Reference	2
1.3	Inertial Frames of Reference or Unaccelerated Frames	3
1.4	Non-inertial Frames of Reference or Accelerated Frames	3
1.5	Is the Earth an Inertial Frame of Reference?	4
1.6	Space–Time Frame	4
1.7	Ether Hypothesis	4
1.8	Michelson–Morley Experiment	5
1.9	Explanations and Interpretation of the Negative Results	8
1.10	Einstein's Postulates of Special Theory of Relativity	9
1.11	Galilean Transformation Equations	9
1.12	Lorentz Transformation Equations of Space and Time	10
1.13	Length Contraction	13
1.14	Time Dilation	14
1.15	Experimental Verification of Time Dilation	15
1.16	Twin Paradox in Special Relativity	16
1.17	Concept of Simultaneity: Relative Character of Time	16
1.18	Velocity Addition	17
1.19	Variation of Mass with Velocity	18
1.20	Einstein's Mass–Energy Relation	21
	Examples	22
1.21	Relativistic Relation between Energy and Momentum	22
1.22	Massless Particles	23
1.23	Space–Time	23
1.24	Space–Time Intervals (Minkowski Space–Time)	24
1.25	Relativistic Doppler Effect in Light	26
1.26	The Expanding Universe	28
1.27	General Theory of Relativity	29
1.28	Principle of Equivalence in General Relativity	29
1.29	Inertial Mass and Gravitational Mass	30

	1.30	Light Deflection in Gravitational Field	30
	1.31	Space–Time Curvature of Space	31
	1.32	The Unification of Mass and Space–Time	31
	1.33	Bending of Light	32
	1.34	Gravitational Lensing	32
	1.35	Light (Photons) and Gravity	33
	1.36	Gravitational Red Shift	34
	1.37	Gravitational Time Dilation	36
	1.38	Perihelion Shift of Mercury	36
	1.39	Light Retardation	36
	1.40	Gravitational Waves	37
	1.41	Black Holes	37
	1.42	Black Hole Detection	39
	1.43	Applications of Relativity	39
		Solved Problems	40
		Previous Year Questions (University Examination)	54
		Multiple Choice Questions	56
		Bibliography	61
Chapter 2	**Quantum Mechanics**		**63**
	2.1	Introduction	63
	2.2	Wave–Particle Duality	64
	2.3	de-Broglie's Hypothesis of Matter Waves	64
	2.4	Properties of Matter Waves	66
	2.5	Wave Packet or Wave Group	66
	2.6	Velocity of de-Broglie Waves	67
	2.7	Relation between Group Velocity and Phase Velocity	68
	2.8	Relation between Group Velocity and Particle Velocity	70
	2.9	Davisson and Germer Experiment	71
	2.10	Heisenberg's Uncertainty Principle	73
	2.11	Derivation of Energy–Time Uncertainty Relation from Position–Momentum Uncertainty Relation	73
	2.12	Derivation of Heisenberg's Uncertainty Principle	74
	2.13	Applications of Heisenberg's Uncertainty Principle	75
	2.14	Wave Function and Its Physical Significance	78
	2.15	Schrödinger's Time-Independent Wave Equation	79
	2.16	Schrödinger's Wave Equation for a Free Particle	80
	2.17	Schrödinger's Time-Dependent Wave Equation	80
	2.18	A Particle in a One-Dimensional Box (Infinite Square Well Potential)	81
	2.19	A Particle in a Three-Dimensional Box	83
	2.20	Basic Postulates of Quantum Mechanics	84
	2.21	Harmonic Oscillator	84
	2.22	Applications of Quantum Mechanics	85
		Solved Problems	86
		Previous Year Questions (University Examination)	92
		Multiple Choice Questions	95
		Bibliography	100

Chapter 3	**Electromagnetic Theory**		**103**
	3.1	Introduction	103
	3.2	Scalar and Vector Fields	104
	3.3	Electric and Magnetic Fields	104
	3.4	The Concept of Gradient	105
	3.5	The Concept of Divergence	106
	3.6	The Concept of Curl	107
	3.7	Laplacian Operator	109
	3.8	Equation of Continuity	109
	3.9	Ampere's Circuital Law	110
	3.10	Maxwell's Equations	112
	3.11	Maxwell's First Equation in Integral Form	115
	3.12	Maxwell's Second Equation in Integral Form	115
	3.13	Maxwell's Third Equation in Integral Form	116
	3.14	Maxwell's Fourth Equation in Integral Form	116
	3.15	Poynting Vector	117
	3.16	Poynting Theorem	117
	3.17	Maxwell's Equation in Free Space	119
	3.18	Maxwell's Equation in Nonconducting Medium	123
	3.19	Maxwell's Equation in Conducting Media	124
	3.20	Skin Depth or Penetration Depth	127
	3.21	Applications of Electromagnetism	128
		Solved Problems	130
		Previous Year Questions (University Examination)	131
		Multiple Choice Questions	133
		Bibliography	138
Chapter 4	**Statistical Mechanics**		**139**
	4.1	Introduction	139
	4.2	Phase Space	139
	4.3	Volume Element of μ-Space	139
	4.4	Number of Accessible Microstates or Phase Cells in the Energy Range E and E+dE	140
	4.5	Evaluation of $\iiint dV_p$	140
	4.6	Density of Microstates	141
	4.7	Ensemble	141
	4.8	Classification of Statistics	142
	4.9	Maxwell–Boltzmann Statistics	143
	4.10	Maxwell–Boltzmann Distribution Law	144
	4.11	Evaluation of β	146
	4.12	Determination of $e^{-\alpha}$	146
	4.13	Evaluation of Integral	147
	4.14	Maxwell–Boltzmann Energy Distribution Function	147
	4.15	Maxwell–Boltzmann Energy Distribution Law	148
	4.16	Maxwell–Boltzmann Speed or Velocity Distribution Law	148
	4.17	Most Probable Speed	149

	4.18	Average Speed	149
	4.19	Root-Mean-Square Speed	150
	4.20	Bose–Einstein Statistics	152
	4.21	Bose–Einstein Distribution Law	152
	4.22	Bose–Einstein Energy Distribution Function	153
	4.23	Bose–Einstein Energy Distribution Law	153
	4.24	Planck's Radiation Formula	154
	4.25	Derivation of Various Laws Related with Black Body	155
	4.26	Fermi–Dirac Statistics	157
	4.27	Fermi–Dirac Distribution	157
	4.28	Fermi–Dirac Energy Distribution Function or Fermi Function (Occupation Index)	159
	4.29	Fermi–Dirac Energy Distribution Law	160
		Solved Problems	163
		Previous Year Questions (University Examination)	167
		Multiple Choice Questions	168
		Bibliography	169
Chapter 5	**Lasers**		**171**
	5.1	Introduction	171
	5.2	Absorption, Spontaneous Emission, and Stimulated Emission of Radiation	171
	5.3	Spontaneous Emission	172
	5.4	Stimulated Emission	172
	5.5	Einstein's A and B Coefficients	172
	5.6	Population Inversion	174
	5.7	Threshold Condition for Laser Action	175
	5.8	Pumping	176
	5.9	Concept of Three- and Four-Level Laser Systems	177
	5.10	Ruby Laser	178
	5.11	Helium-Neon Laser	180
	5.12	Superiority of He-Ne laser over Ruby Laser	180
	5.13	Applications of Lasers	181
		Solved Problems	181
		Previous Year Questions (University Examination)	183
		Multiple Choice Questions	184
		Bibliography	185
Chapter 6	**Dielectric Materials**		**187**
	6.1	Introduction	187
	6.2	Dielectric Constant	187
	6.3	Polar and Nonpolar Molecules	188
	6.4	Dielectric Polarization	189
	6.5	Relation among $\vec{D}$, $\vec{E}$, and $\vec{P}$	190
	6.6	Relation between Dielectric Constant and Electrical Susceptibility	191
	6.7	Polarizability	191
	6.8	Types of Polarization (Polarizability)	192

Contents

6.9	Electronic Polarization	192
6.10	Ionic Polarization	193
6.11	Orientation Polarization	194
6.12	Space-charge Polarization	194
6.13	Total Polarization	194
6.14	Equation of Internal Fields in Liquid and Solids	196
6.15	Clausius–Mossotti Equation	197
6.16	Relation between Dielectric Constant and Refractive Index of Dielectric Material	198
6.17	Frequency Dependence of Dielectric Constant	198
6.18	Dielectric Loss	199
6.19	Dielectric Strength and Dielectric Breakdown	200
6.20	Various Kinds of Dielectric Materials	201
6.21	Hysterisis in Ferroelectric Materials	202
6.22	Applications of Ferroelectric Materials in Devices	202
6.23	Loss Factor and Its Significance	202
6.24	Electrostriction Effect	203
6.25	Direct Piezoelectric Effect and Inverse Piezoelectric Effect	203
6.26	Pyro-electric Material	203
	Solved Problems	204
	Previous Year Questions (University Examination)	205
	Multiple Choice Questions	206
	Bibliography	207

Chapter 7 Semiconducting Materials 209

7.1	Band Theory of Solids	209
7.2	Classification of Solids on the Basis of Band Theory	209
7.3	Fermi Energy	210
7.4	Fermi Energy Level	210
7.5	Density of States	211
7.6	Density of State for Electrons in Conduction Band	211
7.7	Density of State for Holes in Valence Band	211
7.8	Fermi–Dirac Distribution Function	211
7.9	Free Carrier Density (Electron–Hole Concentration)	212
7.10	Fermi Level for Intrinsic Semiconductors	214
7.11	Fermi Level for Extrinsic Semiconductors	215
7.12	Conductivity of Semiconductors	216
7.13	Hall Effect for Conducting Materials (Metals)	218
7.14	Hall Effect in Semiconductors	220
7.15	Applications of Hall Effect	222
7.16	Compound Semiconductors	222
7.17	Applications of Semiconductors	226
	Solved Problems	226
	Previous Year Questions (University Examination)	228
	Multiple Choice Questions	229
	Bibliography	230

Chapter 8	**Nanomaterials**		**233**
	8.1	Nano	233
	8.2	History of Nanotechnology	234
	8.3	Nanoscience	234
	8.4	Nanotechnology	234
	8.5	Nanomaterials	235
	8.6	Types of Nanomaterials	235
	8.7	Properties of Nanomaterials	235
	8.8	Reason behind Property Change at Nanoscale	235
	8.9	Material Fabrication at Nanoscale (Nanomaterials)	236
	8.10	Categories of Nanomaterials	237
	8.11	Buckyballs	239
	8.12	Creation of Buckyballs	239
	8.13	Applications of Buckyballs	241
	8.14	Carbon Nanotubes	241
	8.15	Types of Carbon Nanotubes	242
	8.16	Production (Synthesis) of Carbon Nanotubes	242
	8.17	Structure of Carbon Nanotubes	243
	8.18	Properties of Nanotubes	245
	8.19	Applications of Nanotubes	245
		Previous Year Questions (University Examination)	245
		Multiple Choice Questions	246
		Bibliography	247
Chapter 9	**Superconducting Materials**		**249**
	9.1	Introduction	249
	9.2	Superconductivity	249
	9.3	Superconductivity and Transition Temperature	250
	9.4	Temperature Dependence of Resistivity in Superconducting Materials	250
	9.5	Effect of Critical Magnetic Field	250
	9.6	Critical Current and Current Density	252
	9.7	Meissner Effect (Flux Exclusion)	253
	9.8	Type I and Type II Superconductors	254
	9.9	BCS Theory (Explanation of Superconductivity)	256
	9.10	High-temperature Superconductors	257
	9.11	AC Resistivity	258
	9.12	Entropy	259
	9.13	Specific Heat	260
	9.14	Thermal Conductivity	261
	9.15	Acoustic Attenuation	261
	9.16	The Energy Gap	262
	9.17	Isotope Effect	263
	9.18	Mechanical Effects	264
	9.19	Characteristics of Superconductors	264
	9.20	Organic Superconductor	264
	9.21	One-Dimensional Fabre and Bechgaard Salts	265

Contents xi

	9.22	Two-Dimensional (BEDT-TTF)$_2$X	265
	9.23	Doped Fullerenes	267
	9.24	More Organic Superconductors	267
	9.25	Applications of Superconductors	268
		Solved Problems	269
		Previous Year Questions (University Examination)	270
		Multiple Choice Questions	271
		Bibliography	273
Chapter 10	**Crystal Structure**		**275**
	10.1	Crystallography	275
	10.2	Crystalline Solids	275
	10.3	Amorphous Solids	275
	10.4	Space Lattices	276
	10.5	Basis	276
	10.6	Translational Vectors	276
	10.7	Unit Cell and Lattice Parameter	277
	10.8	Primitive Cell	277
	10.9	Types of Unit Cell	278
	10.10	Seven Crystal Systems and Fourteen Bravais Lattices	278
	10.11	Crystal System Structure	281
	10.12	Some Important Crystal Structure	285
	10.13	Calculation of Lattice Constant	287
	10.14	Lattice Planes in a Crystal	288
	10.15	Miller Indices or Position and Orientation of Lattice Planes	288
	10.16	Interplanar Spacing in the Crystals	290
	10.17	Reciprocal Lattice	290
		Solved Problems	290
		Previous Year Questions (University Examination)	293
		Previous Year Questions (University Examination): Long Questions	294
		Previous Year Questions (University Examination): Numerical Questions	295
		Multiple Choice Questions	295
		Bibliography	296
Chapter 11	**Wave Optics: Interference**		**297**
	11.1	Interference	297
	11.2	Coherent Sources	297
	11.3	Theory of Interference	297
	11.4	Energy Distribution and Conservation of Energy	299
	11.5	Stokes' Treatment	300
	11.6	Interference in Thin Film	300
	11.7	Fringe Width	306
	11.8	Newton's Ring	306
	11.9	Determination of Wavelength of Sodium Light Using Newton's Ring	310

	11.10	Determination of Refractive Index of an Unknown Liquid	311
	11.11	Newton's Ring in Transmitted Light	311
	11.12	Effect of Increasing the Distance between Lens and Plate or Lifting up the Lens from the Flat Surface	314
	11.13	Effect of Placing the Lens on a Silver Glass Plate or Mirror	314
	11.14	Newton's Rings Are Circular, but Air-wedge Fringes Are Straight	314
	11.15	The Effect of Placing the Concave Surface of the Plano-concave Lens Toward the Plane Glass Plate	314
	11.16	Effects of Using a Lens of Small Radius of Curvature	314
		Solved Problems	315
		Previous Year Questions (University Examination): Short Questions	318
		Previous Year Questions (University Examination): Long Questions	319
		Previous Year Questions (University Examination): Numerical Questions	321
		Multiple Choice Questions	324
		Bibliography	324
Chapter 12	**Wave Optics: Diffraction**		**325**
	12.1	Diffraction	325
	12.2	Fresnel and Fraunhofer Diffraction	325
	12.3	Fraunhofer Diffraction due to Single Slit	325
	12.4	Effect of Making Slit Narrower	330
	12.5	Fraunhofer Diffraction due to Double (Two) Slits	330
	12.6	Fraunhofer Diffraction due to Plane Diffraction Grating	332
	12.7	Formation of Spectra with Plane Diffraction Grating	336
	12.8	Absent Spectra for Diffraction Grating	337
	12.9	Maximum Number of Orders Possible in Grating Spectra	337
	12.10	Determination of Wavelength of Light by Using Plane Diffraction Grating	337
	12.11	Grating Element	338
	12.12	Dispersive Power of Plane Transmission Grating	338
	12.13	Resolving Power of an Optical Instrument	339
	12.14	Rayleigh's Criterion for the Limit of Resolution	339
	12.15	Resolving Power of Plane Diffraction Grating	339
		Solved Problems	340
		Previous Year Questions (University Examination): Short Questions	344
		Previous Year Questions (University Examination): Long Questions	345
		Previous Year Questions (University Examination): Numerical Questions	346
		Multiple Choice Questions	348
		Bibliography	349

Chapter 13 Fiber Optics — 351

- 13.1 Introduction — 351
- 13.2 Basic Principle of Optical Fiber — 352
- 13.3 Fiber Classification — 353
- 13.4 Acceptance Angle, Acceptance Cone, and Numerical Aperture of a Fiber — 355
- 13.5 Modes in Optical Fiber — 357
- 13.6 Modal Classification of Optical Fibers — 358
- 13.7 Attenuation in Fibers — 358
- 13.8 Kinds of Attenuation — 358
- 13.9 Merits of Optical Fiber — 359
- 13.10 Optical Fiber Communication — 359
- 13.11 Applications of Optical Fiber — 360
- 13.12 Light-source Materials for Optical Fiber Communication — 361
- 13.13 Semiconductor Materials for Solar Cells — 361
- Solved Problems — 363
- Previous Year Questions (University Examination) — 366
- Multiple Choice Questions — 367
- Bibliography — 368

Chapter 14 Physics Practicals — 369

- List of Experiments — 369
- Experiment 1: Band Gap of Semiconductor — 370
- Experiment 2: Stefan's Law of Black Body Radiation — 374
- Experiment 3: Determination of Thermopower — 378
- Experiment 4: Variation of Magnetic Field — 381
- Experiment 5: Carrey Foster Bridge — 384
- Experiment 6: e/m of Electron Using a Diode (Magnetron Valve) — 388
- Experiment 7: Tangent Galvanometer — 393
- Experiment 8: Newton's Rings — 396
- Experiment 9: Polarimeter — 399
- Experiment 10: Grating Spectrum — 404
- Experiment 11: Charge Sensitivity of a Ballistic Galvanometer — 408
- Experiment 12: Rydberg Constant — 411
- Experiment 13: Planck's Constant — 414
- Previous Year Questions (University Examination) — 416
- Multiple Choice Questions — 418

Index — 421

Preface

Engineering physics plays a crucial role in providing the foundational knowledge necessary for the development of innovative technologies. It is an essential part of the curriculum for students in various streams of science and engineering at the undergraduate level.

The goal of this book is to develop a solid understanding of the basic principles of physics and highlight their relevance to engineering. The content is structured to progressively build the knowledge and skills necessary for further studies in both theoretical and applied sciences. Each chapter begins with the basic concepts and gradually moves to more advanced topics, supported by numerical examples, illustrations, and problem sets that reinforce learning. The problems included are designed to improve the problem-solving skills of students and provide practical insight into the engineering applications of physics.

The manuscript includes 14 chapters that were prepared in accordance with the syllabus taught in various Indian colleges and universities. In addition to core topics, the manuscript also covers advanced topics such as **relativistic mechanics, quantum mechanics, optical fiber, lasers, semiconducting materials, superconducting materials,** and **nanomaterials**. Students who want to pursue higher education and a career in research, as well as instructors who instruct postgraduate courses at universities, will find these topics helpful for building a solid foundational understanding and developing problem-solving abilities. Moreover, each chapter concludes with a set of review questions and problems, along with the answers. This would also serve as a question bank for students preparing for different competitive examinations. They will have an opportunity to study advanced fields and assess their understanding.

The structuring of the book provides in-depth coverage of all topics. Chapter 1 discusses *Relativistic Mechanics*, while Chapter 2 describes *Quantum Mechanics*. Chapter 3 is devoted to *Electromagnetic Theory*, and Chapter 4 focuses on *Statistical Mechanics*. Chapter 5 explores *Laser*, followed by Chapter 6 on *Dielectric Materials*. Chapter 7 describes *Semiconducting Materials*, and Chapter 8 is devoted to *Nanomaterials*. Chapter 9 covers *Superconducting Materials*, and Chapter 10 examines *Crystal Structure*. Chapters 11 and 12 describe *Wave Optics: Interference* and *Wave Optics: Diffraction*, respectively. Chapter 13 is on *Fiber Optics*, and finally, Chapter 14 is devoted to *Physics Practical*.

The manuscript has been organized in such a way that it provides a conceptual link among the different topics covered in the chapters. To enhance clarity and ease of understanding, all mathematical steps have been included along with relevant discussions.

We would like to thank the entire team at Cambridge University Press, especially Karan Gupta, Shweta Pant, Aparna Kumar, Ankush Kumar, and Aniruddha De, for publishing this book in a very short time span. The authors of this book are also thankful to PSIT Group Director, Professor (Dr) Man Mohan Shukla, and PSIT Director, Professor (Dr) Sanjeev K. Bhalla, for their precious academic guidance.

CHAPTER 1

Relativistic Mechanics

1.1 Introduction

Classical mechanics is mainly based on Newton's laws of motion and gravitation. Initially, it was thought that Newton's second law of motion was valid and applicable at all speeds. But new experimental evidence showed that Newton's second law of motion is valid and applicable at low speeds $\left(v < \dfrac{c}{2}\right)$ and invalid when the object is moving at high speeds comparable to the velocity of light $\left(v \geq \dfrac{c}{2}\right)$. This failure of classical mechanics led to the development of the special theory of relativity by young physicist Albert Einstein in 1905, which showed everything in the universe is relative and nothing is absolute. Relativity connects space and time, matter and energy, electricity and magnetism, which are useful and remarkable to our understanding of the physical universe.

The special theory of relativity is applicable to all branches of modern physics, high-energy physics, optics, quantum mechanics, semiconductor devices, atomic theory, nanotechnology, and many other branches of science and technology.

The theory of relativity has two parts: the special theory of relativity and the general theory of relativity. The special theory of relativity deals with the inertial frame of references, while the general theory of relativity deals with the accelerated frame of references. Some common technical terms that are frequently used in relativistic mechanics are as follows:

1. **Particle:** A particle is a tiny bit of matter with almost no linear dimensions and is considered to be located at a single place. Its mass and charge define it. Examples include the electron, proton, and photon, among others.
2. **Event:** An event in relativity is defined as something that happens quickly or instantly at a location in space. It involves both a location and an event's time.
3. **Observer:** An observer is a person who locates, documents, quantifies, and interprets an occurrence. The observer uses his observations to form judgments about the happenings. In today's world, an observer might be a person, a CCTV camera, a webcam attached to a computer, or a mobile camera.

According to the theory of relativity, everything in the universe is relative; nothing is absolute. Position, time, motions, size, and length are all relative concepts. The examples are as follows:

1. Think about two individuals who are facing one another, standing on the opposite banks of a river, and observing a boat passing through the river. The boat is to the left of the second person, whereas it is to the right of the first. This demonstrates unequivocally that position is relative.
2. When we converse with a buddy who is in America, we can see that the time in India is different from the time in America. Thus, time is likewise a relative concept.
3. Think about a person driving along and observing distant, immobile things such as a building or a tree. All of these are traveling in the opposite direction from how the automobile is moving. For a person standing outside the automobile, it is simple to see that the observer of the car is traveling in a certain way, but for the observer within the car, the standing person would appear to be moving in the opposite direction from his own direction. Everything pointed to the motion being relative.

The above examples make it clear that a frame of reference must be established that is common to all for determining whether an object is at rest or in motion.

1.2 Frame of Reference

A frame of reference is a set of coordinate axes that describes the location of an event or particle in two or three dimensions.

A frame of reference must be rigid or stationary. The Cartesian coordinate system, in which the location of the particle is given by its three coordinates x, y, and z along the three perpendicular axes depicted in Figure 1.1, is the most basic frame of reference.

The frame of reference is of two types:

1. Inertial frames of reference or unaccelerated frames
2. Non-inertial frames of reference or accelerated frames

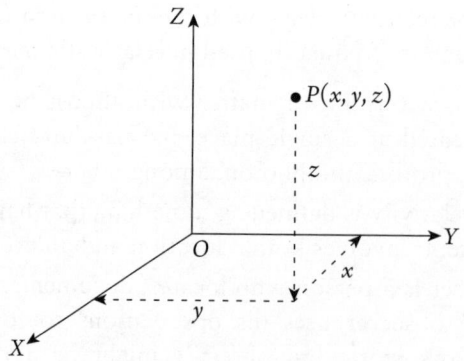

Fig. 1.1 Cartesian system of coordinates as a frame of reference.

1.3 INERTIAL FRAMES OF REFERENCE OR UNACCELERATED FRAMES

The frame of reference in which objects obey Newton's law of inertia and other laws of Newtonian mechanics is known as inertial frames or unaccelerated frames. In other words, all frames of reference that are either immobile with respect to one another or moving at a constant speed are referred to as inertial frames of reference, as shown in Figure 1.2. The value of acceleration is zero for inertial frames of reference; therefore, inertial frames of reference are also known as unaccelerated frames of reference.

In inertial frames, an object that is not being affected by an external force is at rest or travels at a steady speed. Inertial frames are those in which an item is either at rest or moving at a constant speed while not being affected by any external forces.

Inertial frames have the property of being isotropic with respect to mechanical, electrical, electronic, and optical investigations, which means that the rules of physics or the laws of mechanics and optics will be the same for all observers in inertial frames of reference.

1.4 NON-INERTIAL FRAMES OF REFERENCE OR ACCELERATED FRAMES

The frame of reference in which objects do not obey Newton's law of inertia and other laws of Newtonian mechanics is known as non-inertial frames of reference. In other words, non-inertial frames of reference refer to all frames of reference that move at nonuniform velocities. Non-inertial frames of reference are sometimes referred to as accelerated frames of reference since the acceleration is present in them.

The frames of reference with respect to which an unaccelerated object appears accelerated are called non-inertial or accelerated frames of reference. In these frames, Newton's laws are not valid, and an object not acted upon by an external force is accelerated. The simplest example of a non-inertial frame is a rotating merry-go-round shown in Figure 1.3.

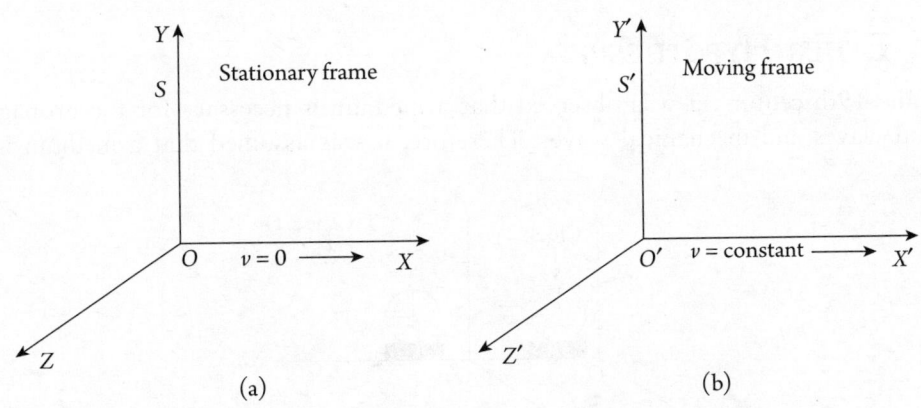

Fig. 1.2 (a) Stationary inertial frame and (b) moving inertial frame.

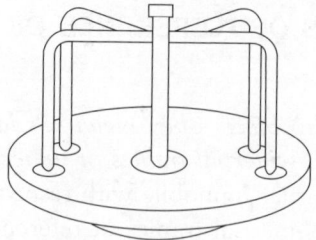

Fig. 1.3 Merry-go-rounds.

1.5 IS THE EARTH AN INERTIAL FRAME OF REFERENCE?

Strictly speaking, the earth is not an inertial frame of reference. The earth is rotating about its own axis and orbiting around the sun. In both these motions, centripetal accelerations are present. So, any frame of reference set up on earth cannot be considered an inertial frame of reference. A centripetal acceleration, whose value at the equator is $\omega^2 R = 3.4$ cm/s^2, is by no means negligible. Therefore, the earth is a non-inertial frame of reference.

> 📝 **Are the Sun and Distant Stars Inertial Frames of Reference?**
>
> No, the sun and distant stars are orbiting around the center of the galaxy, and a small acceleration is present in it; therefore, these massive bodies are also non-inertial frames of reference. Hence, in the universe, all cosmologically massive objects are non-inertial frames of reference.

1.6 SPACE–TIME FRAME

A reference frame with four coordinates x, y, z, and t (time) is referred to as a space time frame. The four axes define a four-dimension continuum called space time.

A space time frame S (stationary) and S' (moving) are shown in Figure 1.4.

1.7 ETHER HYPOTHESIS

Before the 19th century, it was observed that a medium is necessary for the propagation of sound waves and mechanical waves. Therefore, it was assumed that a medium is also

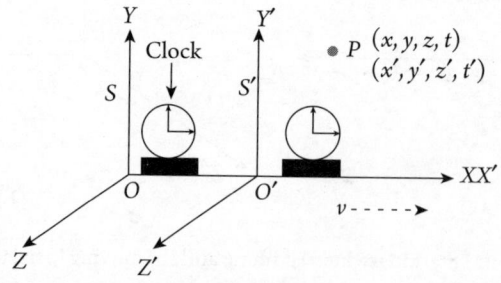

Fig. 1.4 Space time inertial frame S is stationary while S' is moving with constant speed v.

necessary for the propagation of light. A hypothetical medium was called luminiferous ether, or ether wind, or ether current. It was hypothesized that this hypothetical substance, ether, which was rigid, massless, and absolutely transparent, would occupy the whole cosmos, including the vacuum. Hence, the ether medium was supposed to be a stationary frame and could become an absolute inertial frame of reference.

1.8 MICHELSON–MORLEY EXPERIMENT

In 1887, American physicists Albert A. Michelson and Edward W. Morley performed an experiment to detect hypothetical medium ether using an interferometer setup. Michelson received the Noble Prize in Physics for conducting this experiment.

The objective of the Michelson–Morley experiment was to confirm the existence of stationary ether and to measure the absolute velocity of the earth with respect to stationary ether.

The apparatus used by Michelson and Morley is known as an interferometer since it depends on the principle of light.

When light from a monochromatic source (S) is incident parallel to the lens (L), it is divided into two equal-intensity components by the half-silvered plate (A). Plate A receives the transmitted component from mirror M_2, while the reflected component goes to mirror M_1 and is reflected back to it. As a result of the interference between the reflected beams from M_1 and M_2, interference fringes can now be seen via the telescope T (Figure 1.5).

The clear glass plate P is known as a compensating plate, which ensures that both components of light beams pass through the same thicknesses of air and glass, that is, the paths traveled by both beams are equal. The entire equipment is set up to travel with velocity in the direction of the earth's orbit around the sun. $AM_1 = AM_2 = D$ indicates that the distance is equal. The speed of light is $(c - v)$ in the direction that the earth is moving and $(c + v)$ in

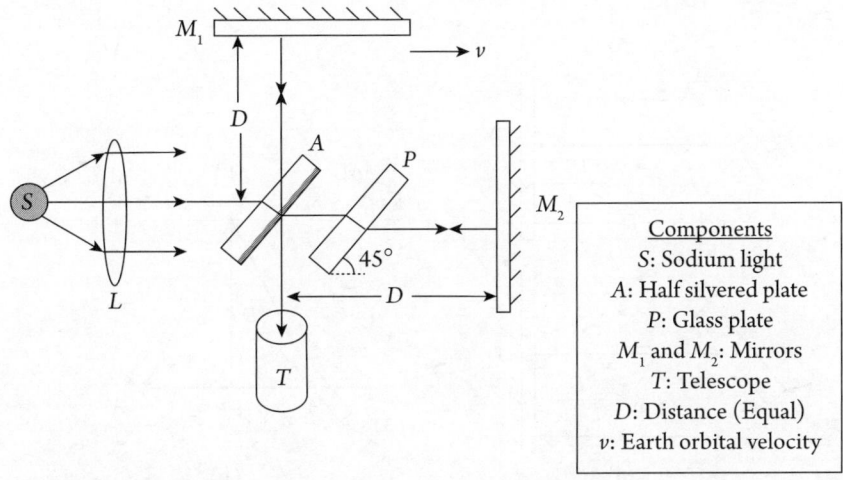

Fig. 1.5 Michelson–Morley experimental setup with monochromatic source S, convex lens L, half-silvered plate A, compensating glass plate P, completely polished mirrors M_1 and M_2, and telescope T.

the reverse direction. Let T_1 be the time taken by the light to travel from A to M_2, and T_2 be the time taken to travel from M_2' to A', as shown in Figure 1.6(a).

Then, the total time (T) taken by the light to travel from A to M_2 and back to plate A is given by

$$T = T_1 + T_2 = \frac{D}{c-v} + \frac{D}{c+v} = \frac{2Dc}{c^2 - v^2}. \tag{1.1}$$

The total distance traveled by light

$$x_1 = T \times c = \frac{2Dc^2}{c^2 - v^2} = \frac{2Dc^2}{c^2\left(1 - \frac{v^2}{c^2}\right)} = 2D\left(1 - \frac{v^2}{c^2}\right)^{-1}.$$

Applying binomial expansion and neglecting higher order terms, we get

$$x_1 \cong 2D\left[1 + \frac{v^2}{c^2}\right]. \tag{1.2}$$

Let the time taken by the light to travel from A to M_1' be T'. During this time, M_1 is shifted to M_1', and A is shifted to R, as shown in Figure 1.6(b). Then, the distance $AR = vT'$.

From the right-angled triangle $\Delta AM_1'R$,

$$\left(AM_1'\right)^2 = D^2 + v^2T'^2$$

or $\quad c^2T'^2 = D^2 + v^2T'^2$

or $$T' = \frac{D}{\sqrt{(c^2 - v^2)}} = \frac{D}{c\left(1 - \frac{v^2}{c^2}\right)^{\frac{1}{2}}} = \frac{D}{c}\left(1 - \frac{v^2}{c^2}\right)^{-\frac{1}{2}}.$$

Using binomial expansion and neglecting higher order terms, we get

$$T' = \frac{D}{c}\left(1 + \frac{1}{2}\frac{v^2}{c^2}\right).$$

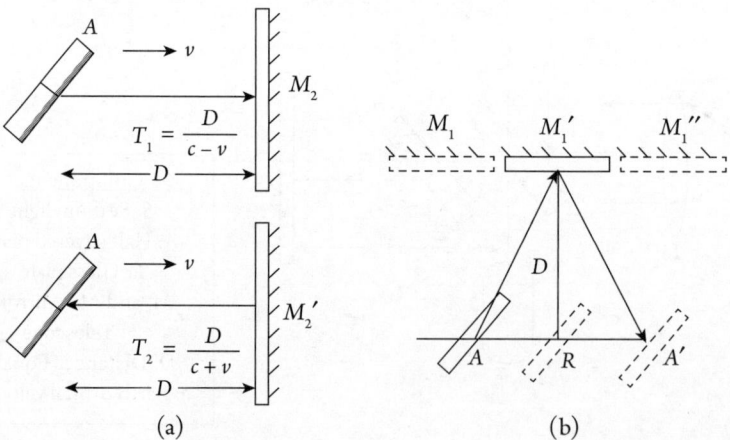

Fig. 1.6 Path traveled by the split light beams in the (a) horizontal direction and (b) vertical direction.

Total time (t) taken by the light in going from A to M_1' and back to plate A' is
$$t = 2T' = \frac{2D}{c}\left[1 + \frac{1}{2}\frac{v^2}{c^2}\right].$$
Total distance traveled is
$$x_2 = c \times t = 2D\left[1 + \frac{1}{2}\frac{v^2}{c^2}\right]. \tag{1.3}$$
From equations (1.2) and (1.3), the path difference can be given by
$$x_1 - x_2 = 2D\left[1 + \frac{v^2}{c^2}\right] - 2D\left[1 + \frac{1}{2}\frac{v^2}{c^2}\right].$$
$$\Delta x = \frac{Dv^2}{c^2} \tag{1.4}$$

If the shifting of n fringes is represented by the path difference in equation (1.4), then we obtain
$$\Delta x = n\lambda,$$
or $\quad n = \dfrac{\Delta x}{\lambda} = \dfrac{Dv^2}{c^2\lambda}. \tag{1.5}$

The apparatus is now rotated through 90°, causing the paths AM_1 and AM_2 to differ by a factor of $\left(\dfrac{Dv^2}{c^2}\right)$. Thus, as the apparatus rotates through 90°, a path difference of the same magnitude is introduced in the opposite direction.

The path difference before the rotation of the apparatus through 90° is
$$\Delta x_1 = \frac{Dv^2}{c^2}.$$
The path difference after the rotation of the apparatus through 90° is
$$\Delta x_2 = -\frac{Dv^2}{c^2}.$$
The total path difference is given by
$$\Delta x_1 - \Delta x_2 = \frac{Dv^2}{c^2} - \left(-\frac{Dv^2}{c^2}\right) = \frac{2Dv^2}{c^2}.$$

So, the total path difference between the two rays becomes $\left(\dfrac{2Dv^2}{c^2}\right)$. Hence, the total fringe shift will be
$$N = \frac{2Dv^2}{c^2\lambda}. \tag{1.6}$$

The interference fringes are observed by the telescope. The circular field of view that is seen from the eyepiece of the telescope with a fixed cross wire is shown in Figure 1.6(c). Figure 1.6(c) illustrates that fringe A shifts to the right side from the vertical cross wire by an amount $\dfrac{Dv^2}{c^2\lambda}$ before rotation of the experiment through 90°, and the same fringe A shifts to the left side from the vertical cross wire by an amount $\dfrac{Dv^2}{c^2\lambda}$ after rotation of the experiment through 90°. Thus, the total amount of shift of A fringe from the right side to the left side is $\dfrac{2Dv^2}{c^2\lambda}$.

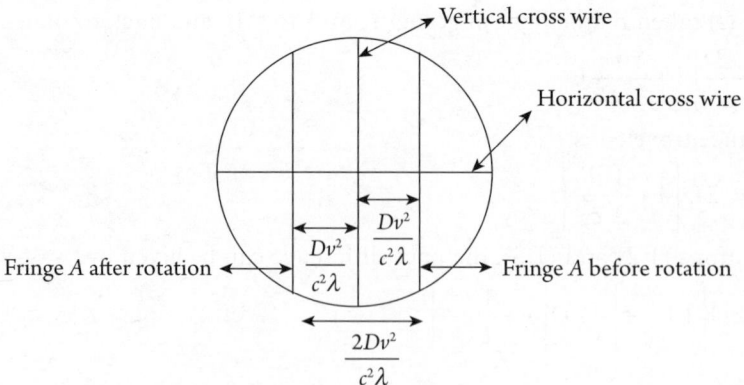

Fig. 1.6 (c) Circular field of view of the telescope.

To get accurate results, the distance D was increased to a value of up to 11 m by the method of multiple reflections using a system of mirrors. On taking the earth's velocity through ether equal to its orbital velocity, that is, $v = 3 \times 10^4$ m/s, the expected fringe shift for visible light $\left(\lambda = 5.5 \times 10^{-7} \text{ m}\right)$ is

$$N = \frac{2Dv^2}{c^2\lambda} = \frac{2 \times 11 \times \left(3 \times 10^4\right)^2}{\left(3 \times 10^8\right)^2 \times 5.5 \times 10^{-7}} = 0.4.$$

The actual shift of the interference pattern observed was almost negligible, indicating no relative velocity between the earth and the ether. Thus, the motion of the earth through the ether could not be detected experimentally. Hence, the hypothesis of the existence of a stationary ether medium was disapproved.

> **Why is the Michelson–Morley Experiment Rotated through 90°?**
>
> Suppose the ether current is parallel to one of the components of two light paths due to the earth's motion, which causes the two beams to have different transit times and the result would be destructive interference at the screen. Now, if the experiment is rotated through 90°, then the ether current is parallel to the other component of two light paths due to the earth's motion, which causes the two beams to have different transit times and the result would be constructive interference at the screen. Consequently, the whole fringe pattern (constructive and destructive both) can be covered at screen by the combination of experiment rotation through 90°.

1.9 Explanations and Interpretation of the Negative Results

The following explanations were provided to explain the Michelson–Morley experiment's negative results:

1. **Ether Drag Hypothesis:** According to Michelson, there is no relative motion between the earth and the ether since the moving earth drags the ether along with it.

2. **Lorentz–Fitzgerald Contraction Hypothesis:** This states that a material body traveling through ether becomes smaller by a factor of $\sqrt{1-\frac{v^2}{c^2}}$ in the direction of motion. The length D in the direction of the motion is therefore reduced to $D\sqrt{1-\frac{v^2}{c^2}}$. No fringe shift should be observed since the Michelson–Morley experiment equalizes the times along the perpendicular direction.
3. **Constancy of Speed of Light:** According to Einstein, the speed of light is invariant, meaning it is constant and unaffected by the velocity of the source, observer, or medium. Therefore, there shouldn't be any predicted fringe shift since the time it takes for the light to travel the two directions in the Michelson–Morley experiment would be the same.

1.10 EINSTEIN'S POSTULATES OF SPECIAL THEORY OF RELATIVITY

- All motion is relative, and the speed of light in free space is the same for all observers.

The two postulates that form the foundation of Einstein's special theory of relativity are listed here:

Postulate 1: The Principle of Equivalence (or Relativity)

It states that the laws of physics are same in all inertial frames of reference moving with the constant velocity (without any acceleration) with respect to one another.

Postulate 2: The Principle of Constancy of the Speed of Light

It states that the speed of light in free space (vacuum) is always the same in all inertial frames of reference and is equal to c. That is, it is unaffected by the source's and the observer's motion, as well as the inertial frames' relative motion.

The theory based on the above two postulates is called the special theory of relativity.

1.11 GALILEAN TRANSFORMATION EQUATIONS

Transformation equations are the equations that describe the connection between the coordinates of two reference systems.

In Newtonian mechanics, the transformation equations are known as Galilean transformations when the speed of frames, objects, or observers is very slow relative to the speed of light.

Consider two inertial frames of reference S and S' with axes (x, y, z) and (x', y', z') respectively. The inertial frame of reference S is stationary while S' is moving with constant velocity along positive x direction, as shown in Figure 1.7. Suppose an event has space time coordinates (x, y, z, t) w.r.t. S frame, while it has space time coordinates (x', y', z', t') w.r.t.

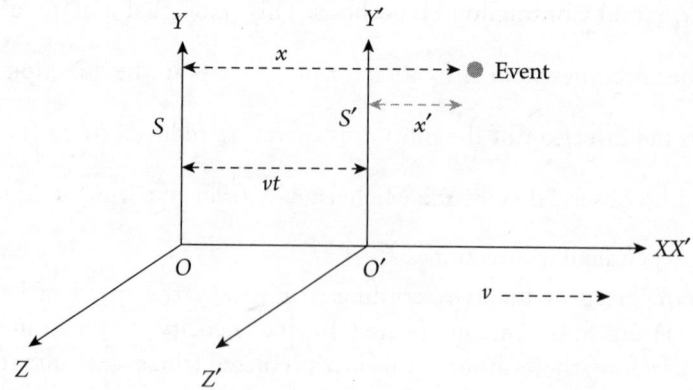

Fig. 1.7 Inertial stationary S frame and motion of S' frame with constant velocity.

S' frame. At any time t, if the x-coordinate of the event exceeds by vt in S' frame, then the Galilean transformation equations are

$$x' = x - vt; \quad y' = y; \quad z' = z; \quad t' = t. \tag{1.7}$$

The inverse Galilean equations can be obtained by interchanging the coordinates and replacing v by $-v$ as

$$x = x' + vt'; \quad y' = y; \quad z = z'; \quad t = t'. \tag{1.8}$$

The length between two points is invariant under Galilean transformation, and is given as

$$L = x_2 - x_1 = x_2' - x_1'.$$

On differentiating equation (1.7) w.r.t. time, we get the velocity as

$$\frac{dx'}{dt} = \frac{dx}{dt} - \frac{vdt}{dt}; \quad \frac{dy'}{dt} = \frac{dy}{dt}; \quad \frac{dz'}{dt} = \frac{dz}{dt}; \quad dt' = dt.$$

$$u_x' = u_x - v; \quad u_y' = u_y; \quad u_z' = u_z. \tag{1.9}$$

Thus, the velocity is not invariant under Galilean transformation.

On differentiating equation (1.9) w.r.t. time, we get the acceleration as

$$\frac{du_x'}{dt} = \frac{du_x}{dt} - 0; \quad \frac{du_y'}{dt} = \frac{du_y}{dt}; \quad \frac{du_z'}{dt} = \frac{du_z}{dt}.$$

$$a_x' = a_x; \quad a_y' = a_y; \quad a_z' = a_z. \tag{1.10}$$

Thus, the acceleration is invariant under Galilean transformation.

1.12 Lorentz Transformation Equations of Space and Time

The equations in relativity physics that relate the space and time coordinates of two coordinate systems moving with uniform velocity relative to one another are called Lorentz transformations.

Relativistic Mechanics

Lorentz transformation equations are applicable for all speeds in the range of $0 \leq v \leq c$.

Let us consider a flashbulb at O' in the frame S' moving at a speed v along XX' axes, as shown in Figure 1.8.

The coordinates (x', y', z', t') show the frame of the flashbulb S', whereas the coordinates (x, y, z, t) indicate the frame of a stationary observer in S. At the point the origins of the two reference frames meet, a flashbulb emits a pulse of light.

We define $t = t' = 0$ as the time at which the flashbulb and these two origins coincide. The fixed point O from where the flash originated acts as the wavefront's origin, as this light signal travels as a spherical wavefront form. A fixed point on the spherical wavefront such as P becomes the fixed point at a later time. Figure 1.8b illustrates this; P is at a distance r from O and r' from O'. The second postulate of Einstein states that for both observers, the speed of light should be C. This means that $r = ct$ gives the distance to the point P on the wavefront as observed by an observer in S, while $r' = ct$ gives the distance to the point P as measured by an observer in S'. Therefore,

$$r = ct; \quad r' = ct'. \tag{1.11}$$

Accepting Einstein's second postulate necessitates that the time t and t' needed for the light to reach P be different.

Now, the radius of a sphere is given by the equation $r^2 = x^2 + y^2 + z^2$, as described by an observer in S. Similarly, the distance r' measured in S' is given by the equation $r'^2 = x'^2 + y'^2 + z'^2$. On substituting the values of r and r' from equation (1.7), we get the expressions of the radius of sphere as follows:

In frame S, $c^2 t^2 = x^2 + y^2 + z^2$. (1.12)

In frame S', $c'^2 t'^2 = x'^2 + y'^2 + z'^2$. (1.13)

The Y and Z coordinates measured in the two frames are always the same because S' moves along the XX' axis. This means that $Y = Y'$ and $Z = Z'$ are unaffected by the motion along X.

Hence, subtracting equation (1.12) from equation (1.11), we have

$$x^2 - x'^2 = c^2 t^2 - c^2 t'^2.$$

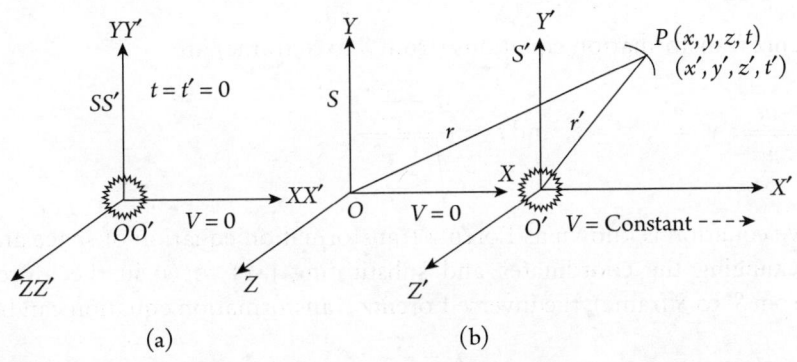

Fig. 1.8 (a) Inertial frames S and S' with flash bulb at O' when time $t = 0$. (b) Inertial frames S and S' with flash bulb at O' when time $t > 0$

$$x^2 - c^2t^2 = x'^2 - c^2t'^2. \tag{1.14}$$

The transformation equations relative to x' and t' can be written as
$$x' = k(x - vt), \tag{1.15}$$
where k is a constant.

Similarly, $t' = a(t - bx)$, \hfill (1.16)

where a and b are constants.

Now, substituting the values of x' and t' in equation (1.13), we get
$$x^2 - c^2t^2 = x'^2 - c^2t'^2,$$
$$x^2 - c^2t^2 = k^2(x - vt)^2 - c^2a^2(t - bx)^2,$$
$$x^2 - c^2t^2 = (k^2 - c^2a^2b^2)x^2 - 2(k^2v - c^2a^2b)xt - \left(a^2 - \frac{k^2v^2}{c^2}\right)c^2t^2. \tag{1.17}$$

Equating the coefficient of corresponding terms in equation (1.17), we get
$$k^2 - c^2a^2b^2 = 1, \tag{1.18}$$
$$k^2v - c^2a^2b = 0, \tag{1.19}$$
$$a^2 - \frac{k^2v^2}{c^2} = 1. \tag{1.20}$$

When we solve these equations for k, a, and b, we obtain
$$k = a = \frac{1}{\sqrt{1 - \frac{v^2}{c^2}}} \tag{1.21}$$

and $b = \dfrac{v}{c^2}.$ \hfill (1.22)

Substituting these values of k, a, and b in equations (1.15) and (1.16), we have
$$x' = \frac{x - vt}{\sqrt{1 - \frac{v^2}{c^2}}} \text{ and } t' = \frac{t - \frac{vx}{c^2}}{\sqrt{1 - \frac{v^2}{c^2}}}.$$

The Lorentz transformation equations (from S to S' frame) are
$$x' = \frac{x - vt}{\sqrt{1 - \frac{v^2}{c^2}}}, \; y' = y, \; z' = z, \text{ and } t' = \frac{t - \frac{vx}{c^2}}{\sqrt{1 - \frac{v^2}{c^2}}}. \tag{1.23}$$

The above equation is known as Lorentz transformation equation of space and time.

By interchanging the coordinates and substituting $(-v)$ for v in the aforementioned equations (from S' to S frame), the inverse Lorentz transformation equation can be achieved:
$$x = \frac{x' + vt'}{\sqrt{1 - \frac{v^2}{c^2}}}, \; y = y', \; z = z', \text{ and } t = \frac{t' + \frac{vx'}{c^2}}{\sqrt{1 - \frac{v^2}{c^2}}}. \tag{1.24}$$

Relativistic Mechanics

The above equation is known as inverse Lorentz transformation equation of space and time.

At low speeds, that is, $v \ll c$ or $v < \dfrac{c}{2}$; $\dfrac{v^2}{c^2} \approx 0$ or $\dfrac{v}{c^2} \approx 0 \Rightarrow \sqrt{1 - \dfrac{v^2}{c^2}} \cong 1$. Therefore, Lorentz transformation equations can become

$x' = x - vt$, $y' = y$, $z' = z$, $t' = t$.

Thus, at low velocity, Lorentz transformation equations are converted to Galilean transformation equations.

1.13 LENGTH CONTRACTION

- Faster means shorter

Consider two frames of reference S and S' whose x-axes meet at $t = 0$. As seen in Figure 1.9, the frame S' is traveling with a constant speed in the direction of positive X relative to the frame S. Take into account a rod AB at rest in S' frame. Assume that x'_1 and x'_2 are the coordinates of the rod's ends at any point in S' frame. The proper length is $l_0 = x'_2 - x'_1$, which is the length determined by an observer at rest with respect to the rod. *Proper length or actual length of an object is defined as the length measured by a scale at rest with respect to an object at rest.*

Suppose x_1 and x_2 be the coordinates of the ends of the rod at the same instant of time in S. *Measured length or relativistic length of an object is defined as the length measured by a frame with respect to an object which is placed in another frame.*

Let the measured length $l = x_2 - x_1$ be the length of rod measured in frame of reference S by an observer.

According to the Lorentz transformation,

$$x'_2 = \dfrac{x_2 - vt}{\sqrt{1 - \dfrac{v^2}{c^2}}}. \tag{1.25}$$

And $$x'_1 = \dfrac{x_1 - vt}{\sqrt{1 - \dfrac{v^2}{c^2}}}. \tag{1.26}$$

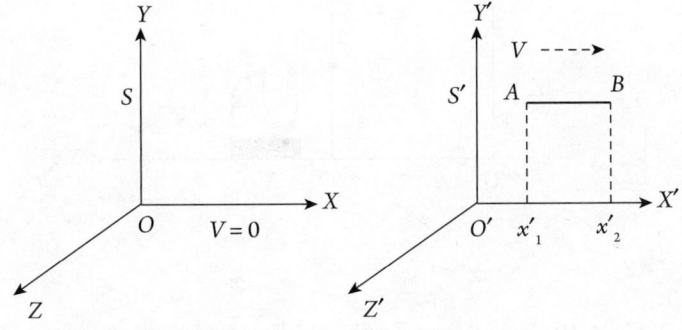

Fig. 1.9 Inertial frames with a rod placed in moving frame.

Subtracting equation (1.26) from equation (1.25), we get

$$x'_2 - x'_1 = \frac{x_2 - x_1}{\sqrt{1 - \frac{v^2}{c^2}}} = \frac{l}{\sqrt{1 - \frac{v^2}{c^2}}},$$

or $l_0 = \dfrac{l}{\sqrt{1 - \dfrac{v^2}{c^2}}}$,

or $l = l_0 \sqrt{1 - \dfrac{v^2}{c^2}}.$ (1.27)

From equation (1.27), it is clear that $l < l_0$. As a result, the observer in S notices that the rod in S' has a shorter length by a factor of $\sqrt{1 - \dfrac{v^2}{c^2}}$. In a perpendicular direction from the motion's direction, there is no contraction. The contraction only becomes noticeable if $v \approx c$. Each observer will notice that the other rod is shorter than the rod of his own system if there are two identical rods at rest, one in the S frame and the other in the S' frame.

1.14 Time Dilation

- A moving clock runs slower than a clock at rest.

Let us consider a clock at the point X' in the moving frame S' shown in Figure 1.10. When an observer in S' finds that the time is t'_1, an observer in S will find it to be t_1, where

$$t_1 = \frac{t'_1 + \dfrac{vx'}{c^2}}{\sqrt{1 - \dfrac{v^2}{c^2}}}.$$

After a time interval of t_0, the observer in the moving system finds that the time is now t'_2 according to his clock. Thus,

$$t_0 = t'_2 - t'_1.$$

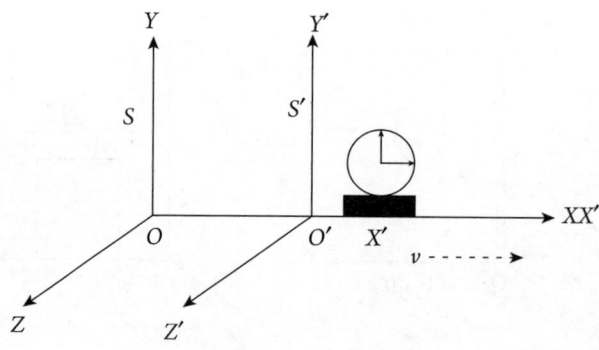

Fig. 1.10 Inertial frames with a clock placed in moving frame.

Relativistic Mechanics

The observer in S, however, measures at the end of the same time interval the time t_2 as

$$t_2 = \frac{t_2' + \frac{vx'}{c^2}}{\sqrt{1 - \frac{v^2}{c^2}}}.$$

Now, the duration of time interval t is

$$t = t_2 - t_1 = \frac{t_2' - t_1'}{\sqrt{1 - \frac{v^2}{c^2}}},$$

$$t = \frac{t_0}{\sqrt{1 - \frac{v^2}{c^2}}}, \quad [t < t_0].$$

Thus, a stationary clock measures a longer time between events occurring in a moving frame of reference by a factor $\frac{1}{\sqrt{1 - \frac{v^2}{c^2}}}$. Thus, a moving clock appears to be slow to a stationary observer. *The time interval between two events that occur at the same position measured by a clock in the inertial frame in which the events occur is called proper time. The time interval between the same two events in one inertial frame (moving/stationary) measured by an observer from another inertial frame (stationary/moving) is known as non-proper or relativistic time.*

1.15 EXPERIMENTAL VERIFICATION OF TIME DILATION

Mesons have proven time dilation to be true. Figure 1.11 illustrates how these are produced at a height of 10 km in the earth's atmosphere by the interaction of photons and are projected at a speed of 2.29×10^8 m/s toward the earth's surface. These mesons decay with an average lifetime 2.0×10^{-6} sec. Thus, mesons can travel a distance

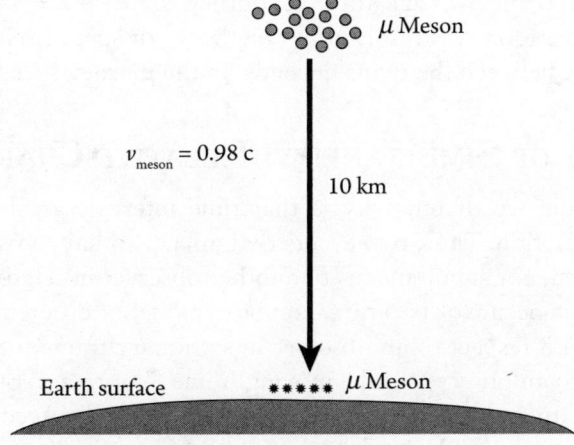

Fig. 1.11 Meson decay.

d = Velocity × decay time = $2.29 \times 10^8 \times 2.0 \times 10^{-6} \approx 0.6$ km.

When compared to the 10 km needed to reach the earth, this distance is quite little. Time dilation makes this conceivable. Consequently, dilated lifetime is provided by

$$t = \frac{t_0}{\sqrt{1-\frac{v^2}{c^2}}} = \frac{2.0 \times 10^{-6}}{\sqrt{1-\left(\frac{2.29 \times 10^8}{3 \times 10^8}\right)^2}} = 3.17 \times 10^{-5} \text{ sec}.$$

In this dilated lifetime, μ mesons can travel

d = Velocity × dilated lifetime = $2.29 \times 10^8 \times 3.17 \times 10^{-5} \approx 9.5$ km.

Thus, time dilation explains the presence of μ mesons on the earth surface. Hence, time dilation is a real effect.

1.16 Twin Paradox in Special Relativity

Consider 30-year-old identical twins A and B. Twin B travels round-trip to a star in a spacecraft with a velocity of $v = 0.99\,c$ in relation to twin A, while twin A stays at rest at the origin O on the earth. O is located 15 light-years from the star. According to Newtonian mechanics, the age of A and B is equal as B finished his journey and reached the earth. But, according to the theory of special relativity, the age of A and B is not equal as B finished his journey and reached the earth. This age difference phenomenon between two twins is named as "twin paradox." On the basis of the theory of special relativity, the twin paradox can be solved. According to A, the time taken by B in the round trip is

$$t_1 = \frac{2 \times 15 \text{ light-years}}{0.99\,c} = \frac{30\,c}{0.99\,c} = 30.3 \text{ years}.$$

Thus, according to A, his own age, as B completes the journey, is $30 + 30.3 = 60.3$ years. According to B, the travel time (proper time period) is as follows:

$$t_2 = 30\sqrt{1-\frac{v^2}{c^2}} = 30\sqrt{1-(0.99)^2} = 4.2 \text{ years}.$$

Thus, according to B, his own age after the journey is $30 + 4.2 = 34.2$ years.

Hence, this twin paradox was resolved by the theory of special relativity, which showed that the age difference between the twins depends on the journey speed.

1.17 Concept of Simultaneity: Relative Character of Time

An interesting consequence of relativity is that time intervals are not the same for two observers in relative motion. Thus, two events that appear to happen simultaneously to one observer are, in general, not simultaneous to another observer in relative motion.

Suppose two events occur (or two time-bombs explode) at different places x_1 and x_2 but at the same time t_0 with respect to an observer in stationary frame (or on the ground). The situation is different to an observer in the moving frame S' or to a pilot of spaceship moving with a velocity v relative to the stationary frame (or ground). According to the Lorentz transformation for time, the explosion at x_1 w.r.t. an observer in S' occurs at

Relativistic Mechanics

$$t'_1 = \frac{t_0 - x_1\left(\frac{v}{c^2}\right)}{\sqrt{1-\frac{v^2}{c^2}}}.$$

And that x_2 occurs at

$$t'_2 = \frac{t_0 - x_2\left(\frac{v}{c^2}\right)}{\sqrt{1-\frac{v^2}{c^2}}}.$$

Hence, the two events (explosion) that occur simultaneously to one observer in the stationary frame are separated to another in the moving frame by a time interval given as

$$t'_2 - t'_1 = \frac{(x_1 - x_2)\left(\frac{v}{c^2}\right)}{\sqrt{1-\frac{v^2}{c^2}}}.$$

Since $x_1 \neq x_2$, $t'_2 \neq t'_1$. Hence, the principle of simultaneity is an absolute concept for two events. It depends upon an observer or a frame of reference. The effect is not due to time dilation. Consequently, there is no such thing as "absolute time," which is the same for all observers. Time is relative and is different for observers in relative motion.

1.18 Velocity Addition

Let's assume that a particle is moving with respect to S and S'. As measured by an observer in S, the three velocity components are

$$u_x = \frac{dx}{dt}, u_y = \frac{dy}{dt}, u_z = \frac{dz}{dt}. \tag{1.28}$$

While, with respect to an observer in S', they are

$$u'_x = \frac{dx'}{dt'}, u'_y = \frac{dy'}{dt'}, u'_z = \frac{dz'}{dt'}. \tag{1.29}$$

By differentiating the inverse Lorentz transformation equation for x, y, z, and t, we obtain

$$dx = \frac{dx' + vdt'}{\sqrt{1-\frac{v^2}{c^2}}}, \, dy = dy', \, dz = dz', \text{ and } dt = \frac{dt' + \frac{vdx'}{c^2}}{\sqrt{1-\frac{v^2}{c^2}}}. \tag{1.30}$$

By substituting the values of dx, dy, dz, and dt in equation (1.28), we have

$$u_x = \frac{dx}{dt} = \frac{dx' + vdt'}{dt' + \frac{vdx'}{c^2}} = \frac{\frac{dx'}{dt'} + v}{1 + \frac{vdx'}{c^2 \, dt'}}$$

or $u_x = \dfrac{u'_x + v}{1 + \dfrac{vu'_x}{c^2}}.$ (1.31)

Similarly, $u_y = \dfrac{u'_y\sqrt{1 - \dfrac{v^2}{c^2}}}{1 + \dfrac{vu'_x}{c^2}}$ and $u_z = \dfrac{u'_z\sqrt{1 - \dfrac{v^2}{c^2}}}{1 + \dfrac{vu'_x}{c^2}}.$

If $u'_x = c$, that is, if light is emitted in the moving reference frame S' in the direction of motion relative to S, the observer in S will measure the velocity

$$u_x = \dfrac{u'_x + v}{1 + \dfrac{vu'_x}{c^2}} = \dfrac{c + v}{1 + \dfrac{vc}{c^2}} = c.$$

Since their relative velocity is always c, regardless of the other object velocity, if one object travels with velocity c in relation to another.

When $u'_x = c = v$, then

$$u_x = \dfrac{c + c}{1 + \dfrac{c^2}{c^2}} = \dfrac{2c}{2} = c.$$

This demonstrates that addition of the speed of light to the speed of light just reproduces the speed of light.

- *It concludes that velocity of light is an absolute constant.*
- It concludes that any light signal cannot travel faster than the velocity of light.

1.19 Variation of Mass with Velocity

- Rest mass is least.

Consider S and S' as two examples of frames of reference. Figure 1.12 depicts the frame S' traveling with a constant velocity v in the direction of positive X with respect to S. Assume that in S', two elastic balls A and B with masses m each approach one another at identical speeds (that is, u and $-u$). They hit to one another and combine into a single body.

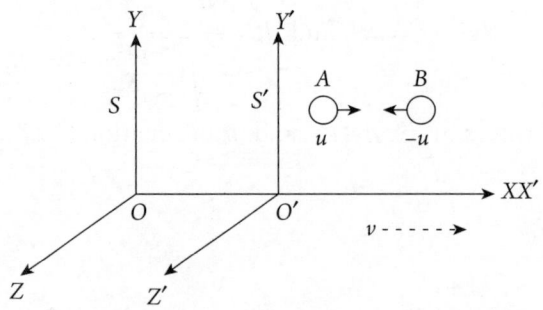

Fig. 1.12 Inertial frames with two similar elastic balls approaching each other placed in moving frame.

Relativistic Mechanics

Considering conservation of momentum,

Momentum of ball A + Momentum of ball B = Momentum of coalesced mass, that is, $mu + (-mu)$ = Momentum of coalesced mass = 0.

In S' frame, the merged mass must thus be at rest. Now think about the collision in terms of frame of reference S. Let the ball's relative speeds to S be u_1 and u_2. Then,

$$u_1 = \frac{u+v}{1+\frac{uv}{c^2}}, \tag{1.32}$$

$$u_2 = \frac{-u+v}{1-\frac{uv}{c^2}}. \tag{1.33}$$

The merged mass's velocity relative to the S frame after its collision is v. Let m_1 be the mass of ball A traveling at speed u_1, and let m_2 be the mass of ball B moving at speed u_2 in the frame of reference S. Considering conservation of momentum, we have

$$m_1 u_1 + m_2 u_2 = (m_1 + m_2)v. \tag{1.34}$$

Substituting for u_1 and u_2 from equations (1.32) and (1.33),

$$m_1 \left[\frac{u+v}{1+\frac{uv}{c^2}}\right] + m_2 \left[\frac{-u+v}{1-\frac{uv}{c^2}}\right] = (m_1 + m_2)v,$$

$$m_1 \left[\frac{u+v}{1+\frac{uv}{c^2}} - v\right] = m_2 \left[v - \frac{(-u+v)}{1-\frac{uv}{c^2}}\right],$$

$$m_1 \left[\frac{u\left(1-\frac{v^2}{c^2}\right)}{1+\frac{uv}{c^2}}\right] = m_2 \left[\frac{u\left(1-\frac{v^2}{c^2}\right)}{1-\frac{uv}{c^2}}\right]$$

or

$$\frac{m_1}{m_2} = \frac{1+\frac{uv}{c^2}}{1-\frac{uv}{c^2}}. \tag{1.35}$$

Let us consider the value of terms,

$$1 - \frac{u_1^2}{c^2} = 1 - \frac{\left(\frac{u+v}{c}\right)^2}{\left(1+\frac{uv}{c^2}\right)^2} = \frac{\left(1-\frac{u^2}{c^2}\right)\left(1-\frac{v^2}{c^2}\right)}{\left(1+\frac{uv}{c^2}\right)^2}. \tag{1.36}$$

Similarly,

$$1 - \frac{u_2^2}{c^2} = \frac{\left(1 - \frac{u^2}{c^2}\right)\left(1 - \frac{v^2}{c^2}\right)}{\left(1 - \frac{uv}{c^2}\right)^2}. \tag{1.37}$$

Dividing equation (1.37) by equation (1.36), we have

$$\frac{1 - \frac{u_2^2}{c^2}}{1 - \frac{u_1^2}{c^2}} = \frac{\left(1 + \frac{uv}{c^2}\right)^2}{\left(1 - \frac{uv}{c^2}\right)^2}.$$

Taking square root on both sides, we get

$$\frac{\sqrt{1 - \frac{u_2^2}{c^2}}}{\sqrt{1 - \frac{u_1^2}{c^2}}} = \frac{\left(1 + \frac{uv}{c^2}\right)}{\left(1 - \frac{uv}{c^2}\right)}. \tag{1.38}$$

From equation (1.35) and equation (1.38), we get

$$\frac{m_1}{m_2} = \frac{\sqrt{1 - \frac{u_2^2}{c^2}}}{\sqrt{1 - \frac{u_1^2}{c^2}}},$$

or $m_1 \sqrt{1 - \frac{u_1^2}{c^2}} = m_2 \sqrt{1 - \frac{u_2^2}{c^2}}.$ (1.39)

From equation (1.39), it is clear that the left-hand side and right-hand side are independent of one another, and this result may be true only if each is a constant.

Therefore, $m_1 \sqrt{1 - \frac{u_1^2}{c^2}} = m_2 \sqrt{1 - \frac{u_2^2}{c^2}} = m_0,$

where m_0 is the rest mass of the body and corresponds to zero velocity.

Thus, $m_1 = \frac{m_0}{\sqrt{1 - \frac{u_1^2}{c^2}}}.$

If m be the mass of the body when it is moving with a velocity v, then

$$m = \frac{m_0}{\sqrt{1 - \frac{v^2}{c^2}}}. \tag{1.40}$$

Equation (1.40) is a relativistic formula for the variation of mass with velocity and m is the relativistic mass or moving mass of the material particle.

At ordinary velocity, that is, when $v \ll c$, $\frac{v^2}{c^2} \approx 0.$

Relativistic Mechanics

Thus, $m = m_0$, that is, relativistic mass is equal to the rest mass of material particle.

If $v = c$, then by equation (1.40), $m = \infty \Rightarrow F = ma = \infty$, which means that an infinite force is required to achieve the velocity of light for any object having some rest mass ($m_0 \neq 0$), which is not possible.

- Thus, it is concluded that *"The material particle cannot have a velocity equal to the velocity of light."*

1.20 EINSTEIN'S MASS–ENERGY RELATION

Consider a particle of mass m acted upon by a force F. Assume that the direction of force is the same as the direction of velocity of the particle. Clearly, the application of the force will increase the energy of the particle.

According to Newton's second law of motion, force acting on a body is defined as the rate of change of its momentum, that is,

$$F = \frac{d}{dt}(mv) = m\frac{dv}{dt} + v\frac{dm}{dt}. \tag{1.41}$$

According to the theory of relativity, both mass and velocity are variable.

Now if this force F displaces the particle by a distance dx, its energy increases by

$$dK = F \cdot dx = m\frac{dv}{dt} \cdot dx + v\frac{dm}{dt} \cdot dx,$$

or $dK = mv\,dv + v^2 dm \quad \left[\because \frac{dx}{dt} = v\right]. \tag{1.42}$

According to the Einstein's relation of relativistic mass,

$$m = \frac{m_0}{\sqrt{1 - \frac{v^2}{c^2}}} = m_0\left(1 - \frac{v^2}{c^2}\right)^{-\frac{1}{2}}. \tag{1.43}$$

Now, differentiating the above equation, we have:

$$dm = m_0\left(\frac{-1}{2}\right)\left(1 - \frac{v^2}{c^2}\right)^{-\frac{3}{2}}\left(\frac{-2v}{c^2}dv\right),$$

or $dm = \frac{mv}{c^2}\left(1 - \frac{v^2}{c^2}\right)^{-1} dv,$

or $dm = \frac{mv\,dv}{c^2\left(1 - \frac{v^2}{c^2}\right)} = \frac{mv\,dv}{c^2 - v^2},$

or $c^2 dm = mv\,dv + v^2 dm. \tag{1.44}$

Comparing equation (1.42) and equation (1.44), we get

$$dK = c^2 dm. \tag{1.45}$$

If the particle is accelerated from rest to a velocity v, let its mass m_0 increase to m.

Integrating, total increase in K.E. $= \int_0^K dk = c^2 \int_{m_0}^m dm,$

or $\quad K = (m - m_0)c^2,$

or $\quad K + m_0 c^2 = mc^2.$ (1.46)

Here, K is the kinetic energy of the particle and the quantity $m_0 c^2$ is the energy associated with the rest mass of the particle. The quantity $m_0 c^2$ is regarded as "internal stored energy or rest energy of the particle."

The sum of the relativistic kinetic energy and the rest energy gives the total energy of the particle. So, the total energy of the particle is given by

$E = mc^2.$ (1.47)

This is the Einstein's mass–energy equivalence relation.

- Total energy and rest energy can be related as

$$E = mc^2 = \frac{m_0}{\sqrt{1 - \frac{v^2}{c^2}}} c^2 = \frac{E_0}{\sqrt{1 - \frac{v^2}{c^2}}}.$$

EXAMPLES

1. **Nuclear Fission:** Total mass of the constituents by which a nucleus is formed is just a little more than the mass of nucleus itself. The difference of mass, known as mass defect, is converted into the binding energy of the nucleus. It is the energy that keeps the nucleus bound. In nuclear fission reactions, the nucleus splits up into two parts and releases a large amount of energy. This principle is used to prepare atom bombs (mass can be converted into energy).

2. **Annihilation of Matter:** If an electron and a positron come close to each other, they annihilate (destroy) each other. High-energy radiation known as γ-radiation of energy equal to the rest mass–energy plus the kinetic energy of the disappeared particles is produced (mass can be converted into energy).

3. **Pair Production:** The process of annihilation of matter is reversible. When high-energy γ-radiation photons are absorbed by a nucleus, the photons disappear and produce electron–positron pair, whose total energy, that is, rest mass–energy and kinetic energy, is equal to the energy disappeared by the γ-radiations, and this phenomenon is known as pair production (energy can be converted into mass).

1.21 RELATIVISTIC RELATION BETWEEN ENERGY AND MOMENTUM

The relativistic total energy of a particle moving with velocity v is given by

$$E = mc^2 = \frac{m_0}{\sqrt{1 - \frac{v^2}{c^2}}} c^2,$$ (1.48)

where m_0 is the rest mass of the particle.

Relativistic Mechanics

The momentum of the particle is $p = mv \Rightarrow v = \dfrac{p}{m}$; equation (1.48) can be written as

$$E = \dfrac{m_0}{\sqrt{1 - \dfrac{p^2}{m^2 c^2}}} c^2 = \dfrac{m_0}{\sqrt{1 - \dfrac{p^2 c^2}{(mc^2)^2}}} c^2 = \dfrac{m_0}{\sqrt{1 - \dfrac{p^2 c^2}{(E)^2}}} c^2.$$

Squaring both sides of the above equation, we get

$$E^2 = \dfrac{m_0^2 c^4}{1 - \dfrac{p^2 c^2}{(E)^2}} \Rightarrow E^2 \left[1 - \dfrac{p^2 c^2}{(E)^2} \right] = m_0^2 c^4.$$

$$E^2 = p^2 c^2 + m_0^2 c^4. \tag{1.49}$$

This is the relation between total energy and momentum of a particle.

1.22 MASSLESS PARTICLES

A particle that has zero rest mass is called a massless particle. In classical physics, the existence of a massless particle is impossible. However, in relativistic mechanics, a particle with zero rest mass can exist.

The relativistic total energy E of a particle of rest mass m_0 in terms of its momentum p may be expressed as $E = \sqrt{m_0^2 c^4 + p^2 c^2}$.

For massless particle, $m_0 = 0$,

$$E = pc \text{ or } p = \dfrac{E}{c}.$$

Since p is also equal to $mv = \dfrac{E}{c}$, $v = c$, that is, the velocity of massless particle is the same as that of light in free space.

The energy of the particle $E = pc = mc^2$, where m is the mass equivalent to energy.

Thus, every massless particle has energy pc and momentum E/c and moves with the velocity of light. The mass of massless particle is equal to E/c^2. It means that a massless particle has mass so long as it is in motion. On being stopped, they cease to exist: they are either absorbed completely or are changed into heat at the surface. Thus, the massless particle has energy and momentum and can exist only when they move with the velocity of light. Photons and neutrinos are best examples of massless particles. In addition, gravitons are also massless particles with zero rest mass.

1.23 SPACE–TIME

As we have seen, space and time are closely entwined in the natural world. It could be necessary for another observer to measure a length that one observer can measure using only a meter stick and a clock. Regarding events as occurring in a four-dimensional space–time in which the typical three coordinates x, y, and z correspond to space and a fourth coordinate ict refers to time, where $i = \sqrt{(-1)}$ is a practical and elegant approach to represent

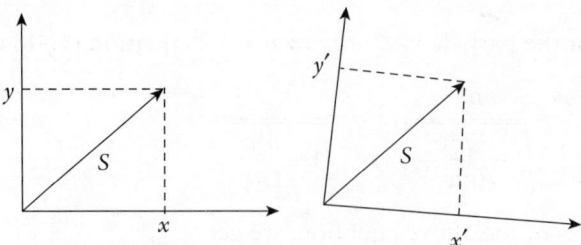

Fig. 1.13 Rotating a two-dimensional coordinate system.

the findings of special relativity. Space–time is equally as easy to deal with mathematically as three-dimensional space, despite the fact that we cannot see it. Instead of merely t, ict was used as the time coordinate since the quantity $S^2 = x^2 + y^2 + z^2 + (ict)^2 = x^2 + y^2 + z^2 - (ct)^2$ is unchanged under a Lorentz transformation. In other words, if an event happens at x, y, z, and t in an inertial frame S and at x', y', z', and t' in an additional inertial frame S', then

$$S^2 = x^2 + y^2 + z^2 - (ct)^2 = x'^2 + y'^2 + z'^2 - (ct')^2.$$

We may think of a Lorentz transformation as simply rotating the coordinate axes x, y, z, and ict in space–time because S^2 is invariant (Figure 1.13).

The four coordinates x, y, z, and ict constitute a vector in space–time, and this vector is unaffected by rotations of the coordinate system or changes in perspective from one inertial frame S to another S'.

The components of a second four-vector, whose magnitude is unaffected by Lorentz transformations, are p_x, p_y, p_z, iE/c. Here, p_x, p_y, p_z are the usual components of the linear momentum of a body whose total energy is E. Hence, the value of $p_x^2 + p_y^2 + p_z^2 - (E/c)^2$ is the same in all inertial frames even though p_x, p_y, p_z and E separately may be different.

The magnetic and electric fields B and E are combined into an invariant quantity termed a tensor in a more complex mathematical description. This method of introducing special relativity into physics has aided in the discovery of novel occurrences and correlations while also deepening our comprehension of fundamental natural principles.

1.24 SPACE–TIME INTERVALS (MINKOWSKI SPACE–TIME)

A combination of three-dimensional space (x, y, z) and time (ct) into four-dimensional space where space–time interval between any two events is independent of the inertial frames of reference in which they are recorded is known as Minkowski space–time.

Figure 1.14 shows two events plotted on the axes x and ct. Event 1 occurs at $x = 0$, $t = 0$ and event 2 occurs at $x = \Delta x$ and $t = \Delta t$. The space–time interval Δs between them is defined by

$$(\Delta S)^2 = (c\Delta t)^2 - (\Delta x)^2.$$

The virtue of this definition is that $(\Delta S)^2$ is invariant under Lorentz transformations. If Δx and Δt are the differences in space and time between two events measured in the S frame and $\Delta x'$ and $\Delta t'$ are the same quantities measured in the S' frame

$$(\Delta S)^2 = (c\Delta t)^2 - (\Delta x)^2 = (c\Delta t')^2 - (\Delta x')^2.$$

Relativistic Mechanics

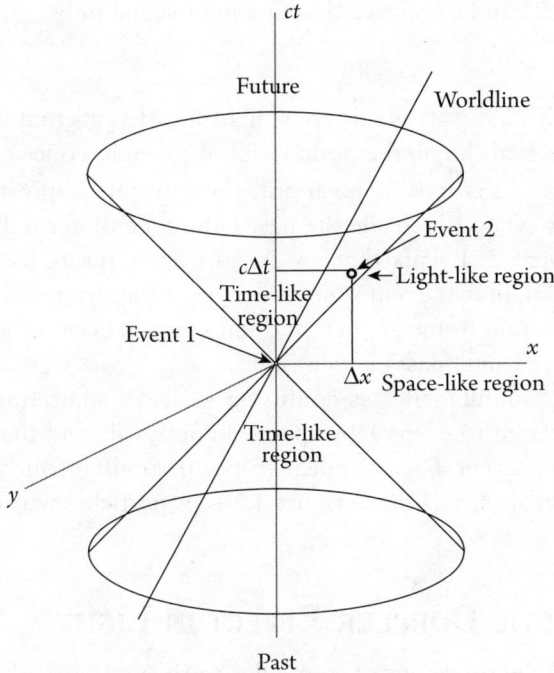

Fig. 1.14 The past and future light cones in space–time of event 1.

Therefore, any inferences we draw from the S frame, where event 1 is at the origin, are valid for every other frame in relative motion with constant speed.

Let us now examine any connections that could exist between events 1 and 2. If a signal that travels slower than the speed of light can connect these occurrences, then event 2 can be causally related to Event 1 in some way.

$c\Delta t > |\Delta x|$ or $(\Delta S)^2 > 0$.

A period when $(\Delta S)^2 > 0$ is regarded as time-like. The light cones in Figure 1.14 that are limited by the line $x = ct$ include every time-like period between events 1 and 2. In the past, light cone are all the events that may have impacted event 1, and in the future, light cone are all the events that event 1 has the potential to influence. (Events connected by time-like intervals do not necessarily have to be related; nonetheless, there is a chance that they may be related.)

On the other hand, the requirement for not being any causal relationship between occurrences 1 and 2 is that

$c\Delta t < |\Delta x|$,

or $(\Delta S)^2 < 0$.

An interval in which $(\Delta S)^2 < 0$ is said to be space-like. Every event that has a space-like gap between it and is connected to event 1 is outside of event 1's light cones, as shown in Figure 1.14. Such events have never interacted with event 1 in the past and can never interact with it in the future; the two events must thus be completely unconnected.

When events 1 and 2 can be connected with a light signal only

$c\Delta t = |\Delta x|$ or $(\Delta S)^2 = 0$.

An interval in which $(\Delta S)^2 = 0$ is said to be light-like. Events that can be connected with event 1 by light-like intervals lie on the boundaries of the light cones.

Due to the fact that $(\Delta S)^2 = 0$ is invariant, these results apply for the light cones of event 2; for instance, if event 2 is inside the past light cone of event 1, event 1 is inside the future light cone of event 2. Events that are in an event's future as seen in one frame of reference S are, in general, in that event's future in every other frame S', and events that are in an event's past as seen in that frame are in that event's past in every other frame S'. Therefore, the definitions of "future" and "past" are constant.

However, the term "simultaneity" is confusing since, in some frames of reference, any occurrences that are related to event 1 by space-like intervals and that are outside the past and future light cones of event 1 may appear to occur simultaneously. The world line of a particle is its passage across space–time (Figure 1.14). A particle's world line must be located within its light cones.

1.25 Relativistic Doppler Effect in Light

The apparent change in the frequency of sound due to the motion of the source of sound and listener (observer) relative to the medium is known as Doppler effect in sound. This Doppler effect draws significant attention in the phenomenon of light motion.

The Doppler effect in light is different from that in sound for the simple reason that, unlike sound, light requires no material medium for its propagation and hence its relative velocity is the same, that is, "c" for all observers regardless of their own state of motion. Finally, any apparent change in the frequency or the wavelength of a light pulse is appreciable only at relativistic speeds, that is, speeds near about "c."

Let us consider two frame of references S and S'. The frame S is stationary while S' is moving with a constant velocity v relative to S in the positive X-direction shown in Figure 1.15.

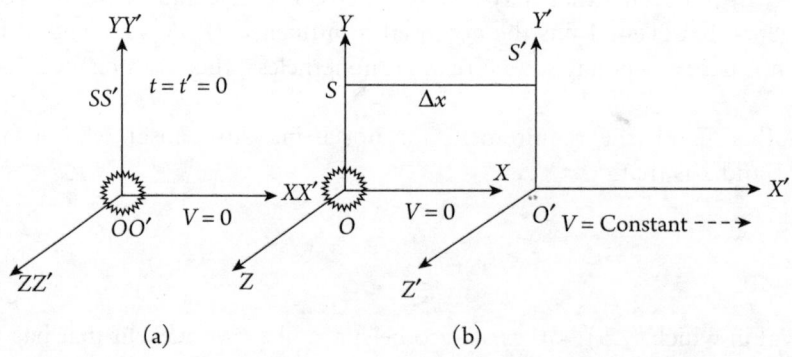

Fig. 1.15 Two inertial frames moving with constant speed to each other.

Relativistic Mechanics

Let two light signals or pulses be emitted from a source placed at the origin O in the frame S at time $t = 0$ and $t = T$, where T is the proper time period of the light pulses. Suppose the interval between the reception of these pulses by an observer at the origin O' in frame S' is $\Delta t'$. Then, clearly $\Delta t'$ is the apparent time interval denoted by T' between the two pulses, as measured in the frame S'.

Since the observer continues to be at O' all the time, the distance $\Delta x'$ covered by him in the frame during the reception of the two pulses is zero. Let us obtain the values of Δx and Δt in frame S, which correspond to those of $\Delta x'$ and $\Delta t'$ in frame S'.

From the inverse Lorentz transformation equation, we have

$$x = \frac{x' + vt'}{\sqrt{1 - \frac{v^2}{c^2}}},$$

or $\Delta x = \dfrac{\Delta x' + v\Delta t'}{\sqrt{1 - \dfrac{v^2}{c^2}}}.$

Since $\Delta x' = 0$, we have

$$\Delta x = \frac{v\Delta t'}{\sqrt{1 - \frac{v^2}{c^2}}} = \frac{vT'}{\sqrt{1 - \frac{v^2}{c^2}}}. \qquad [\because \Delta t' = T']. \qquad (1.50)$$

Similarly, from the inverse Lorentz transformation equation, we have

$$t = \frac{t' + \dfrac{vx'}{c^2}}{\sqrt{1 - \dfrac{v^2}{c^2}}},$$

or $\Delta t = \dfrac{\Delta t' + \dfrac{v\Delta x'}{c^2}}{\sqrt{1 - \dfrac{v^2}{c^2}}}.$

Since $\Delta x' = 0$, we have

$$\Delta t = \frac{\Delta t'}{\sqrt{1 - \frac{v^2}{c^2}}} = \frac{T'}{\sqrt{1 - \frac{v^2}{c^2}}}. \qquad (1.51)$$

It is obvious that both the proper time period T of the pulses and the time taken $(\Delta x/c)$ by the second pulse covers the extra distance Δx in the frame S. Therefore,

$$\Delta t = T + \frac{\Delta x}{c}.$$

On substituting the values of Δx and Δt from equations (1.50) and (1.51), we get

$$\frac{T'}{\sqrt{1 - \frac{v^2}{c^2}}} = T + \frac{vT'}{c\sqrt{1 - \frac{v^2}{c^2}}}$$

or $T = \dfrac{T'}{\sqrt{1 - \dfrac{v^2}{c^2}}}\left(1 - \dfrac{v}{c}\right).$ (1.52)

If ν and ν' are the actual and the observed (or apparent) frequencies of the light pulses, respectively, we have $\nu = 1/T$ and $\nu' = 1/T'$. So, on substituting the values in equation (1.52), we have

$$\dfrac{1}{\nu} = \dfrac{1}{\nu'\sqrt{1 - \dfrac{v^2}{c^2}}}\left(1 - \dfrac{v}{c}\right) \Rightarrow \nu' = \nu\dfrac{\left(1 - \dfrac{v}{c}\right)}{\sqrt{1 - \dfrac{v^2}{c^2}}} = \nu\dfrac{\left(1 - \dfrac{v}{c}\right)}{\sqrt{\left(1 - \dfrac{v}{c}\right)\left(1 + \dfrac{v}{c}\right)}},$$

or $\nu' = \nu\sqrt{\dfrac{1 - \dfrac{v}{c}}{1 + \dfrac{v}{c}}} = \nu\sqrt{\dfrac{1 - \beta}{1 + \beta}}$ $\left[\because \dfrac{v}{c} = \beta\right].$ (1.53)

The above expression is known as relativistic Doppler effect in light and gives the apparent frequency ν' of the light pulses of actual frequency ν as observed in frame S'. In terms of wavelength, the relativistic Doppler effect in light can be written as

$$\dfrac{c}{\lambda'} = \dfrac{c}{\lambda}\sqrt{\dfrac{1 - \dfrac{v}{c}}{1 + \dfrac{v}{c}}} \Rightarrow \lambda' = \lambda\sqrt{\dfrac{1 + \dfrac{v}{c}}{1 - \dfrac{v}{c}}} = \lambda\sqrt{\dfrac{1 + \beta}{1 - \beta}} \quad \left[\because \nu = \dfrac{c}{\lambda}\right].$$

The above expression is known as relativistic Doppler effect in light in terms of apparent wavelength λ'.

1.26 The Expanding Universe

The Doppler effect in light is a significant tool in astronomy. Every element shows its own characteristic lines of definite wavelength in its spectrum. Now, in the spectrum of the light received from the distant stars and galaxies, the characteristic spectral lines of the various known elements present in them are all found to be shifted by various amounts from their normal positions toward the lower frequency or the higher wavelength side, which is termed as red shift of the spectrum.

According to the Doppler effect in light in terms of wavelength $\lambda' = \lambda\sqrt{\dfrac{1 + \dfrac{v}{c}}{1 - \dfrac{v}{c}}}$, that is,

when the source (stars and galaxy) is receding from the observer (earth), there is an apparent increase in wavelength and spectrum shift toward the red end. Thus, the fact that light from the distant stars and galaxies shows the shift of the spectral lines toward the red end means that they are receding from us.

Relativistic Mechanics

Since the shift of the spectral lines toward the red end is due to the recession of the stars and galaxies from the earth, this phenomenon is called recessional red shift, and it gives a strong support to the theory of the expanding universe.

On the basis of a large number of observations made on several galaxies, it is summarized that the motion of astronomical objects due solely to this expansion is known as the Hubble flow and is given by

$v = H_0 D.$

Here, H_0 is the proportionality constant known as Hubble constant, "D" is the proper distance to the galaxy that can change over time, and "v" is the speed of separation. The SI unit of Hubble constant is s^{-1}, and the reciprocal of Hubble constant is known as Hubble time. The Hubble constant can also give the relative rate of expansion. In this form, $H_0 = 7\%/\text{Gyr}$ means that at the current rate of expansion it takes a billion years for an unbound universe to grow by 7%.

1.27 General Theory of Relativity

General relativity is the extension of special relativity. It includes the effects of accelerating objects (non-inertial frames) and their mass on space–time.

As a result, the theory is an explanation of gravity.

It is based on two concepts:

1. The principle of equivalence, which is an extension of Einstein's first postulate of special relativity.
2. The curvature of space–time due to gravity.

1.28 Principle of Equivalence in General Relativity

The principle of equivalence is an experiment in non-inertial reference frames.

Consider an astronaut sitting in a confined space on a rocket placed on the earth, as shown in Figure 1.16. The astronaut is strapped into a chair that is mounted on a weighing scale that indicates a mass M. The astronaut drops a safety manual that falls to the floor.

Now contrast this situation with that of a rocket accelerating through space. The gravitational force of the earth is now negligible. If the acceleration has exactly the same magnitude g on the earth, then the weighing scale indicates the same mass M that it had on the earth, and the safety manual still falls with the same acceleration as measured by the astronaut. The question is: How can the astronaut tell whether the rocket is on the earth or in space?

Hence, the principle of equivalence states that *there is no experiment that can be done in a small confined space that can detect the difference between a uniform gravitational field and an equivalent uniform acceleration.*

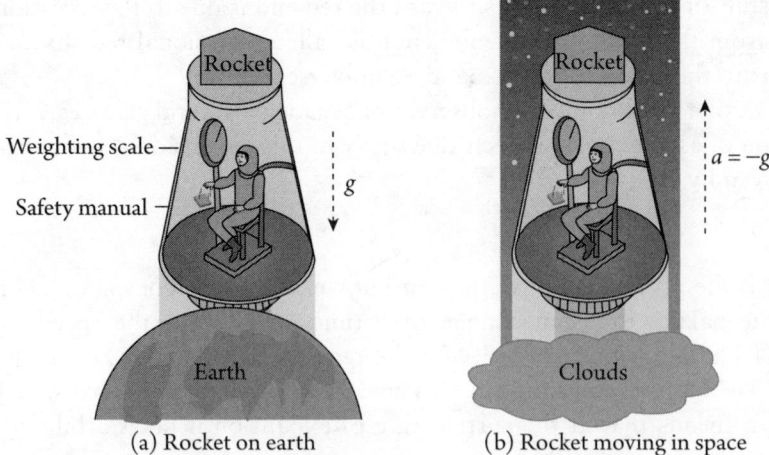

Fig. 1.16 Uniform gravitation or uniform acceleration.

1.29 INERTIAL MASS AND GRAVITATIONAL MASS

Recall from Newton's second law that an object accelerates in reaction to a force according to its inertial mass as

$$\vec{F} = m_I \vec{a}.$$

The inertial mass measures how strongly an object resists a change in its motion. The gravitational mass measures how strongly it attracts other objects.

$$\vec{F} = m_G \vec{g}.$$

For the same force, we get a ratio of masses as

$$\vec{a} = \left(\frac{m_G}{m_I}\right)\vec{g}.$$

Thus, according to the principle of equivalence, the inertial and gravitational masses are equal.

1.30 LIGHT DEFLECTION IN GRAVITATIONAL FIELD

Consider accelerating through a region of space where the gravitational force is negligible. A small window on the rocket allows a beam of starlight to enter the spacecraft. Since the velocity of light is finite, there is a nonzero amount of time for the light to shine across the opposite wall of the spaceship.

During this time, the rocket has accelerated upward. From the point of view of a passenger in the rocket, the light path appears to bend down toward the floor.

The principle of equivalence implies that an observer on the earth watching light pass through the window of a classroom will agree that the light bends toward the ground.

This prediction seems surprising; however, the unification of mass and energy from the special theory of relativity hints that the gravitational force of the earth could act on the effective mass of the light beam, as shown in Figure 1.17.

Relativistic Mechanics

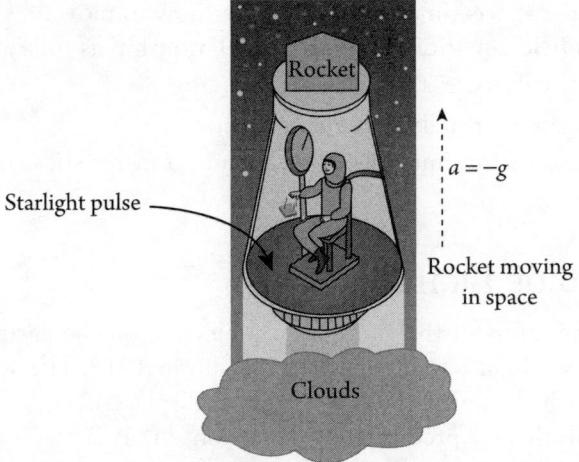

Fig. 1.17 Unification of mass and energy.

1.31 Space–Time Curvature of Space

Light bending for the earth's observer seems to violate the premise that the velocity of light is constant from special relativity. Light traveling at a constant velocity implies that it travels in a straight line.

Einstein recognized that we need to expand our definition of a *straight line*.

The shortest distance between two points on a flat surface appears different than the same distance between points on a sphere. The path on the sphere appears curved. We shall expand our definition of a *straight line* to include any minimized distance between two points.

Thus, if the space–time near the earth is not flat, then the straight line path of light near the earth will appear curved, as shown in Figure 1.18.

1.32 The Unification of Mass and Space–Time

Einstein mandated that the mass of the earth creates a dimple on the space–time surface. In other words, the mass changes the geometry of the space–time.

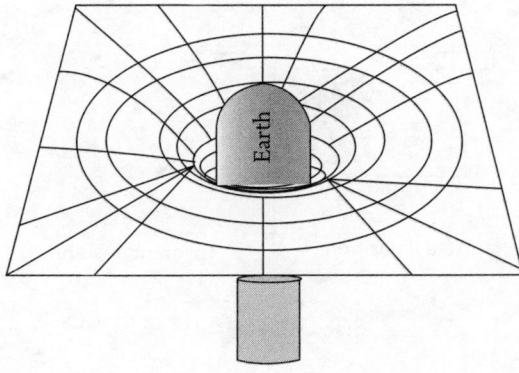

Fig. 1.18 Space–time curvature.

The geometry of the space–time then tells matter how to move.
Einstein's famous field equations sum up this relationship as follows:
- Mass–energy tells space–time how to curve.
- Space–time curvature tells matter how to move.

The result is that standard units of length, such as a meter stick, increase in the vicinity of a mass.

1.33 Bending of Light

During a solar eclipse, most of the sun's light is blocked on the earth, which afforded the opportunity to view starlight passing close to the sun in 1919. The starlight was bent as it passed near the sun, which caused the star to appear displaced.

Einstein's general theory predicted a deflection of 1.75 sec. of arc, and the two measurements found 1.98 ± 0.16 and 1.61 ± 0.40 sec.

Since the eclipse of 1919, many experiments, using both starlight and radio waves from quasars, have confirmed Einstein's predictions about the bending of light with increasingly good accuracy.

1.34 Gravitational Lensing

In 1979, it was revealed that what first seemed to be two neighboring quasars were actually only one, with the light of the single star being diverted by a huge object in between, as shown in Figure 1.19. This was the first detection of gravitational lensing. Other gravitational lenses have since been discovered, and they also affect radio waves coming from far-off sources in addition to light waves.

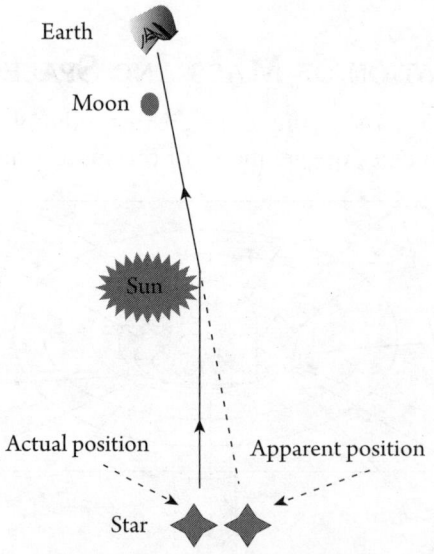

Fig. 1.19 Bending of light beam.

Relativistic Mechanics

> ### 📝 Quasars and Galaxies
>
> A quasar, like a star, may be seen as a sharp point of light with even the most powerful telescope. Quasars, which stand for quasi-stellar radio sources, are strong radio wave emitters in contrast to stars. There appear to be many more quasars than the hundreds that have already been found. The energy output of a typical quasar, despite its smaller size than the solar system, may be thousands of times greater than that of the Milky Way galaxy as a whole. The majority of scientists think that every quasar contains a black hole with a mass of at least 100 million suns. The substance of the neighboring stars is compressed and heated as they are drawn inward into the black hole, resulting in the radiation that can be seen. A star may release ten times as much energy while being devoured as it would have if it had lived a normal life. A quasar appears to be able to maintain its measured rates on a diet of a few stars every year. Quasars could actually be the centers of freshly created galaxies. Did quasar phases formerly occur in all galaxies?
>
> There is evidence that all galaxies, including the Milky Way, have enormous black holes at their centers; thus, it is impossible to determine for sure at this time.
>
> The gravitational red shift and black holes are two further effects of the interplay between gravity and light; more information on these topics is provided below.

1.35 Light (Photons) and Gravity

- Photons act as though they have gravitational mass even though they don't have any at rest.

The curvature of space-time around a mass causes gravity to have an impact on light. The mass of the photons is provided by

$$m = \frac{p}{v} = \frac{h\nu}{c^2}.$$

According to the principle of equivalence in general relativity, gravitational mass is always equal to inertial mass, so a photon of frequency ν must act gravitationally like a particle of mass $\frac{h\nu}{c^2}$.

Thus, a photon that falls through a height H can manifest the increase of mgH in its energy by an increase in frequency from ν to ν', as shown in Figure 1.20.

Final photon energy = Initial photon energy + Increase in energy

$$h\nu' = h\nu + mgH.$$

On substituting the value of photon mass, we get

$$h\nu' = h\nu + \left(\frac{h\nu}{c^2}\right)gH,$$

$$h\nu' = h\nu\left(1 + \frac{gH}{c^2}\right).$$

Thus, the above expression is the energy of photon after falling through the height H. It is concluded that the energy or frequency of the photon that moves toward earth increases; therefore, the energy or frequency of a photon moving away from earth should decrease.

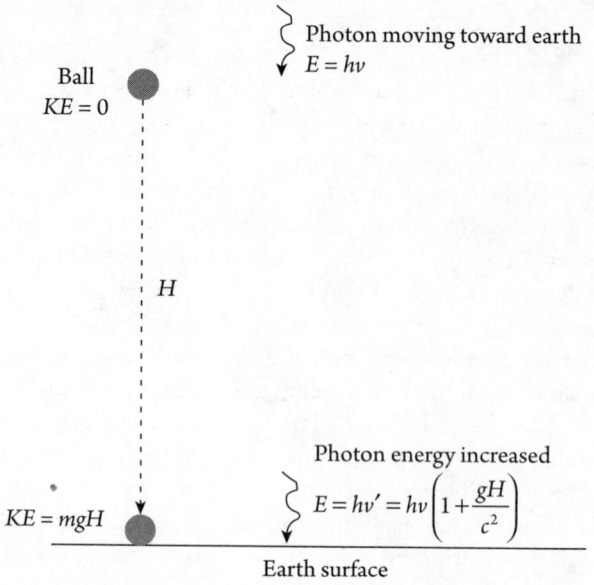

Fig. 1.20 Variation in total energy of falling ball and photon through a height H.

1.36 GRAVITATIONAL RED SHIFT

The gravitational fields of many stars are quite powerful, in contrast to the weak gravitational field of the earth. Assume a star with mass M and radius R emits a photon with initial frequency v, as shown in Figure 1.21. A mass's potential energy at the star's surface is

$$P.E. = -\frac{GMm}{R}.$$

The force between M and m is attracting; hence, the negative sign is necessary. On a star's surface, a photon of mass $\left(\frac{hv}{c^2}\right)$ has the potential energy given by

$$P.E. = -\frac{GMhv}{c^2 R}.$$

Thus, the total energy is the sum of potential energy, and its quantum energy hv is

$$E = hv - \frac{GMhv}{c^2 R} = hv\left(1 - \frac{GM}{c^2 R}\right).$$

The photon is outside of the star's gravitational field at a greater distance from the star, as at the earth, but its overall energy is constant, that is,

$$E = hv'.$$

Here, v' is the frequency of the arriving photon. The potential energy of the photon in the earth's gravitational field is negligible compared with that in the star's field. Hence,

$$hv' = hv\left(1 - \frac{GM}{c^2 R}\right),$$

$$\frac{v'}{v} = \left(1 - \frac{GM}{c^2 R}\right).$$

Relativistic Mechanics

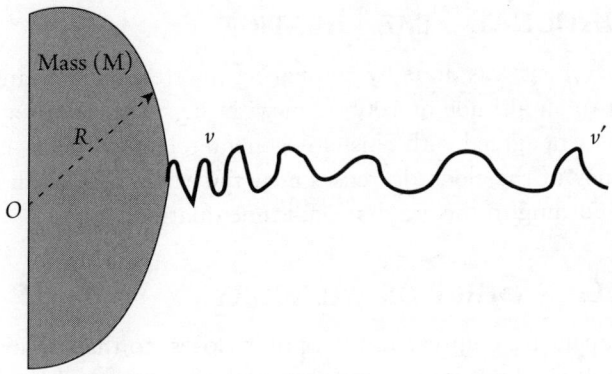

Fig. 1.21 The red shift of emitted photons

The relative frequency change is given by
$$\frac{\Delta v'}{v} = \frac{v - v'}{v} = 1 - \frac{v'}{v} = \frac{GM}{c^2 R}.$$

The photon in the lower frequency at the earth corresponds to its loss in energy as it leaves the field of the star.

The result is a photon that is displaced toward the red end of the spectrum in the visible area, a process known as gravitational red shift.

Due to the apparent recession of distant galaxies from the earth, which appears to be caused by a general expansion of the universe, the gravitational red shift differs from the Doppler red shift seen in their spectra.

An experiment conducted in a tall tower measured the "blue-shift" change in frequency of a light pulse sent down the tower. The energy gained when traveling downward a distance H is mgH. If v is the energy frequency of light at the top and v' is the frequency at the bottom, energy conservation gives $hv = hv' + mgH$.

The effective mass of light is $m = \frac{E}{c^2} = \frac{hv}{c^2}$.

This yields the ratio of frequency shift to the frequency as
$$\frac{\nabla v}{v} = \frac{GM}{c^2 R},$$

or, in general, $\frac{\nabla v}{v} = -\frac{GM}{c^2}\left(\frac{1}{R_1} - \frac{1}{R_2}\right).$

Using gamma rays, the frequency ratio was observed to be
$$\frac{\nabla v}{v} \approx 10^{-15}.$$

The M/R ratio is often too low for gravitational red shift to be seen for most stars, including the sun. Although it has been spotted, it is barely on the edge of measurement for a type of stars known as white dwarfs. A typical white dwarf is around the size of the earth and has the mass of the sun. It is an ancient star that is very tiny because its core is made up of atoms whose electron structures have collapsed.

1.37 GRAVITATIONAL TIME DILATION

A very accurate experiment was done by comparing the frequency of an atomic clock flown on a Scout D rocket to an altitude of 10000 km with the frequency of a similar clock on the ground. The measurement agreed with Einstein's general relativity theory is within 0.02% only.

Since the frequency of the clock decreases near the earth, a clock in a gravitational field runs more slowly according to the gravitational time dilation.

1.38 PERIHELION SHIFT OF MERCURY

The orbits of the planets are ellipses, and the point closest to the sun in a planetary orbit is called the perihelion. It has been known for hundreds of years that Mercury's orbit precesses about the sun, as shown in Figure 1.22. Accounting for the perturbations of the other planets left 43 sec. of arc per century that was previously unexplained by classical physics. The curvature of space–time explained by general relativity accounted for the 43 sec. of arc shift in the orbit of Mercury.

1.39 LIGHT RETARDATION

As light passes by a massive object, the path taken by the light is longer because of the space–time curvature, as shown in Figure 1.23.

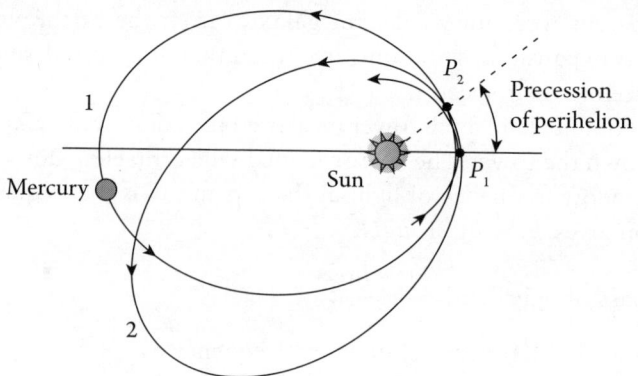

Fig. 1.22 Precession of perihelion of Mercury.

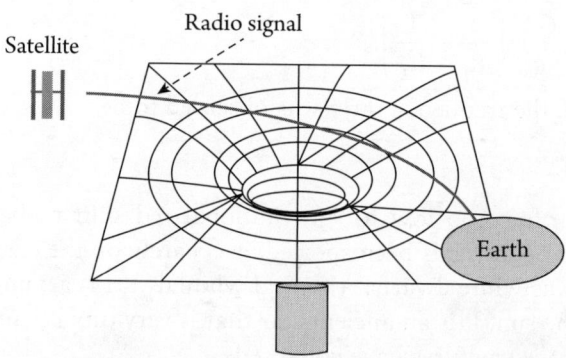

Fig. 1.23 Space time curvature.

Relativistic Mechanics

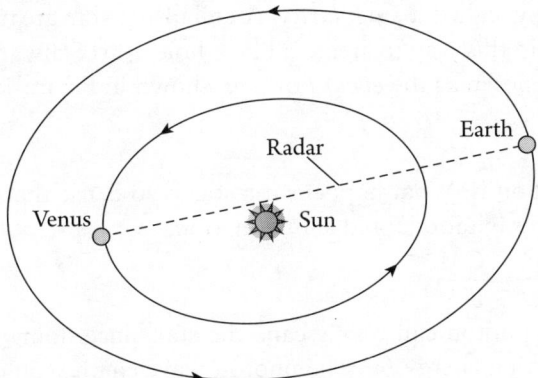

Fig. 1.24 Superior conjunction position.

The longer path causes a time delay for a light pulse traveling close to the sun. This effect was measured by sending a radar wave to Venus, where it was reflected back to the earth. The position of Venus had to be in the "superior conjunction" position on the other side of the sun from the earth, shown in Figure 1.24. The signal passed near the sun experienced a time delay of about 200 microsec. This was in excellent agreement with the general theory of relativity.

1.40 Gravitational Waves

When a charge accelerates, the electric field surrounding the charge redistributes itself. This change in the electric field produces an electromagnetic wave, which is easily detected. In much the same way, an accelerated mass should also produce gravitational waves.

Gravitational waves carry energy and momentum, travel at the speed of light, and are characterized by frequency and wavelength.

As gravitational waves pass through space–time, they cause small ripples. The stretching and shrinking are of the order of 1 part in 10^{21} even due to a strong gravitational wave source.

Due to their small magnitude, gravitational waves would be difficult to detect. Large astronomical events could create measurable space–time waves such as the collapse of a neutron star, a black hole, or the Big Bang.

This effect has been likened to noticing a single grain of sand added to all the beaches of Long Island, New York.

1.41 Black Holes

When a star is burning, the heat produced by the thermonuclear reactions pushes out the star's matter and balances the force of gravity. When the star's fuel is depleted, no heat is left to counteract the force of gravity, which becomes dominant. The star's mass collapses into an incredibly dense ball that could wrap space–time enough to not allow light to escape.

The point at the center is called a *singularity*. A collapsing star greater than 3 solar masses will distort space–time in this way to create a black hole. Karl Schwarzschild determined the radius of a black hole known as the *event horizon*, shown in Figure 1.25.

$$r_s = \frac{2GM}{c^2}.$$

An interesting question is: What happens if a star is so dense that $GM/c^2R \geq 1$? If this is the case, then from the gravitational red shift equation,

$$\frac{\Delta v'}{v} = \frac{v - v'}{v} = 1 - \frac{v'}{v} = \frac{GM}{c^2R}.$$

We can see that no photon can ever escape the star, since doing so would need energy above the photon's starting energy $h\nu$. The photon wavelength would have effectively been extended to infinity by the red shift at that point. Because it cannot emit light, a star of this type would be invisible, just as a black hole is invisible in space.

In a situation in which gravitational energy is comparable with total energy, as for a photon in a black hole, general relativity must be applied in detail. The correct criterion for a star to be a black hole turns out to be $GM/c^2R \geq \frac{1}{2}$.

General relativity must be applied in detail in cases when gravitational energy is similar to total energy, such as for a photon in a black hole. It turns out that $GM/c^2R \geq \frac{1}{2}$ is the proper criterion for determining if a star is a black hole. For a body of mass M, the Schwarzschild radius R_s is defined as

$$R_s = \frac{2GM}{c^2}.$$

If all of a body's mass ($M > 3M_{sun}$) is contained within a sphere of this radius, the object is a black hole. The event horizon of a black hole is its perimeter. Nothing can ever escape a black hole because, for a star with the mass of the sun, the escape velocity is equal to the speed of light at the Schwarzschild radius, which is 3 km—a quarter of a million times smaller than the sun's present radius. Anything that comes close to a black hole will be pulled into it and will never leave again.

What makes a black hole observable, and how can one be found? A black hole that is a part of a double-star system will be detected by the gravitational pull it exerts on the companion star; the two stars orbit each other. Additionally, the other star's stuff will be drawn in by

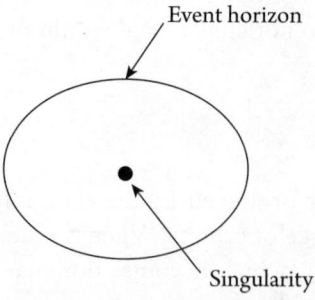

Fig. 1.25 Schematic of singularity and event horizon.

Relativistic Mechanics

the black hole's powerful gravitational field. This matter will be squeezed and heated to extremely high temperatures, which will cause it to release a lot of x-rays. Cygnus X-1 is one of several unseen objects that scientists think are black holes based on this evidence; it has a mass that is around eight times that of the sun and a radius that may be as small as 10 km. A black hole's x-ray emission area should reach several hundred kilometers into space. Only extremely massive stars become black holes.

White dwarfs $(M < 1.4 M_{sun})$ and neutron stars $(1.4 M_{sun} < M \leq 3 M_{sun})$ are the evolution of lighter stars, and as their names imply, they are primarily composed of neutrons. However, as time passes, both white dwarfs and neutron stars' powerful gravitational fields continue to draw in an increasing amount of cosmic dust and gas. They too will develop into black holes once they have accumulated sufficient mass. Black holes may make up the entirety of the universe if it exists for a long-enough time. Galaxies' centers are thought to contain black holes as well. Once more, the movements of neighboring bodies and the quantity and kind of radiation released provide hints. Stars near a galactic center have been discovered to travel so quickly that they could only be kept in their orbits by the gravitational pull of a massive object. How enormous? Up to a billion times the mass of the sun. Additionally, radiation emanates from galactic centers in such abundance that only black holes might be to blame, as is the case with black holes that were previously stars.

1.42 BLACK HOLE DETECTION

Since light can't escape, they must be detected indirectly, especially when severe red shifting of light occurs. Hawking radiation results from particle–antiparticle pairs created near the event horizon. One member slips into the singularity as the other escapes. Antiparticles that escape will radiate as they combine with matter. Energy expended to pair production at the event horizon decreases the total mass–energy of the black hole. Hawking calculated the blackbody temperature of the black hole to be

$$T = \frac{\hbar c^3}{8 \pi k G M}.$$

The power radiated is

$$P(T) = 4 \pi \sigma r_s^2 \left(\frac{\hbar c^3}{8 \pi k G M} \right)^4.$$

This result is used to detect a black hole by its Hawking radiation.

Mass falling into a black hole would create a rotating accretion disk. Internal friction would create heat and emit x-rays.

1.43 APPLICATIONS OF RELATIVITY

1. The popular engineering application of relativity is satellite global positioning systems, which uses general relativity to predict that the satellite's precision clocks run 45 microsec per day faster than those on the earth due to the lower gravity in space.
2. Relativity is used to predict things such as the existence of black holes, light bending due to gravity, and the behavior of planets in their orbits.

3. Relativity is used in television or LED TV due to fast-moving electrons or photons, which are incident on the screen to produce moving frame pictures.
4. Relativity is used in radar guns in military due to fast-moving short square pulses and calculations of distance of any object from the radar guns.
5. Since electrons accelerated at 200 kV run with a velocity, that is 0.7 times of light velocity (c), the scattering and imaging theories of transmission electron microscopy are considered under the special theory of relativity.
6. Supersonic aeroplanes that move with ultra-high speed close to the speed of light; the physical parameters like energy, mass, and time will be changed based on the relativistic effect of motion at the ultra-high speed, which is described by relativity theory.
7. Einstein showed that magnetism is a purely relativistic effect. Magnetic fields are a relativistic correction that an observer observes when charges move relative to him.
8. General relativity is employed to calculate and predict exactly how anything with mass will warp space–time, and how the gravity caused by that warped space–time will affect not just the matter but also the space, time, and light around it.
9. In physics, the twin paradox is a thought experiment in special relativity involving identical twins, one of whom makes a journey into space in a high-speed rocket and returns home to find that the twin who remained on the earth has aged more. This paradox was explained by time dilation (theory of special relativity).
10. Relativity is also utilized in quantum mechanics, quantum field theory, and nano-technology.

Solved Problems

Ex. 1: What will be fringe shift according to the ether theory in the Michelson–Morley experiment if the effective path length of each path is 7 m and light has 7000 Å wavelength? The velocity of the earth is 3×10^4 m/s.

Solution:
According to the ether theory in the Michelson–Morley experiment, the expected fringe shift is given by
$$N = \frac{2Dv^2}{c^2\lambda}.$$
Here, $D = 7$ m, $v = 3 \times 10^4$ m/s, and $\lambda = 7000$ Å $= 7 \times 10^{-7}$ m,
$$N = \frac{2 \times 7 (3 \times 10^4)^2}{(3 \times 10^8)^2 \times 7 \times 10^{-7}} = 0.2.$$
The fringe shift is 0.2 according to the Michelson–Morley experiment.

Ex. 2: In the Michelson–Morley experiment, the effective path length of paths of two beams is 11 m each. The wavelength of the light used is 6000 Å. If the expected fringe shift is 0.4 fringes, then calculate the velocity of the earth relative to ether.

Relativistic Mechanics

Solution:

The expected fringe shift is given by

$$N = \frac{2Dv^2}{c^2\lambda} \Rightarrow v = c\sqrt{\frac{\lambda N}{2D}}.$$

Here, $D = 11$ m, $\lambda = 6000$ Å $= 6 \times 10^{-7}$ m, and $N = 0.4$,

$$v = (3 \times 10^8)^2 \sqrt{\frac{6 \times 10^{-7} \times 0.4}{2 \times 11}} = 3.13 \times 10^4 \text{ m/s}.$$

The velocity of the earth relative to ether is 3.13×10^4 m/s.

Ex. 3: Calculate the expected fringe shift in the Michelson–Morley experiment if the distance of each path is 11 m and the wavelength of light is 5.6×10^{-7} m. The experimental setup was not rotated through 90°. The linear velocity of the earth may be taken as 30 Km/s.

Solution:

When the setup is not rotated through 90°. Then, the expected fringe shift is given by

$$n = \frac{Dv^2}{c^2\lambda}.$$

Here, $D = 11$ m, $v = 3 \times 10^4$ m/s, and $\lambda = 5.6 \times 10^{-7}$ m,

$$n = \frac{11 \times (3 \times 10^4)^2}{(3 \times 10^8)^2 \times 5.6 \times 10^{-7}} = 0.196.$$

The expected fringe shift is 0.196 in the Michelson–Morley experiment.

Ex. 4: Prove that $x^2 + y^2 + z^2 - c^2t^2$ is invariant under Lorentz transformation.

or

Show that space–time interval between two events remain invariant under Lorentz transformation.

Solution:

The Lorentz transformation equations are

$$x = \frac{x' + vt'}{\sqrt{1 - \frac{v^2}{c^2}}}, \ y = y', \ z = z', \text{ and } t = \frac{t' + \frac{vx'}{c^2}}{\sqrt{1 - \frac{v^2}{c^2}}}. \tag{1.54}$$

The given equation is

$$x^2 + y^2 + z^2 - c^2t^2.$$

Put the values of x, y, z, and t from equation (1.54), we get

$$\left[\frac{x' + vt'}{\sqrt{1 - \frac{v^2}{c^2}}}\right]^2 + y'^2 + z'^2 - c^2 \left[\frac{t' + \frac{vx'}{c^2}}{\sqrt{1 - \frac{v^2}{c^2}}}\right]^2$$

or $\dfrac{1}{1 - \frac{v^2}{c^2}}\left[x'^2 + v^2t'^2 + 2vt'x' - c^2t'^2 - \dfrac{v^2x'^2}{c^2} - 2vt'x'\right] + y'^2 + z'^2,$

or $\dfrac{1}{1-\dfrac{v^2}{c^2}}\left[x'^2 + v^2 t'^2 - c^2 t'^2 - \dfrac{v^2 x'^2}{c^2}\right] + y'^2 + z'^2$,

or $\dfrac{1}{1-\dfrac{v^2}{c^2}}\left[x'^2\left(1-\dfrac{v^2}{c^2}\right) - c^2 t'^2\left(1-\dfrac{v^2}{c^2}\right)\right] + y'^2 + z'^2$,

or $\left[x'^2 - c^2 t'^2\right] + y'^2 + z'^2$,

or $x'^2 + y'^2 + z'^2 - c^2 t'^2$.

Thus, $x^2 + y^2 + z^2 - c^2 t^2 = x'^2 + y'^2 + z'^2 - c^2 t'^2$.

Hence, $x^2 + y^2 + z^2 - c^2 t^2$ is invariant under Lorentz transformation.

Ex. 5: In an inertial frame S, two lights (red and blue) are separated by a distance $\Delta x = 2.45$ km, with red light at the larger value of x. The blue light flashes, and 5.35 microsec later the red light flashes. Frame S' is moving in the direction of increasing x with speed of 0.85 c. What is the distance between the two flashes and the time between them as measured in S'?

Solution:

According to Lorentz transformation equations,

$$x_1' = \dfrac{x_1 - vt_1}{\sqrt{1-\dfrac{v^2}{c^2}}} \text{ and } x_2' = \dfrac{x_2 - vt_2}{\sqrt{1-\dfrac{v^2}{c^2}}},$$

$$t_1' = \dfrac{t_1 - \dfrac{vx_1}{c^2}}{\sqrt{1-\dfrac{v^2}{c^2}}} \text{ and } t_2' = \dfrac{t_2 - \dfrac{vx_2}{c^2}}{\sqrt{1-\dfrac{v^2}{c^2}}}.$$

The distance between the red light and blue light with respect to an observer in moving frame S' is

$$\Delta x' = x_2' - x_1' = \dfrac{(x_2 - x_1) - v(t_2 - t_1)}{\sqrt{1-\dfrac{v^2}{c^2}}},$$

$$\Delta x' = \dfrac{\Delta x - v\Delta t}{\sqrt{1-\dfrac{v^2}{c^2}}}.$$

Here, $\Delta x = 2.45$ km $= 2.45 \times 10^3$ m, $\Delta t = 5.35 \times 10^{-6}$ sec, and $c = 3 \times 10^8$ m. After substituting these values in the above equation, we get

$\Delta x' = 20.78 \times 10^2$ m or 2.08 km.

Similarly, the time between the red light and blue light is given as

$$\Delta t' = \dfrac{(t_2 - t_1) - v\left(\dfrac{x_2 - x_1}{c^2}\right)}{\sqrt{1-\dfrac{v^2}{c^2}}} = \dfrac{(\Delta t) - v\left(\dfrac{\Delta x}{c^2}\right)}{\sqrt{1-\dfrac{v^2}{c^2}}}.$$

Relativistic Mechanics

Since $v = 0.85\,c$, putting the values, we have

$\Delta t' = -1.35 \times 10^{-5}$ sec. or -0.135 microsec.

Thus, the distance between the two flashes is 2.08 km, and the time between them is -0.135 microsec as measured in S'. The negative sign in time difference indicates that in S' frame, an observer observes that red light flashes at first, then after 0.135 microsec, the blue light flashes.

Ex. 6: A bar that is 1 m long and situated along the X' axis travels away from a stationary observer at a speed of 0.75 c. How long is the bar according to the observer's measurements who is standing still?

Solution:

According to the length contraction formula,

$$l = l_0 \sqrt{1 - \frac{v^2}{c^2}}.$$

Given $l_0 = 1$ m and $v = 0.75\,c$,

then $l = 1\sqrt{1 - \frac{(0.75\,c)^2}{c^2}} = 0.66$ m.

By a stationary observer, the length of the bar is 0.66 m.

Ex. 7: At what speed would a rocket be moved to reduce its length to 99% of its full length with respect to an observer?

Solution:

Length contraction is given by

$$l = l_0 \sqrt{1 - \frac{v^2}{c^2}}.$$

Given $l = 99 \frac{l_0}{100} \Rightarrow \frac{l}{l_0} = \frac{99}{100}$,

$$\frac{l}{l_0} = \sqrt{1 - \frac{v^2}{c^2}} \Rightarrow \frac{99}{100} = \sqrt{1 - \frac{v^2}{c^2}}.$$

or $v^2 = 0.0199 \times (3.0 \times 10^8)^2$,

or $v = 42.3 \times 10^6 \frac{m}{s}$.

The velocity of the rocket should be 42.3×10^6 m/s.

Ex. 8: Determine the percentage of contraction for a rod traveling at 0.8 c in a direction 60° degrees inclined to its own length.

Solution:

Consider l_0 be the rod length in the frame S, and S' is the moving frame with velocity 0.8 c in a direction making an angle 60° with x-axis. The l_0 has horizontal and vertical components $l_0 \cos 60°$ and $l_0 \sin 60°$, respectively, as shown in Figure 1.26.

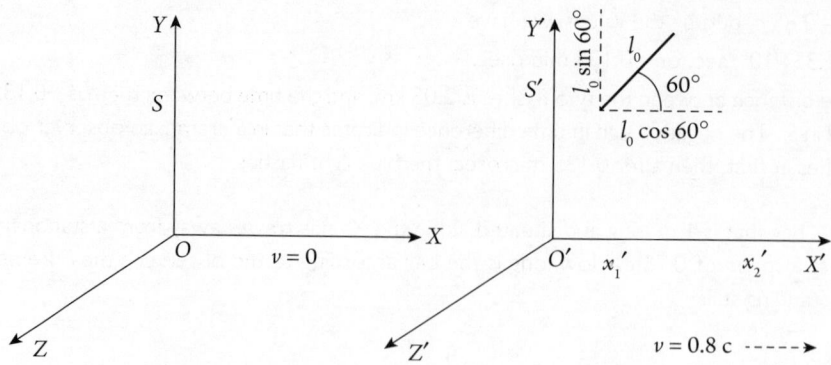

Fig. 1.26 Systematic visualization of Example 8.

Now, length of the rod along the direction of motion

$$= l_0 \cos 60° \sqrt{1 - \frac{(0.8c)^2}{c^2}} = \frac{l_0}{2} \times 0.6 = 0.3 l_0.$$

Length of the rod perpendicular to the direction of motion

$$= l_0 \sin 60° = \frac{\sqrt{3}}{2} l_0.$$

Length of the moving rod,

$$l = \left[(0.3 l_0)^2 + \left(\frac{l_0 \sqrt{3}}{2} \right)^2 \right]^{\frac{1}{2}} = 0.916 l_0.$$

Percentage contraction $= \dfrac{l_0 - 0916 l_0}{l_0} \times 100 = 8.4\%.$

The percentage contraction of a rod is 8.4% to its own length.

Ex. 9: Find the volume of a cube whose proper edge lengths are l_0 while one of its edges is moving at a velocity of v.

Solution:

The edge of the cube along which it is traveling will experience a contraction in length. The lengths of the other edges that are parallel to the direction of motion won't change. If l_0 is the proper edge length of the cube, then the length of the edge along which it is moving will be $l_0 \sqrt{1 - \dfrac{v^2}{c^2}}$. Since the cube's other two edges are unaffected and continue to measure l_0, the volume of the moving cube,

$$V' = l_0 \sqrt{1 - \frac{v^2}{c^2}} (l_0) \times (l_0),$$

$$V' = l_0^3 \sqrt{1 - \frac{v^2}{c^2}}.$$

Thus, the volume of the cube is $l_0^3 \sqrt{1 - \dfrac{v^2}{c^2}}.$

Relativistic Mechanics

Ex. 10: When a reference frame S is at rest, a circular lamina travels with its plane parallel to the X–Y plane. Calculate the velocity at which the object's surface area would seem to be cut in half to a viewer in frame S assuming that the object is moving down the axis of X (or Y).

Solution:
Let's assume that the circular lamina is moving with velocity v along the axis of X of the reference frame S. As a result, the radius R along this axis will now become R', but the radius perpendicular to it will not change. Since the major and minor axes of the circular lamina are identical to R and R' along the axes of Y and X, respectively, it will look as though it has taken on an elliptical form.

Thus, the surface area will appear to be $\pi R R'$ while the original surface area was πR^2. Now, we have $\pi R R' = \frac{1}{2}\pi R^2$; hence, $R' = \frac{R}{2}$. We know that $R' = R\sqrt{1-\frac{v^2}{c^2}}$, where v is the velocity of the circular lamina with respect to the reference frame S.

$$R' = R\sqrt{1-\frac{v^2}{c^2}}, \text{ or } \frac{1}{4} = 1-\frac{v^2}{c^2},$$

$$\frac{v^2}{c^2} = \frac{3}{4} \Rightarrow \frac{v}{c} = \frac{\sqrt{3}}{2} \Rightarrow v = \left(\frac{\sqrt{3}}{2}\right)c,$$

$v = 0.866 c.$

The velocity of the circular lamina relative to frame S is 0.866 c.

Ex. 11: Calculate the length and orientation of a rod of length 5 m in a frame of reference moving with a velocity of 0.6 c in the direction making an angle 30° with the rod.

Solution:
Given that S' is the frame that is traveling at a speed of 0.6 c in a direction that forms a 30° angle with the x-axis. Let l_0 be the length of the rod in the frame in which it is at rest. $l_0 \cos 30°$ and $l_0 \sin 30°$, respectively, are the parts of l_0 that are parallel to and perpendicular to the direction of motion. Given $l_0 = 5$ m.

Now, length of the rod along the direction of motion (X-axis),

$$l_x = l_0 \cos 30° \sqrt{1-\frac{(0.6c)^2}{c^2}} = 5 \times \frac{\sqrt{3}}{2} \times 0.8 = 3.464 \text{ m}.$$

Length of the rod perpendicular to the direction of motion (Y-axis),

$$l_y = l_0 \sin 30° = 5 \times \frac{1}{2} = 2.5 \text{ m}.$$

Length of the moving rod,

$$l = \sqrt{(l_x)^2 + (l_y)^2} = \sqrt{(3.464)^2 + (2.5)^2} = 4.27 \text{ m}.$$

The orientation of the rod,

$$\tan\theta = \frac{l_y}{l_x} = \frac{2.5}{3.464} = 0.72,$$

$\theta = \tan^{-1} 0.72 = 35°45' = 35.8°.$

The length of a rod is 4.27 m and the orientation is 35.8°.

Ex. 12: Calculate the percentage contraction in the length of the rod in a frame of reference moving with velocity 0.8 c in a direction (a) parallel to its length and (b) at an angle of 30° with its length. What is the orientation of the moving frame of reference in case (b)?

Solution:

(a) Let L_0 be the length of the rod, placed along the axis of x in a reference frame S at rest. Then, its length in a frame S', moving with velocity 0.8 c relative to S in a direction parallel to its length, that is, along the axis of x is given by

$$l = l_0\sqrt{1-\frac{v^2}{c^2}} = l_0\sqrt{1-\frac{(0.8c)^2}{c^2}} = l_0\sqrt{1-0.64} = 0.60l_0.$$

The percentage contraction produced in the length of the rod

$$= \frac{l_0 - 0.60l_0}{l_0} \times 100 = 40\%.$$

(b) In this case, the component of the length of the rod along its direction of motion

$$= l_0 \cos 30° = \frac{\sqrt{3}}{2} l_0.$$

And, the component of its length, perpendicular to the direction is

$$= l_0 \sin 30° = 0.5 l_0.$$

Only the former component undergoes a change in length and not the latter (being perpendicular to the direction of motion).

If l'_0 be the length of the former component in the moving frame S', we have

$$l'_0 = \frac{\sqrt{3}}{2} l_0 \sqrt{1-\frac{(0.8c)^2}{c^2}} = 0.52 l_0.$$

Since the other component remains unchanged at $0.5 l_0$, we have the total length of the rod in frame S', say, $l' = \sqrt{(0.52 l_0)^2 + (0.5 l_0)^2} = 0.7228 l_0.$

∴ Percentage contraction produced in the length of the rod $= \frac{0.7228 l_0}{l_0} \times 100 = 22.72\%.$

If θ be the angle that the length of the rod appears to make with the direction of velocity v of frame S', we have

$$\tan\theta = \frac{\text{Length component in } \perp}{\text{Length component in } \parallel} = \frac{0.5 l_0}{0.52 l_0} = 0.96,$$

$$\theta = \tan^{-1}(0.96) = 43°5'.$$

Thus, the rod makes an angle of 43°5' with its direction of motion.

Ex. 13: A rocket ship is 50 m long. To a ground observer, it seems to be 49.5 m long when it is in flight. Identify the rocket's speed.

Solution:

If l_0 is the proper length, then according to the length contraction formula

$$l = l_0\sqrt{1-\frac{v^2}{c^2}} \Rightarrow v = c\sqrt{1-\frac{l^2}{l_0^2}},$$

Relativistic Mechanics

$$v = 3 \times 10^8 \sqrt{1 - \frac{49.5^2}{50^2}} = 4.23 \times \frac{10^7 \, m}{s}.$$

The speed of the rocket is 4.23×10^7 m/s.

Ex. 14: How quickly should a clock be moved such that it seems to lose one minute every hour?

Solution:

According to the time dilation formula,

$$t = \frac{t_0}{\sqrt{1 - \frac{v^2}{c^2}}}.$$

Here, $t_0 = 59$ min. is proper time, and $t = 60$ min is dilated time,

$$v = c\sqrt{1 - \frac{t_0^2}{t^2}} = 3 \times 10^8 \sqrt{1 - \frac{59^2}{60^2}} = 5.45 \times 10^7 \, \frac{m}{s}.$$

The speed of the clock is 5.45×10^7 m/s.

Ex. 15: A certain particle called meson has a lifetime 2×10^{-6} s. What is the mean lifetime when the particle is traveling with a speed of 0.9 c? How far does it go during one mean life?

Solution:

According to the time dilation formula,

$$t = \frac{t_0}{\sqrt{1 - \frac{v^2}{c^2}}}.$$

Here, $t_0 = 2 \times 10^{-6}$ sec, $v = 0.9$ c, put in above, we get

$$t = \frac{2 \times 10^{-6}}{\sqrt{1 - \frac{(0.9c)^2}{c^2}}} = 4.587 \times 10^{-6} \text{ sec.}$$

Hence, mean lifetime of a meson $= 4.587 \times 10^{-6}$ sec.

The distance traveled by the meson in one life is

$x = v \times t = 0.9 \times 3.0 \times 10^8 \times 4.587 \times 10^{-6} = 1238.49$ m.

Ex. 16: A π^+ meson's appropriate life is 2.5×10^{-8} sec. If a beam of these mesons of velocity 0.8 c is produced, calculate the distance the beam can travel before the flux of the meson beam is reduced to $1/e^2$ times the initial flux.

Solution:

If t_0 be the proper lifetime of π^+ mesons in its own frame of reference and the observed lifetime is given by

$$t = \frac{t_0}{\sqrt{1 - \frac{v^2}{c^2}}} = \frac{2.5 \times 10^{-8}}{\sqrt{1 - \frac{(0.8c)^2}{c^2}}} = 4.167 \times 10^{-8} \text{ sec.}$$

If φ_0 be the initial flux and φ is the flux after time t, then

$$\varphi = \varphi_0 e^{-\frac{T}{t}} \text{ or } \frac{\varphi_0}{\varphi} = e^{\frac{T}{t}}.$$

According to the given problem, $\varphi = \frac{\varphi_0}{e^2}$,

$$e^{\frac{T}{t}} = e^2 \text{ or } \frac{T}{t} = 2 \Rightarrow T = 2t = 2 \times 4.167 \times 10^{-8} = 8.334 \times 10^{-8} \text{ s}.$$

The distance traveled by the meson beam is produced before the flux is reduced to $1/e^2$ times the initial flux. This is given by

$$d = vt = 0.8 \times 3.0 \times 10^8 \times 8.334 \times 10^{-8} = 20 \text{ m}.$$

Ex. 17: At a speed of 0.8 c, a man departs from the earth on a rocket ship and travels four light years to the nearest star. How much younger will he be than his identical brother who chose to stay behind when he returns?

Solution:
According to the time dilation formula,

$$t = \frac{t_0}{\sqrt{1-\frac{v^2}{c^2}}} \Rightarrow t_0 = t\sqrt{1-\frac{v^2}{c^2}}.$$

The distance to the star is 4 light years, and the speed is 0.8 c. Therefore, the time interval for the person remaining on the earth is

$$t = \frac{2 \times 4 ly}{0.8c} = 10 \text{ years}.$$

However, according to the astronaut's clock,

$$t_0 = 10\sqrt{1-\frac{(0.8c)^2}{c^2}} = 10 \times 0.6 = 6 \text{ years}.$$

Therefore, $10 - 6 = 4$ years.
Thus, his twin brother is 4 years younger.

Ex. 18: A wristwatch, keeping correct time on earth, is worn by the pilot of a spaceship. How much will it appear to lose per day with respect to an observer on the earth when the spaceship leaves the earth with a constant velocity of 10^{-7} m/s.

Solution:
According to the time dilation formula,

$$t = \frac{t_0}{\sqrt{1-\frac{v^2}{c^2}}} \Rightarrow t_0 = t\sqrt{1-\frac{v^2}{c^2}}.$$

Here, $t = 24$ hours, $v = 10^{-7}$ m/s, and $c = 3.0 \times 10^8$ m/s.

$$t_0 = 24\sqrt{1-\frac{10^{-14}}{c^2}} = 23.986 \text{ hours}.$$

Hence, loss of time per day = $24 - 23.986 = 0.014$ hours = 50.4 sec.

Relativistic Mechanics

Ex. 19: Two particles came toward each other with speed 0.7 c with respect to the laboratory. What is their relative speed?

Solution:
According to the velocity addition theorem,
$$u = \frac{u'+v}{1+\frac{u'v}{c^2}} = \frac{0.7c+0.7c}{1+\frac{(0.7c \times 0.7c)}{c^2}},$$

$$u = \frac{1.4c}{1.49} = 0.9396c.$$

Their relative speed is 0.9396 c.

Ex. 20: A particle has a velocity $\vec{u'} = 3i + 4j + 12k$ m/s in a coordinate system while traveling down the positive x-axis direction at a speed of 0.8 c in relation to the laboratory. In the lab frame, find the value velocity "u."

Solution:
According to the velocity addition theorem,
$$u_x = \frac{u'_x + v}{1+\frac{u'_x v}{c^2}} = \frac{3+0.8c}{1+\frac{0.8c \times 3}{c^2}} = 2.4 \times 10^{-8}\frac{m}{s},$$

$$u_y = \frac{u'_y \sqrt{1-\frac{v^2}{c^2}}}{1+\frac{u'_x v}{c^2}} = \frac{4\sqrt{1-\frac{0.8c^2}{c^2}}}{1+\frac{0.8c \times 3}{c^2}} = 2.4 \, m/s,$$

$$u_z = \frac{u'_z \sqrt{1-\frac{v^2}{c^2}}}{1+\frac{u'_x v}{c^2}} = \frac{12\sqrt{1-\frac{0.8c^2}{c^2}}}{1+\frac{0.8c \times 3}{c^2}} = 7.2 \frac{m}{s}.$$

Therefore, $\vec{u}$ in laboratory frame is given by
$$\vec{u} = u_x i + u_y j + u_z k,$$
$$\vec{u} = (2.4 \times 10^{-8} i + 2.4 j + 7.2 k) m/s.$$

Thus, $\vec{u}$ in laboratory frame is $(2.4 \times 10^{-8} i + 2.4 j + 7.2 k) m/s$.

Ex. 21: The mass of a body will be 2.25 times its rest mass at what speed?

Solution:
According to the mass variation formula,
$$m = \frac{m_0}{\sqrt{1-\frac{v^2}{c^2}}}.$$

Here, $m = 2.25\, m_0$ $\quad \therefore 2.25\, m_0 = \dfrac{m_0}{\sqrt{1-\dfrac{v^2}{c^2}}}$,

$v = c\sqrt{1-\dfrac{1^2}{2.25^2}} = 2.68 \times \dfrac{10^8\, m}{s}$.

Thus, the speed is found to be $2.68 \times 10^8\, m/s$.

Ex. 22: A person observes two men, each of rest mass 60 kg, moving toward each other, each with velocity of 0.5 c. What is the mass of one man as observed by the other?

Solution:
According to the velocity addition theorem,

$$u = \dfrac{u' + v}{1 + \dfrac{u'v}{c^2}} = \dfrac{0.5\,c + 0.5\,c}{1 + \dfrac{(0.5\,c \times 0.5\,c)}{c^2}} = \dfrac{1}{1.25}c = 0.8\,c.$$

The mass of one man as observed by the other is

$$m = \dfrac{m_0}{\sqrt{1-\dfrac{v^2}{c^2}}} = \dfrac{60}{\sqrt{1-\dfrac{(0.8)^2}{c^2}}} = \dfrac{60}{0.6} = 100\, kg.$$

Hence, $m = 100\, kg$.

Ex. 23: On the earth, a man weighs 50 kg. His mass, as determined by a witness on the earth, is 50.5 kg when he is aboard a rocket ship in flight. How fast is the rocket moving?

Solution:
According to the relativistic mass with velocity formula,

$$m = \dfrac{m_0}{\sqrt{1-\dfrac{v^2}{c^2}}} \text{ or } v = c\sqrt{1-\dfrac{m_0^2}{m^2}}.$$

Here, $m_0 = 50\, kg$, $m = 50.5\, kg$, and $c = 3 \times 10^8$,

$$v = 3.0 \times 10^8 \sqrt{1-\dfrac{50^2}{50.5^2}} = 4.23 \times \dfrac{10^7\, m}{s}.$$

Ex. 24: A proton has a rest mass of 1.67×10^{-27} kg. At what velocity will its mass double compared to its rest mass?

Solution:
According to the relativistic mass with velocity formula,

$$m = \dfrac{m_0}{\sqrt{1-\dfrac{v^2}{c^2}}} \text{ or } v = c\sqrt{1-\dfrac{m_0^2}{m^2}},$$

$$v = 3.0 \times 10^8 \sqrt{1-\dfrac{m_0^2}{4m_0^2}} = 2.6 \times 10^8\, m/s.$$

Relativistic Mechanics

Ex. 25: When a meter stick's mass is 3/2 times that of its rest mass, what is the length of the stick when it moves parallel to its length?

Solution:

In the given problem, the mass of the rod is 3/2 times its rest mass, that is,

$$m = \frac{3}{2}m_0 \quad \text{or} \quad \frac{m_0}{\sqrt{1-\frac{v^2}{c^2}}} = \frac{3}{2}m_0,$$

$$\sqrt{1-\frac{v^2}{c^2}} = \frac{2}{3} = 0.667.$$

The length of the meter stick moving parallel to its length, according to the length contraction formula, is

$$l = l_0\sqrt{1-\frac{v^2}{c^2}}.$$

Here, $l_0 = 1$ m,

$l = 1 \times 0.667 = 0.667$ m.

The length of a meter stick is 0.667 m.

Ex. 26: What speed must an electron travel for its mass to be equal to the proton's rest mass?

Solution:

Considering the relationship between the fluctuation of mass and velocity,

$$m = \frac{m_0}{\sqrt{1-\frac{v^2}{c^2}}} \quad \text{or} \quad v = c\sqrt{1-\frac{m_0^2}{m^2}}.$$

Here, $m_0 = m_e = 9.11 \times 10^{-31}$ kg, $m = m_p = 1.67 \times 10^{-27}$ kg, and $c = 3 \times 10^8$ m/s,

$$v = 3.0 \times 10^8 \sqrt{1-\frac{(9.11\times 10^{-31})^2}{(1.67\times 10^{-27})^2}} = 2.99 \times 10^8 \text{ m/s}.$$

Ex. 27: Calculate the rest mass, relativistic mass, and momentum of a photon of energy 5eV.

Solution:

Energy of the photon is given as

$E = 5$ eV $= 5 \times 1.6 \times 10^{-19}$ J $= 8.0 \times 10^{-19}$ J.

The momentum of photon is

$$p = \frac{E}{c} = \frac{8.0 \times 10^{-19}}{3 \times 10^8} = 2.67 \times 10^{-27} \text{ kgm/s}.$$

Therefore, the relativistic mass of photon is

$$m = \frac{p}{c} = \frac{2.67 \times 10^{-27}}{3 \times 10^8} = 8.9 \times 10^{-36} \text{ kg}.$$

According to the variation of relativistic mass with velocity formula,

$$m = \frac{m_0}{\sqrt{1 - \frac{v^2}{c^2}}} \text{ or } m_0 = m\sqrt{1 - \frac{v^2}{c^2}}.$$

The velocity of photon, $v = c$.
Therefore, the rest mass of photon, $m_0 = 0$.

Ex. 28: Calculate the 2 MeV electron's mass and speed.

Solution:

In relativistic mechanics, the energy of a particle is expressed as

$$E = mc^2 \text{ or } m = \frac{E}{c^2}.$$

Here, $E = 2$ MeV $= 2 \times 10^6 \times 1.6 \times 10^{-19}$ J and $c = 3 \times 10^8$ m/s.

$$m = \frac{2 \times 10^6 \times 1.6 \times 10^{-19}}{3 \times 10^8} = 3.55 \times 10^{-30} \text{ kg}.$$

Using the formula for the variation of mass with velocity,

$$m = \frac{m_0}{\sqrt{1 - \frac{v^2}{c^2}}} \text{ or } v = c\sqrt{1 - \frac{m_0^2}{m^2}},$$

$$v = 3 \times 10^8 \sqrt{1 - \frac{(9.11 \times 10^{-31})^2}{(3.55 \times 10^{-30})^2}} = 2.90 \times \frac{10^8 \text{ m}}{\text{s}}.$$

Ex. 29: What is the particle's velocity if the total energy of the particle is precisely three times that of the rest energy?

Solution:

According to Einstein mass–energy relation,
$E = mc^2$.

Here, total energy = 3 × rest energy; therefore,
$E = 3 m_0 c^2 \Rightarrow mc^2 = 3 m_0 c^2 \Rightarrow m = 3 m_0$.

Using the formula for the variation of mass with velocity,

$$m = \frac{m_0}{\sqrt{1 - \frac{v^2}{c^2}}} \text{ or } v = c\sqrt{1 - \frac{m_0^2}{m^2}},$$

$$v = 3 \times 10^8 \sqrt{1 - \frac{1}{9}} = 2.8 \times \frac{10^8 \text{ m}}{\text{s}}.$$

Ex. 30: The total energy of a moving meson is exactly twice its rest energy. Find the speed of meson.

Solution:

According to the mass–energy relation, the total energy of a moving meson is
$E = mc^2$.

Relativistic Mechanics

According to the given problem, $E = 2 \times$ rest energy $= 2\, m_0 c^2$,

$$2\, m_0 c^2 = mc^2 \text{ or } 2\, m_0 = \frac{m_0}{\sqrt{1-\frac{v^2}{c^2}}},$$

$$v = 3 \times 10^8 \sqrt{1-\frac{1}{4}} = 2.598 \times \frac{10^8\, m}{s}.$$

Ex. 31: Determine a body's velocity if its kinetic energy is two times that of its rest mass–energy.

Solution:

The relativistic kinetic energy can be expressed as

$$K = (m-m_0)c^2 \text{ or } mc^2 = K + m_0 c^2,$$

$$K = 2\, m_0 c^2 \quad \therefore mc^2 = 3\, m_0 c^2 \text{ or } m = 3\, m_0.$$

Using the formula for the variation of mass with velocity,

$$m = \frac{m_0}{\sqrt{1-\frac{v^2}{c^2}}} \text{ or } v = c\sqrt{1-\frac{m_0^2}{m^2}},$$

$$v = 3 \times 10^8 \sqrt{1-\frac{1}{9}} = 2.829 \times 10^8\, \frac{m}{s}.$$

Ex. 32: Find out the velocity of a particle if its kinetic energy is three times the rest energy.

Solution:

The relativistic kinetic energy can be expressed as

$$K = (m-m_0)c^2 \text{ or } mc^2 = K + m_0 c^2,$$

$$K = 3\, m_0 c^2 \quad \therefore mc^2 = 4\, m_0 c^2 \text{ or } m = 4\, m_0.$$

According to the variation of mass with velocity formula,

$$m = \frac{m_0}{\sqrt{1-\frac{v^2}{c^2}}} \text{ or } v = c\sqrt{1-\frac{m_0^2}{m^2}},$$

$$v = 3 \times 10^8 \sqrt{1-\frac{1}{16}} = 2.884 \times 10^8\, \frac{m}{s}.$$

Ex. 33: A moving electron has a mass that is 11 times that of its rest mass. Find the momentum and kinetic energy of it.

Solution:

The relativistic kinetic energy can be expressed as

$$K = (m-m_0)c^2 \text{ or } K = (11 m_0 - m_0)c^2 \quad [\because m = 11 m_0],$$

$K = 10\,m_0 c^2 = 10 \times 9.11 \times 10^{-31} (3 \times 10^8)^2 = 5.17 \times 10^6$ eV.

The momentum p of the particle is given by $p = mv$.

According to the variation of mass with velocity formula,

$$m = \frac{m_0}{\sqrt{1 - \frac{v^2}{c^2}}} \text{ or } v = c\sqrt{1 - \frac{m_0^2}{m^2}},$$

$$v = 3 \times 10^8 \sqrt{1 - \frac{m_0^2}{(11 m_0)^2}} = 2.99 \times 10^8 \text{ m/s},$$

$p = 11 m_0 \times v = 11 \times 9.11 \times 10^{-31} \times 2.99 \times 10^8$,

$p = 2.99 \times 10^{-21}$ kgm/s.

Ex. 34: Determine how much work is required to raise an electron's speed from 0.6 c to 0.8 c. Considering that an electron's rest energy is 0.5 MeV.

Solution:

The relativistic kinetic energy can be expressed as

$$K = (m - m_0)c^2 = c^2 \left[\frac{m_0}{\sqrt{1 - \frac{v^2}{c^2}}} - m_0 \right] = c^2 m_0 \left[\frac{1}{\sqrt{1 - \frac{v^2}{c^2}}} - 1 \right].$$

Initial kinetic energy of an electron, when its velocity is 0.6 c, is given by

$$K_1 = c^2 m_0 \left[\frac{1}{\sqrt{1 - \frac{(0.6c)^2}{c^2}}} - 1 \right] = c^2 m_0 [1.25 - 1] = 0.5 \times 10^6 \times 0.25 = 1.25 \times 10^5 \text{ eV}.$$

Final kinetic energy of an electron, when its velocity increases to 0.8 c, is given by

$$K_2 = c^2 m_0 \left[\frac{1}{\sqrt{1 - \frac{(0.8c)^2}{c^2}}} - 1 \right] = c^2 m_0 [1.67 - 1] = 0.5 \times 10^6 \times 0.67 = 3.35 \times 10^5 \text{ eV}.$$

Therefore, the amount of work to be done to increase the speed of an electron from 0.6 to 0.8 c is

$\Delta K = K_2 - K_1 = 3.35 \times 10^5 - 1.25 \times 10^5 = 2.1 \times 10^5$ eV,

$\Delta K = 2.1 \times 10^5 \times 1.6 \times 10^{-19} = 3.36 \times 10^{-14}$ J.

Previous Year Questions (University Examination)

1. What does "frame of reference" mean? Compare and contrast inertial and non-inertial frames of reference. Is the earth a gravitational frame?

Relativistic Mechanics

2. Write down the postulate of the theory of special relativity, and discuss them briefly.
3. What were Michelson and Morley trying to achieve when they conducted the experiment? How did they explain their experimental results?
4. Describe the Michelson and Morley experiment with neat experimental setup, and deduce the total fringe shift expression.
5. Derive the Galilean transformation equations. State the fundamental postulates of the theory of special relativity.
6. Derive Lorentz transformation equations. Show that Lorentz transformation equations are reduced to Galilean transformation equations at very low speeds.
7. Apply Lorentz transformation equations to derive expression for length contraction and time dilation.
8. What do you understand by time dilation? Show that time dilation is a real effect by giving experimental evidence.
9. Show that any signal cannot travel faster than the velocity of light by applying relativistic velocity addition theorem.
10. Obtain the relativistic formula for the addition of velocities and also show that the velocity of light is an absolute constant independent of the frame of reference.
11. Deduce an expression for the variation of mass with velocity at relativistic speed.
12. Establish Einstein's mass–energy relation mathematically. Explain the physical significance of this relation. Mention the nuclear phenomena supporting this relation.
13. What will be the expected fringe shift on the basis of stationary ether hypothesis in Michelson–Morley experiment if the effective path length of each path is 7 m and wavelength of light used is 7000 Å?
14. Calculate the percentage contraction in the length of a rod in a frame of reference moving with velocity 0.8 c in a direction parallel to its length.
15. How fast would a rocket have to go relative to an observer for its length to be contracted to 99% of its length at rest?
16. Calculate the percentage contraction of a rod moving with a velocity of 0.8 c in a direction inclined at 60° to its own length.
17. Calculate the length and orientation of a rod of length 2 m in a frame of reference that is moving with 0.6 c velocity in a direction making an angle of 30° with the rod.
18. At what speed will the mass of a body be 2.25 times its rest mass?
19. The rest mass of proton is 1.67×10^{-27} kg. At what speed will its mass be double its rest mass?
20. What is the length of a 1 m rod moving parallel to its length when its mass is 1.5 times of its rest mass?
21. A clock measures the proper time. With what velocity should it travel relative to an observer so that it appears to go slow by 30 sec in a day?
22. How rapidly must an electron travel in order for its mass to match the proton's rest mass?
23. Determine the 2 Mev electron's mass and speed.
24. A particle with rest mass m_0 travels at a speed of $c\sqrt{2}$. Calculate the object's mass, momentum, total energy, and kinetic energy.
25. A moving electron has a mass that is 11 times that of its rest mass. Find the momentum and kinetic energy of it.

26. Determine a body's velocity if its kinetic energy is two times that of its rest mass–energy.
27. Calculate the rest mass, relativistic mass, and momentum of a photon of energy 5 eV.
28. Show that the relativistic kinetic energy of the particle of rest mass m_0 and moving with velocity v is given by $K = m_0 c^2 \left[\left(1 - \dfrac{v^2}{c^2}\right)^{-1/2} - 1 \right]$.
29. Prove that the circle in frame S appears to be an ellipse in frame S', which is traveling with a velocity v in relation to frame S.
30. Show that the relativistic form of Newton's second law, when $\vec{F}$ is parallel to $\vec{v}$, is given by $\vec{F} = m_0 \dfrac{d\vec{v}}{dt} \left(1 - \dfrac{v^2}{c^2}\right)^{-1/2}$.

Multiple Choice Questions

1. Newtonian mechanics is applicable for
   (a) All velocities
   (b) Velocities less than half the velocity of light
   (c) Velocities approaching the velocity of light
   (d) None of the above

2. Relativistic mechanics is applicable for
   (a) All velocities
   (b) Velocities less than half the velocity of light
   (c) Velocities approaching the velocity of light
   (d) None of the above

3. Einstein proposed the theory of relativity in
   (a) 1900
   (b) 1903
   (c) 1905
   (d) 1915

4. An inertial frame of reference is one which
   (a) Does not accelerate
   (b) Remains at absolute rest
   (c) Remains at absolute motion
   (d) Attached to an observer

5. A non-inertial frame of reference is one which
   (a) Obeys the law of inertia or other Newtonian laws
   (b) Does not obey the law of inertia or other Newtonian laws
   (c) Accelerates an object which is not accelerated
   (d) Both (b) and (c)

6. Is the earth an inertial frame of reference?
   (a) Yes, the earth is moving with constant speed around the sun
   (b) No, the earth is rotating and revolving, which produces a centripetal acceleration
   (c) Yes, the law of inertia holds good on the earth
   (d) Both (a) and (c)

Relativistic Mechanics

7. What was an objective of the Michelson–Morley experiment?
   (a) To confirm the existence of non-inertial frame and measure the relative velocity of the earth
   (b) To confirm the existence of stationary ether and measure the relative velocity of the earth
   (c) To verify length contraction and constancy of speed of light
   (d) Both (a) and (b)

8. What was an outcome of the Michelson–Morley experiment?
   (a) Ether hypothesis was disapproved; there is no absolute frame
   (b) Ether hypothesis was approved; there is an absolute frame
   (c) Relativity is correct for high speed only
   (d) Relativity is correct for low speed only

9. Michelson–Morley performed an experiment to
   (a) Find the velocity of light
   (b) Find the velocity of earth
   (c) Confirm the existence of ether
   (d) None of the above

10. Explanations of negative results of the Michelson–Morley experiment are
    (a) Ether drag hypothesis
    (b) Lorentz–Fitzgerald contraction hypothesis
    (c) Constancy of speed of light
    (d) All of the above

11. Einstein's postulate of special theory of relativity is
    (a) Principle of equivalence
    (b) Constancy of speed of light
    (c) Time dilation
    (d) Both (a) and (b)

12. Speed of light in vacuum depends upon
    (a) Velocity of source
    (b) Velocity of observer
    (c) Velocity of both source and observer
    (d) None of the above

13. Galilean transformations are applicable
    (a) For relativistic motion
    (b) For non-relativistic motion
    (c) For both (a) and (b)
    (d) None of these

14. Which of the following is not invariant under Galilean transformations?
    (a) Space intervals
    (b) Time intervals
    (c) Mass
    (d) Momentum

15. Lorentz transformations are applicable
    (a) For relativistic motion
    (b) For non-relativistic motion
    (c) For both (a) and (b)
    (d) None of these

16. Which of the following is invariant under Lorentz transformations?
    (a) Space coordinates
    (b) Time intervals
    (c) Both space coordinates and time intervals
    (d) Space–time interval

17. If l is the measured length and l_0 is the proper length, then the length contraction formula according to relativity is given by

(a) $l = l_0\sqrt{1 - \dfrac{v^2}{c^2}}$

(b) $l_0 = l\sqrt{1 - \dfrac{v^2}{c^2}}$

(c) $l = \dfrac{l_0}{\sqrt{1 - \dfrac{v^2}{c^2}}}$

(d) None of the above

18. Relativity states that the length of a moving rod
    (a) Has length equal to its rest length
    (b) Has length greater than its rest length
    (c) Has length smaller than its rest length
    (d) None of the above

19. A circle in x–y plane moves along the x-direction with constant velocity w.r.t. an observer. The observer will observe
    (a) A circle with smaller diameter
    (b) A circle with larger diameter
    (c) Ellipse with major axis along the x-direction
    (d) Ellipse with major axis along the y-direction

20. According to the relativity, a moving clock appears to be
    (a) Fast
    (b) Slow
    (c) Normal
    (d) None of the above

21. If t is the measured time interval and t_0 is the proper time interval, then time dilation formula according to relativity is given by

    (a) $t = t_0\sqrt{1 - \dfrac{v^2}{c^2}}$

    (b) $t_0 = t\sqrt{1 - \dfrac{v^2}{c^2}}$

    (c) $t_0 = \dfrac{t}{\sqrt{1 - \dfrac{v^2}{c^2}}}$

    (d) None of the above

22. The presence of μ mesons in the atmosphere of the earth provided the experimental verification of
    (a) Length contraction
    (b) Time dilation
    (c) Galilean transformation
    (d) Michelson–Morley experiment

23. The relativistic velocity addition theorem is consistent with
    (a) First postulate of special theory of relativity
    (b) Second postulate of special theory of relativity
    (c) Einstein's mass–energy relation
    (d) None of these

24. No signal can travel greater than velocity of light is verified by
    (a) Time dilation
    (b) Mass variation with velocity
    (c) Velocity addition theorem
    (d) Einstein's mass–energy relation

25. No material particle can travel equal to the velocity of light is verified by
    (a) Time dilation
    (b) Mass variation with velocity
    (c) Velocity addition theorem
    (d) Einstein's mass–energy relation

Relativistic Mechanics

26. If m is the relativistic mass and m_0 is the rest mass, then mass variation with velocity formula according to relativity is given by
    (a) $m = m_0 \sqrt{1 - \dfrac{v^2}{c^2}}$
    (b) $m_0 = m \sqrt{1 - \dfrac{v^2}{c^2}}$
    (c) $m_0 = \dfrac{m}{\sqrt{1 - \dfrac{v^2}{c^2}}}$
    (d) None of the above

27. The force required for a material particle having mass m to increase its speed equal to the velocity of light is
    (a) Infinite
    (b) Zero
    (c) 1000 N
    (d) 150 N

28. Einstein's mass–energy relation is
    (a) $E = m_0 c^2$
    (b) $\dfrac{E}{c^2} = m$
    (c) $p = mc$
    (d) None of the above

29. Which one of the following is an example of mass–energy relation?
    (a) Pair production
    (b) Presence of μ mesons in the atmosphere
    (c) Both (a) and (b)
    (d) None of the above

30. Pair production is an example of
    (a) Mass converted into energy
    (b) Energy converted into mass
    (c) Energy converted into photons
    (d) None of the above

31. The rest mass of photons is
    (a) Infinity
    (b) Zero
    (c) E/c^2
    (d) Ec^2

32. The velocity of photons is
    (a) Infinite
    (b) Zero
    (c) C
    (d) 0.84 c

33. The massless particles are
    (a) Photons
    (b) Gravitons
    (c) Neutrinos
    (d) All of the above

34. The relativistic momentum of massless particles is
    (a) 0
    (b) E/c^2
    (c) E/c
    (d) None of the above

35. The relativistic mass of massless particles is
    (a) 0
    (b) Greater than 0
    (c) Less than 0
    (d) None of the above

36. The relativistic relation between energy and momentum is
    (a) $E = pc + m_0 c^4$
    (b) $E^2 = p^2 c^2 + m_0^2 c^4$
    (c) $p = mv$
    (d) None of the above

37. What will be fringe shift according to the ether theory in the Michelson–Morley experiment if the effective path length of each path is 7 m and light has 7000 Å wavelength? The velocity of the earth is 3×10^4 m/s.
    (a) 0.4
    (b) 0.3
    (c) 0.2
    (d) 0.1

38. What is the percentage contraction of a rod moving with a velocity of 0.8 c in a direction inclined at 60° to its own length?
    (a) 3.4%
    (b) 4.3%
    (c) 8.4%
    (d) 0%

39. At what speed should a clock be moved so that it may appear to lose 1 minute in each hour?
    (a) 5.45×10^7 m/s
    (b) 4.45×10^7 m/s
    (c) 3.45×10^7 m/s
    (d) 2.45×10^7 m/s

40. Two particles came toward each other with speed 0.7 c with respect to the laboratory. What is their relative speed?
    (a) 0.8516 c
    (b) 0.9396 c
    (c) 0.2016 c
    (d) 0.7065 c

41. At what speed will the mass of a body be 2.25 times its rest mass?
    (a) 5.58×10^8 m/s
    (b) 3.38×10^8 m/s
    (c) 2.68×10^8 m/s
    (d) 1.58×10^8 m/s

42. What is the length of one meter stick moving parallel to its length when its mass is 3/2 times of its rest mass?
    (a) 0.667 m
    (b) 0.432 m
    (c) 1.211 m
    (d) 0 m

43. Find out the velocity of a particle if its kinetic energy is three times the rest energy.
    (a) 3.811×10^5 m/s
    (b) 2.118×10^6 m/s
    (c) 1.511×10^7 m/s
    (d) 2.884×10^8 m/s

44. The rest mass of an electron is m_0. What would be its mass if it moves with velocity 0.6 c?
    (a) $\frac{3}{2}m_0$
    (b) $\frac{4}{3}m_0$
    (c) $\frac{5}{4}m_0$
    (d) $\frac{6}{5}m_0$

45. The rest mass of an electron is m_0. What would be its kinetic energy if it moves with velocity 0.6 c?
    (a) $\frac{1}{2}m_0c^2$
    (b) $\frac{2}{3}m_0c^2$
    (c) $\frac{1}{4}m_0c^2$
    (d) $\frac{3}{4}m_0c^2$

46. The kinetic energy of a body is twice its rest mass–energy. Then the ratio of relativistic mass to rest mass of the body is
    (a) 2
    (b) 3
    (c) 1/2
    (d) 1/3

47. The kinetic energy of a particle is double of its rest mass–energy. Then the speed of the particle in terms of speed of light c is
    (a) c
    (b) $c/2$
    (c) $2c/3$
    (d) $2\sqrt{2}/3$

48. An electron is chased by a photon. The speed of the electron is 0.9 c. Their relative velocity is
    (a) 0.1 c
    (b) 0.9 c
    (c) c
    (d) None of the above

49. An observer moves with a speed $c/2$ toward a stationary source of light; the speed of light appears to the observer to be
    (a) c
    (b) $2c$
    (c) $c/2$
    (d) $3c/2$

50. At what speed would the mass of an electron be double its rest mass?
    (a) 1.3×10^8 ms^{-1}
    (b) 2.6×10^8 ms^{-1}
    (c) 1.9×10^8 ms^{-1}
    (d) 2.25×10^8 ms^{-1}

BIBLIOGRAPHY

1. Rindler, W. (1982). *Introduction to Special Relativity*. Oxford University Press.
2. Griffiths, Jerry B. and Podolsky, J. (2009). *Exact Space-Times in Einstein's General Relativity*. Cambridge University Press.
3. Plebanski, J. and Krasinski, A. (2006). *An Introduction to General Relativity and Cosmology*. Cambridge University Press.
4. Ashtekar, A., Berger, B. K., Isenberg, J., and MacCallum, M., eds. (2015). *General Relativity and Gravitation: A Centennial Perspective*. Cambridge University Press.
5. Gourgoulhon, E. (2013). *Special Relativity in General Frames, from Particles to Astrophysics*. Springer.
6. Rindler, W. (1979). *Essential Relativity*. Springer-Verlag.
7. Choquet-Bruhat, Y. (2015). *Introduction to General Relativity, Black Holes & Cosmology*. Oxford University Press.
8. Stewart, J. (1990). *Advanced General Relativity*. Cambridge University Press.
9. Wald, R. (1984). *General Relativity*. University Chicago Press.
10. Carmeli, M. (1977). *Group Theory and General Relativity*. McGraw-Hill.

Keys

1. (b)	2. (a)	3. (c)	4. (a)	5. (d)	6. (b)	7. (b)	8. (a)	9. (c)	10. (d)
11. (d)	12. (d)	13. (b)	14. (d)	15. (c)	16. (d)	17. (a)	18. (c)	19. (d)	20. (b)
21. (b)	22. (b)	23. (b)	24. (c)	25. (b)	26. (b)	27. (a)	28. (b)	29. (a)	30. (a)
31. (b)	32. (c)	33. (d)	34. (c)	35. (b)	36. (b)	37. (c)	38. (c)	39. (a)	40. (b)
41. (c)	42. (a)	43. (d)	44. (c)	45. (c)	46. (b)	47. (d)	48. (c)	49. (a)	50. (b)

CHAPTER 2

Quantum Mechanics

2.1 Introduction

Numerous microscopic events, including atomic stability, blackbody radiation, the photoelectric effect, and atomic spectroscopy, could not be explained by classical physics. When Max Planck presented the idea of the quantum of energy in 1900, it marked the first significant advancement. Only after positing that the energy exchange between radiation and its surroundings occurs in discrete, or quantized, amounts was he able to replicate the experimental findings in his attempts to understand the phenomenon of blackbody radiation. He claimed that an electromagnetic wave of frequency v and matter can only exchange energy in integer multiples of h, or what he termed a quantum's energy, where h is a fundamental constant known as Planck's constant. The concept of quantizing electromagnetic radiation proved to have far-reaching effects.

Blackbody radiation was correctly explained by Planck's hypothesis, which inspired fresh thinking and set off a wave of new findings that provided answers to the most pressing issues of the day.

Planck's quantum idea received a potent reinforcement from Einstein in 1905. Einstein realized that Planck's theory of the quantization of electromagnetic waves must also apply to light when attempting to comprehend the photoelectric effect. So, adopting Planck's methodology, he proposed that light itself is composed of discrete energy units (or minuscule particles) called photons, each of which has energy hv, which corresponds to the light's frequency. The photoelectric issue, which had persisted since Hertz's initial experimental discovery in 1887, finally found an elegant and accurate explanation, thanks to Einstein's introduction of the photon concept.

Niels Bohr was responsible for another fundamental discovery. Bohr presented his model of the hydrogen atom in 1913, immediately following Rutherford's experimental finding of the atomic nucleus in 1911. This model combined Rutherford's atomic model, Planck's quantum theory, and Einstein's photons. In this work, he made the case that atoms can only exist in discrete energy states and that their interactions with radiation, such as their emission or absorption of radiation, only occur in discrete amounts of hv as a consequence of the transitions between these discrete energy states. For a number of open issues, including atomic stability and atomic spectroscopy, this study offered a satisfactory explanation.

Then, in 1923, Compton made a significant finding that provided the clearest evidence for the corpuscular nature of light. He demonstrated that X-ray photons act like particles with momenta h/c by scattering X-rays with electrons. The theoretical underpinnings and definitive experimental proof for the particle aspect of waves—that is, the idea that waves show particle behavior at the microscopic scale—came from this sequence of discoveries credited to Planck, Einstein, Bohr, and Compton. At this size, classical physics fails conceptually and qualitatively in addition to quantitatively.

In 1923, de-Broglie proposed that material particles themselves show wave-like behavior in addition to radiation's particle-like behavior, a powerful new idea that classical physics could not reconcile. By demonstrating that material particles like electrons can be used to create interference patterns, a characteristic of waves, Davisson and Germer empirically supported this idea in 1927.

2.2 WAVE–PARTICLE DUALITY

Light displays its wave nature in optical phenomena like interference, diffraction, and polarization. The wave theory of light can be used to describe these phenomena. However, the photoelectric effect, Compton effect, radiation absorption, and radiation emission cannot be explained by the wave theory of light. The quantum theory of light can be used to describe these phenomena. According to Einstein, light's energy is concentrated into smaller areas known as photons or energy particles. As a result, light exhibits wave-like behavior in some experiments and particle-like behavior in others. This is referred to as light's dual nature, and the characteristic of this dual nature is referred to as wave–particle duality.

2.3 de-BROGLIE'S HYPOTHESIS OF MATTER WAVES

A moving microscopic particle behaves as wave.

According to de-Broglie hypothesis, *A moving microscopic particle is associated with a wave known as de-Broglie wave or matter wave.* The wavelength of de-Broglie waves is given by

$$\lambda = \frac{h}{mv} = \frac{h}{p} \tag{2.1}$$

where h is the Plank's constant (6.63×10^{-34} J.s), m is the mass of the particle in kg, v is the velocity of the particle in m/s, $p = |\vec{P}|$ is the momentum magnitude (scalar quantity) of the particle in kg × m/s, and λ is the de-Broglie wavelength (scalar quantity) of the particle in meters.

2.3.1 The Expression for the Wavelength of de-Broglie Waves

According to the Planck's theory of radiation, the energy of photons is given by

$$E = h\nu \quad \left[\because c = \nu\lambda \Rightarrow \nu = \frac{c}{\lambda}\right].$$

$$E = \frac{hc}{\lambda} \tag{2.2}$$

Quantum Mechanics

According to the Einstein's mass–energy relation,

$$E = mc^2 \tag{2.3}$$

On substituting the value of E from equation (2.3) in equation (2.2), we get

$$m = \frac{h}{c\lambda} \Rightarrow \lambda = \frac{h}{mc} \tag{2.4}$$

Where mC is the momentum of photons. If a particle of mass m moves with the velocity v, then the de-Broglie wavelength is given by

$$\lambda = \frac{h}{mv} = \frac{h}{p}. \tag{2.5}$$

2.3.2 de-Broglie Wavelength for a Free Particle in Terms of Its Kinetic Energy

1. If a particle has kinetic energy E and moves with the velocity v having mass m, then

$$E = \frac{1}{2}mv^2 = \frac{1}{2}mv^2 \times \frac{m}{m} = \frac{1}{2}\frac{m^2v^2}{m} = \frac{1}{2}\frac{p^2}{m}.$$

$$p = \sqrt{2mE}.$$

Therefore, the de-Broglie wave is given by

$$\lambda = \frac{h}{\sqrt{2mE}}.$$

2. If q is the charge of the particle that is accelerated by a potential difference of V volts, then its kinetic energy is given by

$$E = qV.$$

Therefore, the de-Broglie wavelength is given by

$$\lambda = \frac{h}{\sqrt{2mqV}}.$$

3. If a material particle is in thermal equilibrium at temperature T, then the kinetic energy is given by

$$E = \frac{3}{2}kT.$$

Here, k is the Boltzmann constant (1.38×10^{-23} J/K) and T is the absolute temperature. Therefore, the de-Broglie wavelength is given by

$$\lambda = \frac{h}{\sqrt{3mkT}}.$$

4. If a particle moves with high velocity, which is comparable to the velocity of light, then the mass of the particle is given by

$$m = \frac{m_0}{\sqrt{1 - \frac{v^2}{c^2}}}.$$

Therefore, the de-Broglie wavelength is given by

$$\lambda = \frac{h\sqrt{1-\frac{v^2}{c^2}}}{m_0 v}.$$

5. If an electron is accelerated by a potential of V volts, then the de-Broglie wavelength is given by

$$\lambda = \frac{h}{\sqrt{2mqV}} = \frac{6.63 \times 10^{-34}}{\sqrt{2 \times 9.1 \times 10^{-31} \times 1.6 \times 10^{-19} \times V}},$$

$$\lambda = \frac{12.27 \times 10^{-10}}{\sqrt{V}} m = \frac{12.27}{\sqrt{V}} \text{Å}.$$

2.4 Properties of Matter Waves

The properties of matter waves or de-Broglie's are as follows:

1. Lighter the particle, longer will be the de-Broglie's wavelength.
2. If the velocity is smaller, then the de-Broglie's wavelength will be larger.
3. When velocity $v = 0$, $\lambda = \infty$; this is the indeterminate case. Thus, de-Broglie's wavelength is associated with a moving particle only. Matter waves are not electromagnetic waves, but they are a new kind of wave.
4. Heavier the particle, smaller will be the de-Broglie's wavelength.
5. If the velocity is higher, the de-Broglie's wavelength will be smaller.

2.5 Wave Packet or Wave Group

A moving microscopic particle is not equivalent to a single wave, but it is equivalent to a group of waves or wave group.

A wave packet is constructed by a bunch of waves that slightly differs in velocity and propagation constant. The phase of these waves is such that they interfere in the space to produce a resulting wave with the varying envelope. Therefore, a wave packet has a beginning and an end, as shown in Figure 2.1.

The velocity with which the wave packet moves known as group velocity and it is represented by v_g. The expression for group velocity is given by

$$v_g = \frac{d\omega}{dk}.$$

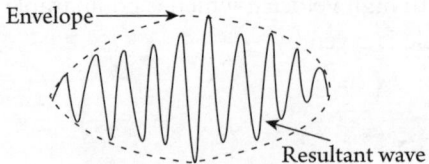

Fig. 2.1 Wave packet.

Quantum Mechanics

The average velocity of the individual monochromatic wave with which a wave packet is constructed is known as phase velocity or wave velocity, and it is denoted by v_p. The expression for phase velocity is given by

$$v_p = \frac{\omega}{k}.$$

2.6 Velocity of de-Broglie Waves

According to de-Broglie wavelength,

$$\lambda = \frac{h}{mv}. \tag{2.6}$$

According to the Plank's theory of radiation, the energy of a photon is given by

$$E = h\nu \Rightarrow \nu = \frac{E}{h}. \tag{2.7}$$

The velocity of de-Broglie waves is given by

$$v_p = \nu\lambda.$$

From equations (2.6) and (2.7),

$$v_p = \frac{E}{h} \times \frac{h}{mv},$$

$$v_p = \frac{E}{mv} \quad \left[\because E = mc^2\right],$$

$$v_p = \frac{mc^2}{mv} \Rightarrow v_p = \frac{c^2}{v}.$$

According to the above expression, the velocity of the particle v is always less than the velocity of light c. Therefore, $v_p > c$, which is not possible according to the relativistic mass variation with velocity. This contradiction was resolved by Schrödinger by postulating that a material particle is not equivalent to a single wave train. But it is equivalent to a group of wave or wave group.

2.6.1 Expression for Group Velocity and Phase Velocity

Let us consider two monochromatic waves are associated with the moving microscopic particle, that slightly differ in angular velocity and propagation constant. The displacement of two waves is given by

$$Y_1 = a\sin(\omega_1 t - k_1 x), \tag{2.8}$$

$$Y_2 = a\sin(\omega_2 t - k_2 x). \tag{2.9}$$

Since these two waves with equal amplitudes in a same medium interfere with each other to produce a wave packet. According to the superposition principle,

$$Y = Y_1 + Y_2,$$

$$Y = a\sin(\omega_1 t - k_1 x) + a\sin(\omega_2 t - k_2 x),$$

$$[\text{Since, } \sin A + \sin B = 2\sin\frac{A+B}{2}\cos\frac{A-B}{2}],$$

$$Y = 2a\left[\cos\left\{\frac{(\omega_1-\omega_2)t}{2} - \frac{(k_1-k_2)x}{2}\right\} \times \sin\left\{\frac{(\omega_1+\omega_2)t}{2} - \frac{(k_1+k_2)x}{2}\right\}\right],$$

$$Y = 2a\left[\cos\left\{\frac{\Delta\omega t}{2} - \frac{\Delta k x}{2}\right\} \times \sin\{\omega t - kx\}\right],$$

$$Y = A\sin(\omega t - kx)$$

Where $A = 2a\left[\cos\left\{\frac{\Delta\omega t}{2} - \frac{\Delta k x}{2}\right\}\right]$ is the modified amplitude of the wave packet, which is modulated both in space and time by a very slowly varying envelope with frequency $\frac{\Delta\omega t}{2}$ and propagation constant $\frac{\Delta k x}{2}$ and has a maximum value $2a$.

For a constant phase, we have
$\omega t - kx = constant.$
On differentiating the above equation w.r.t. x, we get
$$\omega - k\frac{dx}{dt} = 0,$$
$$\frac{dx}{dt} = \frac{\omega}{k},$$
$$v_p = \frac{\omega}{k}.$$

This is an expression of phase velocity or wave velocity of wave packet.
Similarly, for constant amplitude of wave packet, we have
$$\frac{\Delta\omega t}{2} - \frac{\Delta k x}{2} = constant.$$
On differentiating the above equation w.r.t. x, we get
$$\frac{dx}{dt} = \frac{\Delta\omega}{\Delta k},$$
$$v_g = \frac{\Delta\omega}{\Delta k}.$$

For limiting case, it will become
$$v_g = \frac{d\omega}{dk}.$$

This is an expression of group velocity of the wave packet. A bunch of waves that slightly differs in velocity and propagation constant generates a wave packet shown in Figure 2.2.

2.7 RELATION BETWEEN GROUP VELOCITY AND PHASE VELOCITY

We know that the phase velocity is given by

$$v_p = \frac{\omega}{k} \Rightarrow \omega = v_p.k. \tag{2.10}$$

Since we have a relation,

Quantum Mechanics

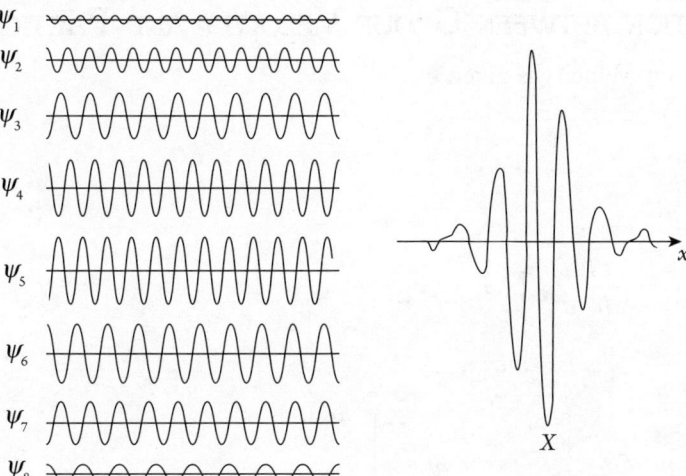

Fig. 2.2 A bunch of waves that slightly differs in velocity and propagation constant produces a wave packet.

$$v_g = \frac{d\omega}{dk} = \frac{d\omega}{d\left(\frac{2\pi}{\lambda}\right)} \quad \left[\because k = \frac{2\pi}{\lambda}\right],$$

$$V_g = \frac{d\omega}{-d\lambda \cdot \left(\frac{2\pi}{\lambda^2}\right)} = -\frac{\lambda^2}{2\pi}\frac{d\omega}{d\lambda},$$

On substituting the value of ω from equation (2.10),

$$v_g = \frac{-\lambda^2}{2\pi}\frac{d}{d\lambda}(v_p \cdot k). \tag{2.11}$$

$$v_g = \frac{-\lambda^2}{2\pi}\frac{d}{d\lambda}\left(v_p \cdot \frac{2\pi}{\lambda}\right),$$

$$v_g = -\lambda^2 \cdot \left[\frac{1}{\lambda}\frac{dv_p}{d\lambda} - \frac{1}{\lambda^2}v_p\right],$$

$$v_g = v_p - \lambda \cdot \frac{dv_p}{d\lambda}.$$

This is the relation between group velocity and phase velocity in dispersive medium.

(When wave velocity is dependent on frequency, the medium is known as dispersive medium.)

For non-dispersive medium, the wave velocity is independent of frequency. Then,

$$\frac{dv_p}{d\lambda} = 0,$$

$$v_g = v_p.$$

For non-dispersive medium, group velocity is equal to phase velocity.

2.8 RELATION BETWEEN GROUP VELOCITY AND PARTICLE VELOCITY

We know that group velocity is given by

$$v_g = \frac{d\omega}{dk} = \frac{\frac{d\omega}{dv}}{\frac{dk}{dv}}. \tag{2.12}$$

Since $\omega = 2\pi v \Rightarrow \frac{2\pi E}{h}$ $\quad [\because E = hv],$

$$\omega = \frac{2\pi mc^2}{h} \quad [\because E = mc^2],$$

$$\omega = \frac{2\pi c^2}{h} \cdot \frac{m_0}{\sqrt{1-\frac{v^2}{c^2}}} \quad \left[\because m = \frac{m_0}{\sqrt{1-\frac{v^2}{c^2}}}\right],$$

$$\omega = \frac{2\pi c^2 m_0}{h}\left(1-\frac{v^2}{c^2}\right)^{-\frac{1}{2}}. \tag{2.13}$$

On differentiating the above equation w.r.t. v, we get

$$\frac{d\omega}{dv} = \frac{2\pi m_0 c^2}{h}\left[-\frac{1}{2}\left(1-\frac{v^2}{c^2}\right)^{-\frac{3}{2}} \times -\frac{2v}{c^2}\right],$$

$$\frac{d\omega}{dv} = \frac{2\pi m_0 v}{h\left(1-\frac{v^2}{c^2}\right)^{\frac{3}{2}}}. \tag{2.14}$$

Since the propagation vector is given by

$$k = \frac{2\pi}{\lambda} = \frac{2\pi mv}{h} \quad \left[\because \lambda = \frac{h}{mv}\right],$$

$$k = \frac{2\pi m_0}{h}\left[v\left(1-\frac{v^2}{c^2}\right)^{-\frac{1}{2}}\right] \quad \left[\because m = \frac{m_0}{\sqrt{1-\frac{v^2}{c^2}}}\right].$$

On differentiating the above equation w.r.t. v, we get

$$\frac{dk}{dv} = \frac{2\pi m_0}{h}\left[-\frac{1}{2}v\left(1-\frac{v^2}{c^2}\right)^{-\frac{3}{2}} \times -\frac{2v}{c^2} + \left(1-\frac{v^2}{c^2}\right)^{-\frac{1}{2}}\right],$$

$$\frac{dk}{dv} = \frac{2\pi m_0}{h}\left(1-\frac{v^2}{c^2}\right)^{-\frac{3}{2}}\left[\frac{v^2}{c^2} + \left(1-\frac{v^2}{c^2}\right)\right],$$

$$\frac{dk}{dv} = \frac{2\pi m_0}{h}\left(1-\frac{v^2}{c^2}\right)^{-\frac{3}{2}},$$

$$\frac{dk}{dv} = \frac{2\pi m_0}{h\left(1-\frac{v^2}{c^2}\right)^{\frac{3}{2}}}. \tag{2.15}$$

On substituting the values of equations (2.14) and (2.15) in equation (2.12),

$$v_g = \frac{\left\{\dfrac{2\pi m_0 v}{h\left(1-\dfrac{v^2}{c^2}\right)^{\frac{3}{2}}}\right\}}{\left\{\dfrac{2\pi m_0}{h\left(1-\dfrac{v^2}{c^2}\right)^{\frac{3}{2}}}\right\}} = v,$$

$$v_g = v.$$

Thus, the velocity of microscopic particle is equal to the velocity of wave packet that is, the particle must be located inside the wave packet.

2.9 DAVISSON AND GERMER EXPERIMENT

The first evidence that the stream of material particle shows wave-like properties was given by the Davisson and Germer experiment in 1927 during his diffraction experiment with slow electrons.

The Davisson and Germer experiment confirms the existence of wave associated with electron by detecting de-Broglie waves but also succeeded in measuring their wavelength.

The Davisson and Germer apparatus is shown in Figure 2.3. Electrons from the heated filament are accelerated through a variable potential "V" and emerge from the electron gun "G." The electron beam falls normally on a Nickel single-target crystal.

The electrons are diffracted or scattered from the crystal in all directions because the Nickel single-target crystal acts as a three-dimensional diffraction grating, that is, the spacing between two consecutive Nickel atoms is considered as slit width. The intensity of the diffracted or scattered beam of electrons in different directions is measured by a Faraday cylinder (Detector), which is connected to a galvanometer "G" and can move on a circular scale. The whole apparatus is enclosed in an evacuated chamber.

The current, which is the measure of intensity of diffracted beam or diffracted electron beams, is plotted against the diffraction angle φ for each accelerating potential, as shown in Figure 2.4.

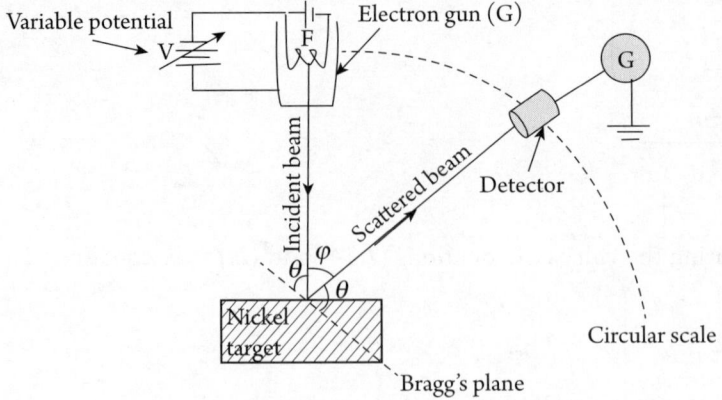

Fig. 2.3 Davisson and Germer experimental setup.

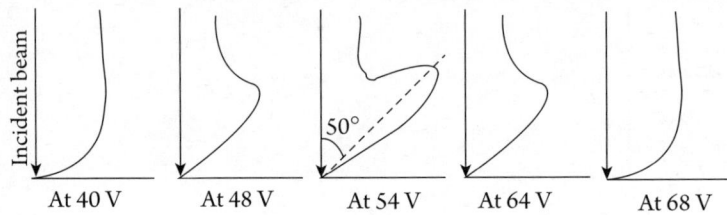

Fig. 2.4 Intensity of diffracted beam plotted against the diffraction angle φ.

From the above curves as the potential difference is increased, the peak starts shifting upward and becomes most prominent. The peak is most significant for 54 V accelerating potential and at diffraction angle 50°. Beyond 54 V, the peak gradually diminishes and becomes insignificant. Thus, this is evidence that electrons were diffracted by a target and verified the existence of an electron wave.

2.9.1 Calculations of the Wavelength

The corresponding angle (Figure 2.5) of incidence relative to the family of Bragg's planes is given by

$$\theta = \frac{180° - 50°}{2} = 65°.$$

According to the Bragg's equation for the maxima, we have

$$2d \sin \theta = n\lambda,$$

where $n = 1$, $d = 0.091$ nm, and $\theta = 65°$; on substituting these values, we get

$$\lambda = 1.65 \text{ Å} \tag{2.16}$$

According to the de-Broglie's wavelength for an electron, we have

$$\lambda = \frac{12.27}{\sqrt{V}} \text{ Å} = \frac{12.27}{\sqrt{54}} \text{ Å},$$

$$\lambda = 1.67 \text{ Å}. \tag{2.17}$$

Quantum Mechanics

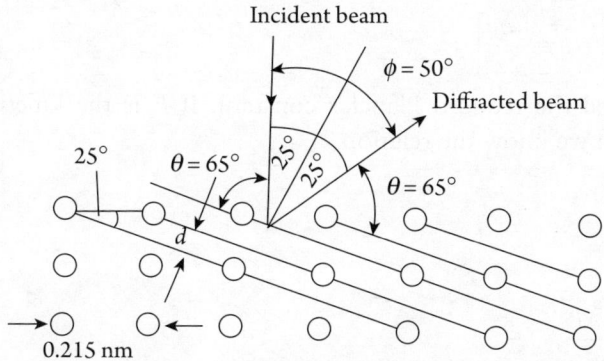

Fig. 2.5 The corresponding angles of incidence in target plate.

Thus, from equations (2.16) and (2.17), it is clear that the de-Broglie wavelength of an electron wave by hypothesis and experiment is almost the same. Therefore, de-Broglie hypothesis is verified by Davisson and Germer experiment.

2.10 HEISENBERG'S UNCERTAINTY PRINCIPLE

Heisenberg states that, *It is not possible to measure simultaneously both the position and momentum (or velocity) of a microscopic particle with an absolute accuracy or certainty.*

Mathematically, it is expressed as

$$\Delta x \cdot \Delta p \geq \frac{h}{4\pi}. \tag{2.18}$$

Where Δx is uncertainty in position, Δp is uncertainty in momentum, and h is the Planck's constant.

We know that $p = mv$ in terms of uncertainty can be written as
$\Delta p = m\Delta v$,

$$\Delta x \cdot (m\Delta v) \geq \frac{h}{4\pi},$$

$$\Delta x \cdot \Delta v \geq \frac{h}{4\pi m}. \tag{2.19}$$

So, the position of an electron or any microscopic particle and its velocity cannot be measured simultaneously with accuracy.

2.11 DERIVATION OF ENERGY–TIME UNCERTAINTY RELATION FROM POSITION–MOMENTUM UNCERTAINTY RELATION

We know that

$$\Delta x \cdot \Delta p \geq \frac{h}{4\pi},$$

or $\Delta x \cdot \Delta p \geq \dfrac{\hbar}{2}. \quad \left[\because \hbar \cong \dfrac{h}{2\pi} \right]$ (2.20)

Where $\hbar$ is called the reduced Planck's constant. If E is the kinetic energy of a free-moving particle, then we know the relation

$$E = \frac{1}{2}mv^2 = \frac{P^2}{2m},$$

$$\Delta E = \frac{2P\Delta P}{2m},$$

$$\Delta E = \frac{P\Delta P}{m} = \frac{mv\Delta P}{m},$$

$$\Delta E = v \times \Delta P = \frac{\Delta x}{\Delta t} \times \Delta P \quad \left[\because v = \frac{\Delta x}{\Delta t} \right],$$

$$\Delta E \times \Delta t = \Delta x \times \Delta P.$$

From equation (2.20), we can write

$$\Delta E \times \Delta t \geq \frac{\hbar}{2}.$$

This is the energy–time uncertainty relation.

2.12 Derivation of Heisenberg's Uncertainty Principle

The relationship between the distance Δx and the wave number spread Δk depends upon the shape of the wave group. The minimum value of product $(\Delta x \times \Delta k)$ occurs when the envelope of the group has a familiar bell shape of Gaussian function. If Δx and Δk are taken as the standard deviations of the respective functions $\psi(x)$ and $g(x)$, the minimum value of the product is $\Delta x \times \Delta k = \dfrac{1}{2}$. Since wave groups in general do not have Gaussian form, it is more realistic to express the relationship between Δx and Δk as

$$\Delta x \Delta k \geq \frac{1}{2}. \tag{2.21}$$

The de-Broglie wavelength of a particle of momentum p is $\lambda = \dfrac{h}{p}$, and the corresponding wave number is

$$k = \frac{2\pi}{\lambda} = \frac{2\pi p}{h}.$$

Hence, if an uncertainty Δk in the wave number of the de-Broglie waves associated with the particle results in an uncertainty Δp in the particle's momentum, then

$$\Delta k = \frac{2\pi \Delta p}{h}. \tag{2.22}$$

On substituting the value of Δk from equation (2.22) to equation (2.21), we get

$$\Delta x \times \frac{2\pi \Delta p}{h} \geq \frac{1}{2},$$

or $\Delta x \times \Delta p \geq \dfrac{h}{4\pi}.$ (2.23)

Thus, the above expression is a Heisenberg's uncertainty relation.

2.13 Applications of Heisenberg's Uncertainty Principle

1. **Non-existence of Electrons Inside an Atomic Nucleus**

   On the basis of Heisenberg's uncertainty principle, it can be shown that electrons cannot exist (reside) inside the nucleus. The radius of the atomic nucleus is of the order of 10^{-15} m. Therefore, if $\Delta x = 10^{-15}$ m, then the velocity of the electron is given by

   $$\Delta x \times \Delta v \geq \frac{h}{4\pi m},$$

   $$\Delta v \geq \frac{h}{4\pi m \Delta x},$$

   $$\Delta v \geq \frac{6.6 \times 10^{-34}}{4 \times 3.14 \times 10^{-15} \times 9.1 \times 10^{-31}}.$$

   $$\Delta v \geq 5.77 \times 10^{10} \text{ m/s}.$$

   Thus, the value of uncertainty in the velocity of an electron is very high; that is, it is greater than the velocity of light, which is not possible by relativistic mechanics; therefore, electrons cannot exist inside the nucleus.

2. **The Radius of Bohr's Orbit**

   If Δp and Δx are the uncertainty in the momentum and position of an electron in the orbit, then

   $$\Delta x \times \Delta p \geq \hbar \quad \left[\because \frac{h}{4\pi} = \frac{h}{2\pi} = \hbar\right],$$

   $$\Delta p \geq \frac{\hbar}{\Delta x}. \tag{2.24}$$

   The kinetic energy of an electron can be written as

   $$\Delta K = \frac{(\Delta p)^2}{2m},$$

   $$\Delta K = \frac{\hbar^2}{2m(\Delta x)^2}. \tag{2.25}$$

   The potential energy of an electron can be written as

   $$V = -\frac{Ze^2}{x},$$

   $$\Delta V = -\frac{Ze^2}{\Delta x}.$$

   So, the uncertainty in the total energy of an electron is given as

   $$\Delta E = \frac{\hbar^2}{2m(\Delta x)^2} - \frac{Ze^2}{\Delta x}. \tag{2.26}$$

   The uncertainty in the energy will be minimum if

   $$\frac{d\Delta E}{d\Delta x} = 0.$$

On differentiating equation (2.26), we get
$$\frac{d\Delta E}{d\Delta x} = -\frac{2\hbar^2}{2m(\Delta x)^3} + \frac{Ze^2}{(\Delta x)^2} = 0,$$

$$\frac{\hbar^2}{m(\Delta x)^3} = \frac{Ze^2}{(\Delta x)^2},$$

$$\Delta x = \frac{\hbar^2}{mZe^2}.$$

Therefore, the radius of Bohr's first orbit is given by
$$\Delta x = r = \frac{\hbar^2}{mZe^2} = \frac{h^2}{4\pi^2 mZe^2}.$$

This is an expression for the radius of Bohr's first orbit.

3. **Binding Energy of an Electron in an Atom**

Each atom consists of an electron moving in a certain definite orbit around a positively charged nucleus. The uncertainty in position Δx of an electron is of the order of $2a$, where a is the radius of an orbit. The corresponding uncertainty in the momentum is given by
$$\Delta x \times \Delta p \geq \frac{h}{4\pi},$$

$$\Delta p \approx \frac{h}{2\pi 2a}. \tag{2.27}$$

The non-relativistic kinetic energy of an electron is given as
$$K \approx \frac{p^2}{2m} \approx \frac{h^2}{32\pi^2 ma^2}. \qquad [\because \Delta p \approx p] \tag{2.28}$$

The potential energy in the electrostatic field of the nucleus with atomic weight Z is given by
$$V = -\frac{Ze^2}{4\pi\varepsilon_0 a}.$$

Thus, the total energy of an electron in the orbit is given by
$E = K + V,$

$$E = \frac{h^2}{32\pi^2 ma^2} - \frac{Ze^2}{4\pi\varepsilon_0 a},$$

$$E = \frac{(6.6 \times 10^{-34})^2}{32 \times (3.14)^2 \times a^2 \times 9.1 \times 10^{-31}} - \frac{Z(1.6 \times 10^{-19})^2}{4\pi \times 8.86 \times 10^{-12} a},$$

$$E = \frac{10^{-20}}{a^2} - \frac{15 \times 10^{-10} Z}{a} \text{ eV}.$$

On taking the radius of the orbit $a \approx 10^{-10}$ m, we have
$E = 1 - 15Z$ eV.

For hydrogen atom, that is, $Z = 1$,
$E = -14$ eV ≈ -13.6 eV.

It is well known that the binding energy of the outermost electron in hydrogen is −13.6 eV. It is the same as obtained by the above expression. Hence, the binding energy derived from the uncertainty principle is comparable with magnitudes.

4. **Zero-point energy for harmonic oscillator**

   According to the quantum mechanics description of a simple harmonic oscillator, the lowest energy of a simple harmonic oscillator is not zero but $\frac{1}{2}h\upsilon$, known as zero-point energy of the harmonic oscillator. This zero-point energy of an oscillator can be obtained by the uncertainty principle.

   If Δp and Δx are the uncertainty in the momentum and position of a particle of mass executing simple harmonic motion along the x-axis, then we have

   $$\Delta x \times \Delta p \geq \frac{\hbar}{2},$$

   $$\Delta p \cong \frac{\hbar}{2\Delta x}. \qquad (2.29)$$

   The kinetic energy of the particle of mass m can be written as

   $$\Delta K = \frac{(\Delta p)^2}{2m}.$$

   The potential energy is given by

   $$U = \frac{1}{2}k(\Delta x)^2.$$

   Here, k is the force constant.

   Thus, the total energy of the system = K.E. + U.

   $$E = \frac{(\Delta p)^2}{2m} + \frac{1}{2}k(\Delta x)^2. \qquad (2.30)$$

   On substituting the value of Δp from equation (2.29) in equation (2.30), we get

   $$E = \frac{\left(\frac{\hbar}{2\Delta x}\right)^2}{2m} + \frac{1}{2}k(\Delta x)^2 = \frac{\hbar^2}{8m(\Delta x)^2} + \frac{1}{2}k(\Delta x)^2. \qquad (2.31)$$

   The energy would be minimum if $\frac{\partial E}{\partial(\Delta x)} = 0$.

   Differentiating equation (2.31) w.r.t. Δx, we get

   $$\frac{\partial E}{\partial(\Delta x)} = -\frac{\hbar^2}{4m(\Delta x)^3} + \frac{1}{2}k \cdot 2\Delta x = 0.$$

   or $-\frac{\hbar^2}{4m(\Delta x)^3} + k\Delta x = 0 \Rightarrow \Delta x = \left(\frac{\hbar^2}{4mk}\right)^{1/4}.$

   On substituting the value of Δx in equation (2.31), we get

   $$E_{min} = \frac{\hbar^2}{8m}\left(\frac{4mk}{\hbar^2}\right)^{1/2} + \frac{1}{2}k\left(\frac{\hbar^2}{4mk}\right)^{1/2},$$

$$E_{min} = \frac{\hbar}{4}\left(\frac{k}{m}\right)^{1/2} + \frac{\hbar}{4}\left(\frac{k}{m}\right)^{1/2} = \frac{\hbar}{2}\left(\frac{k}{m}\right)^{1/2},$$

$$E_{min} = \frac{\hbar}{2}\left(\frac{k}{m}\right)^{1/2} = \frac{1}{2}\hbar\omega = \frac{1}{2}h\upsilon \quad \left[\because \left(\frac{k}{m}\right)^{\frac{1}{2}} = \omega\right].$$

This is known as zero-point energy of the harmonic oscillator.

5. **Finite width of spectral lines**

Heisenberg's uncertainty principle for energy E and time t is expressed as
$\Delta E \times \Delta t \geq \hbar$.

Since the lifetime of an electron in an excited state is finite (10^{-8} sec), therefore the energy level must have finite width. The energy spread of the excited energy levels is given by

$$\Delta E \cong \frac{\hbar}{\Delta t} \cong \frac{1.054 \times 10^{-34}}{10^{-8}} = 1.054 \times 10^{-24} \text{ J}.$$

This indicates that the radiation emitted during the jump of electrons from higher to lower energy levels is not truly monochromatic. Hence, spectral lines can never be sharp, but they have a finite width.

2.14 WAVE FUNCTION AND ITS PHYSICAL SIGNIFICANCE

The quantity whose variations make a de-Broglie wave is called wave function ψ (Psi).

The probability amplitude of matter waves at a given place in space (x, y, z) at a given instant of time (t) is characterized by a wave function ψ (x, y, z, t). The wave function is either real or imaginary.

According to Max Born, *the wave function itself has no physical significance but the square of its absolute magnitude $|\psi|^2 = \psi\psi^*$ (ψ^* is the complex conjugate of ψ) gives the probability of finding the particle at a particular space and time.*

$$|\psi|^2 = \psi\psi^*.$$

This expression is also known as probability density.

Since the particle is found to be somewhere in space, therefore, the probability of finding a particle in a volume dv, that is, $dv = dx\, dy\, dz$, is given by

$$\int_{-\infty}^{+\infty} |\psi|^2 \, dv = 1.$$

A wave function satisfying the above relation is called the normalized wave function.

The wave function ψ must fulfil the following conditions:
- It must be finite everywhere.
- It must be single valued.
- It must be continuous.

Quantum Mechanics

2.15 SCHRÖDINGER'S TIME-INDEPENDENT WAVE EQUATION

Consider a system of stationary waves associated with a moving particle. If the position of the particle is described by (x, y, z) and ψ is the periodic displacement of the wave, then the motion of the wave in the differential form is given by

$$\frac{\partial^2 \psi}{\partial x^2} + \frac{\partial^2 \psi}{\partial y^2} + \frac{\partial^2 \psi}{\partial z^2} - \frac{1}{u^2}\frac{\partial^2 \psi}{\partial t^2} = 0. \tag{2.32}$$

Here, u is the wave velocity of the matter waves.

The solution of the above differential equation is given by

$$\psi(x,y,z,t) = \psi_0(x,y,z)e^{-i\omega t},$$

$$\psi = \psi_0 e^{-i\omega t}. \tag{2.33}$$

On differentiating the above equation twice w.r.t. time, we get

$$\frac{\partial^2 \psi}{\partial t^2} = \psi_0 (-i\omega)^2 e^{-i\omega t},$$

$$\frac{\partial^2 \psi}{\partial t^2} = -\omega^2 \psi. \tag{2.34}$$

On substituting the value from equation (2.34) in equation (2.32), we get

$$\frac{\partial^2 \psi}{\partial x^2} + \frac{\partial^2 \psi}{\partial y^2} + \frac{\partial^2 \psi}{\partial z^2} + \frac{\omega^2 \psi}{u^2} = 0,$$

$$\omega = 2\pi\nu = \frac{2\pi u}{\lambda} \Rightarrow \frac{\omega}{u} = \frac{2\pi}{\lambda},$$

$$\frac{\partial^2 \psi}{\partial x^2} + \frac{\partial^2 \psi}{\partial y^2} + \frac{\partial^2 \psi}{\partial z^2} + \frac{4\pi^2 \psi}{\lambda^2} = 0.$$

Since $\nabla^2 = \frac{\partial^2}{\partial x^2} + \frac{\partial^2}{\partial y^2} + \frac{\partial^2}{\partial z^2}$ is known as the Laplacian operator,

$$\nabla^2 \psi + \frac{4\pi^2 \psi}{\lambda^2} = 0.$$

Since ψ represents the quantum mechanical wave function of matter waves, it must satisfy de-Broglie's equation.

$$\lambda = \frac{h}{mv} = \frac{h}{P},$$

$$\nabla^2 \psi + \frac{4\pi^2 P^2 \psi}{h^2} = 0.$$

If E is the total energy and V is the potential energy, then the kinetic energy of the particle is given as

$$E - V = \frac{1}{2}mv^2 = \frac{P^2}{2m},$$

$$P^2 = 2m(E-V).$$

On substituting the value of P^2 in the above equations,

$$\nabla^2 \psi + \frac{8\pi^2 m(E-V)\psi}{h^2} = 0 \text{ or, } \nabla^2 \psi + \frac{2m(E-V)\psi}{\hbar^2} = 0. \tag{2.35}$$

This is an expression of Schrödinger's time-independent wave equation.

2.16 Schrödinger's Wave Equation for a Free Particle

For a free moving particle, the potential energy of a particle is zero. Therefore, Schrödinger's equation for a free particle is given by

$$\nabla^2 \psi + \frac{8\pi^2 m(E-0)\psi}{h^2} = 0 \Rightarrow \nabla^2 \psi + \frac{8\pi^2 mE\psi}{h^2} = 0. \tag{2.36}$$

2.17 Schrödinger's Time-Dependent Wave Equation

In order to obtain the time-dependent wave equation, we eliminate the energy term from Schrödinger's time-independent wave equation (2.35). For it, let us differentiate equation (2.33) w.r.t. time; we get

$$\psi = \psi_0 e^{-iwt},$$

$$\frac{\partial \psi}{\partial t} = \psi_0(-iw)e^{-iwt},$$

$$\frac{\partial \psi}{\partial t} = (-iw)\psi,$$

$$\frac{\partial \psi}{\partial t} = (-i2\pi v)\psi,$$

$$\frac{\partial \psi}{\partial t} = \left(-i2\pi \frac{E}{h}\right)\psi \quad [\because E = hv],$$

$$\frac{\partial \psi}{\partial t} = \left(-i\frac{E}{\hbar}\right)\psi,$$

$$-\frac{\hbar}{i}\frac{\partial \psi}{\partial t} = E\psi.$$

On multiplying by i to the numerator and denominator to the LHS., we have

$$i\hbar \frac{\partial \psi}{\partial t} = E\psi.$$

According to Schrödinger's time-independent wave equation, we have

$$\nabla^2 \psi + \frac{2m(E-V)\psi}{\hbar^2} = 0,$$

$$\nabla^2 \psi + \frac{2m}{\hbar^2}\left(i\hbar \frac{\partial \psi}{\partial t} - V\psi\right) = 0.$$

On multiplying the above equation by $\frac{\hbar^2}{2m}$ to both sides, we get

$$\frac{\hbar^2}{2m}\nabla^2\psi + \left(i\hbar\frac{\partial\psi}{\partial t} - V\psi\right) = 0,$$

$$i\hbar\frac{\partial\psi}{\partial t} = \left(-\frac{\hbar^2}{2m}\nabla^2 + V\right)\psi \tag{2.37}$$

The above expression is Schrödinger's time-dependent wave equation. This equation can be modified in terms of the Hamiltonian operator (H) as follows:

$$H = -\frac{\hbar^2}{2m}\nabla^2 + V.$$

So, the above equation can be written as

$$E\psi = H\psi.$$

The Hamiltonian operator (H) thus coincides with the energy operator. Schrödinger's time-dependent wave equation is also known as energy conservation law of quantum mechanics.

2.18 A Particle in a One-Dimensional Box (Infinite Square Well Potential)

Consider a particle moving inside a box along x direction. The particle is bouncing back and forth between the walls of the box, as shown in Figure 2.6. The box has potential barriers at $x = 0$ and at $x = L$. In terms of boundary conditions, the potential function is given by

$$V(x) = \begin{cases} V = 0 \text{ for } 0 < x < L \\ V = \infty \text{ for } x \leq 0 \\ V = \infty \text{ for } x \geq L \end{cases}.$$

The wave function $\psi = 0$ outside the box and it exists only within the box. Schrödinger's time-independent wave equation for free particle in a one-dimensional box is given by

$$\frac{\partial^2\psi}{\partial x^2} + \frac{8\pi^2 mE\psi}{h^2} = 0 \tag{2.38}$$

Putting the value $\frac{8\pi^2 mE}{h^2} = K^2$, we have

$$\frac{\partial^2\psi}{\partial x^2} + K^2\psi = 0. \tag{2.39}$$

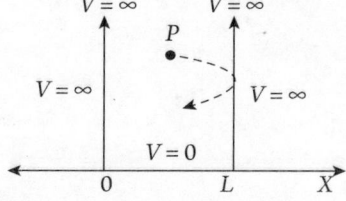

Fig. 2.6 Potential function.

The general solution of the above differential equation is given by

$$\psi = Ce^{ikx} + De^{-ikx},$$
$$\psi = C(\cos kx + i\sin kx) + D(\cos kx - i\sin kx),$$
$$\psi = (C+D)\cos kx + i(C-D)\sin kx,$$
$$\psi = A\sin(Kx) + B\cos(Kx), \tag{2.40}$$

Where $A = i(C-D)$, $B = (C+D)$ and K are constants.

On applying the boundary conditions, we get
$\psi = 0$ at $x = 0 \Rightarrow B = 0$,
$\psi = 0$ at $x = L \Rightarrow 0 = A\sin(KL)$.

Since $A \neq 0 \Rightarrow \sin(KL) = 0$ or $KL = n\pi$

or $K = \dfrac{n\pi}{L}$. [where $n = 0, 1, 2, 3, \ldots$].

On substituting the value of B and K in equation (2.40), we get

$$\psi_n = A\sin\left(\frac{n\pi}{L}x\right). \tag{2.41}$$

Since $\dfrac{8\pi^2 mE}{h^2} = K^2$

or $E = \dfrac{n^2\pi^2 h^2}{L^2 8\pi^2 m},$

$$E_n = \frac{n^2 h^2}{8mL^2}, \tag{2.42}$$

Where $n = 1, 2, 3, 4, \ldots$. For each value of n, there is an energy level, and the corresponding wave function is given by equation (2.41). Each value of E_n is called eigen value, and the corresponding wave function ψ_n is called eigen function.

2.18.1 Wave Function

The particle is certainly found within the box. Therefore, the normalized wave function of the particle is given by

$$\int_0^L |\psi|^2\, dx = 1.$$

On substituting the value of ψ from equation (2.41), we get

$$\int_0^L A^2 \sin^2\left(\frac{n\pi}{L}x\right) dx = 1,$$

$$\frac{A^2}{2}\int_0^L \left\{1 - \cos\left(\frac{2n\pi}{L}x\right)\right\} dx = 1,$$

Quantum Mechanics

$$\frac{A^2}{2}\left\{(x)_0^L - \left(\sin\frac{2n\pi}{L}x\right)_0^L\right\} = 1,$$

$$\frac{A^2}{2} \times L = 1,$$

$$A = \sqrt{\frac{2}{L}}.$$

Therefore, the wave function associated with the particle within the box is given by

$$\psi_n = \sqrt{\frac{2}{L}} \sin\left(\frac{n\pi}{L}x\right). \tag{2.43}$$

The normalized wave functions ψ_1, ψ_2, ψ_3, and so on are plotted against energy levels E_1, E_2, E_3, and so on, as shown in Figure 2.7.

2.19 A Particle in a Three-Dimensional Box

In terms of boundary conditions, the potential function for a three-dimension box of sides L_x, L_y and L_z is given by

$$V(x,y,z) = \begin{cases} V = 0 \text{ for } 0 < x < L_x, 0 < y < L_y, 0 < z < L_z \\ \quad V = \infty \text{ for } x \leq 0 \\ \quad V = \infty \text{ for } x \geq L \end{cases}.$$

The eigen function and eigen energy are given by

$$\psi_n = \sqrt{\frac{8}{L_x L_y L_z}} \sin\left(\frac{n_x \pi}{L_x}x\right) \cdot \sin\left(\frac{n_y \pi}{L_y}y\right) \cdot \sin\left(\frac{n_z \pi}{L_z}z\right), \tag{2.44}$$

$$E_n = \frac{h^2}{8m}\left[\frac{n_x^2}{L_x^2} + \frac{n_y^2}{L_y^2} + \frac{n_z^2}{L_z^2}\right]. \tag{2.45}$$

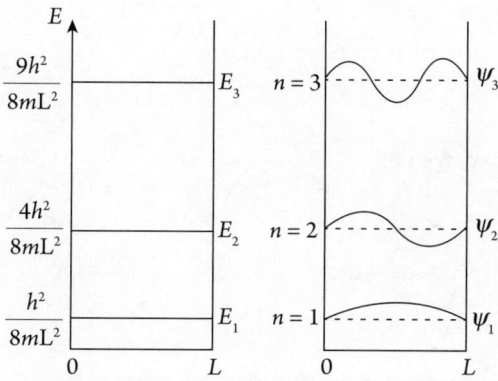

Fig. 2.7 Normalized wave function ψ_1, ψ_2, ψ_3 plotted against energy levels E_1, E_2, E_3.

2.20 Basic Postulates of Quantum Mechanics

The three basic postulates of quantum mechanics are as follows:
(a) Each dynamical variable relating to the motion of a particle can be represented by a linear operator.
(b) A linear eigenvalue equation can be always linked with each operator.
(c) In general, when a measurement of dynamical quantity "a" is made on a particle for which the wave function is ψ, we get the different values of "a" during different trials. This is due to the uncertainty principle. The most probable value of a is given by

$$\langle a \rangle = \int_{-\infty}^{\infty} \psi^* \widehat{A} \psi \, dV,$$

where $\widehat{A}$ is the operator associated with the quantity "a" and ψ^* is the complex conjugate of ψ. The quantity $\langle a \rangle$ is called the expectation value of $\widehat{A}$.

2.21 Harmonic Oscillator

Harmonic motion occurs when a system vibrates about an equilibrium configuration. A body suspended by a spring or floating in a liquid, a diatomic molecule, an atom in a crystal lattice all executes harmonic motion.

Let us consider a particle executing simple harmonic motion; then, the restoring force "F" on the particle of mass m is proportional to the displacement of the particle x from its equilibrium position and in the opposite direction. Thus,

$$F = -kx. \tag{2.46}$$

The potential energy of the particle is given by

$$V(x) = -\int_0^x F(x)\,dx = k\int_0^x x\,dx = \frac{1}{2}kx^2. \tag{2.47}$$

Substituting this value for the potential energy in Schrödinger's equation, we get

$$\frac{\partial^2 \psi}{\partial x^2} + \frac{8\pi^2 m}{h^2}\left[E - \frac{1}{2}kx^2\right]\psi = 0. \tag{2.48}$$

Let us assume the solution of the above equation is

$$\psi(x) = Ae^{-bx^2}.$$

Differentiating it w.r.t. x, we get

$$\frac{\partial \psi}{\partial x} = -2Abxe^{-bx^2}.$$

Differentiating it again w.r.t. x, we get

$$\frac{\partial^2 \psi}{\partial x^2} = -2Abe^{-bx^2} + 4Ab^2x^2e^{-bx^2}.$$

Substituting these values in equation (2.44), we get

$$-2Abe^{-bx^2} + 4Ab^2x^2e^{-bx^2} + \frac{8\pi^2 m}{h^2}\left[E - \frac{1}{2}kx^2\right]Ae^{-bx^2} = 0.$$

Removing the common factor Ae^{-bx^2}, we have

$$-2b + 4b^2 x^2 + \frac{8\pi^2 m}{h^2}\left[E - \frac{1}{2}kx^2\right] = 0,$$

$$\left\{\frac{8\pi^2 mE}{h^2} - 2b\right\} + \left\{4b^2 - \frac{4\pi^2 mk}{h^2}\right\}x^2 = 0. \tag{2.49}$$

The two sets of quantity within the brackets in equation (2.49) must be separately zero because we have to get a solution, which should be valid for all values of x.

$$\frac{8\pi^2 mE}{h^2} - 2b = 0 \Rightarrow E = \frac{bh^2}{4\pi^2 m}. \tag{2.50}$$

$$4b^2 - \frac{4\pi^2 mk}{h^2} = 0 \Rightarrow b = \frac{\pi\sqrt{mk}}{h}. \tag{2.51}$$

Substituting the value of b in equation (2.50), we get

$$E = \frac{\pi\sqrt{mk}}{h} \times \frac{h^2}{4\pi^2 m} = \frac{h}{4\pi}\sqrt{\frac{k}{m}}. \tag{2.52}$$

The classical frequency υ is given as

$$\upsilon = \frac{1}{2\pi}\sqrt{\frac{k}{m}}$$

$$E = \frac{1}{2}h\upsilon.$$

This is the ground energy or zero-point energy for the harmonic oscillator.

The general solution of equation (2.48) is of the form

$$\psi(x) = Af_n(x)e^{-bx^2},$$

where $f_n(x)$ is a polynomial in which the highest power of x is x^n.

The solution leads to the energy values,

$$E_n = \left(n + \frac{1}{2}\right)h\upsilon.$$

The above expression gives the eigen values of the harmonic oscillator by substituting the values of $n = 0, 1, 2, 3, \ldots$. The wave function and probable density of ground state and first three excited states for the harmonic oscillator are shown in Figure 2.8.

2.22 Applications of Quantum Mechanics

Important applications of quantum mechanics are in the following areas:

1. Quantum chemistry
2. Quantum optics
3. Quantum computing
4. Superconducting magnets
5. Light-emitting diodes

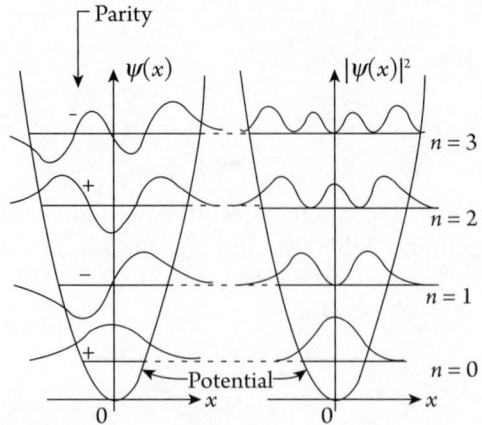

Fig. 2.8 Wave function and probable density of ground state and first three excited states.

6. Optical amplifiers
7. Lasers
8. Transistors
9. Semiconductors
10. Microprocessors
11. Medical and research imaging
12. Electron microscopy

Solved Problems

Ex. 1: Find the de-Broglie wavelength for an electron of energy V eV.

Solution:

The de-Broglie wavelength for an electron is given by

$$\lambda = \frac{h}{\sqrt{2mqV}}.$$

Here, $m = 9.1 \times 10^{-31}$ Kg, $h = 6.63 \times 10^{-34}$ J-sec, and $q = 1.6 \times 10^{-19}$ C,

$$\lambda = \frac{6.63 \times 10^{-34}}{\sqrt{2 \times 9.1 \times 10^{-31} \times 1.6 \times 10^{-19} \times V}}$$

or $\lambda = \dfrac{12.27}{\sqrt{V}}$ Å

Ex. 2: Find the de-Broglie wavelength of a 15 keV electron.

Solution:

The de-Broglie wavelength for an electron is given by,

$$\lambda = \frac{h}{\sqrt{2mE}}$$

Quantum Mechanics

Here, $m = 9.1 \times 10^{-31}$ kg, $h = 6.63 \times 10^{-34}$ J-sec, and
$E = 15 \times 10^3 \times 1.6 \times 10^{-19} = 24 \times 10^{-16}$ J

$$\lambda = \frac{6.63 \times 10^{-34}}{\sqrt{2 \times 9.1 \times 10^{-31} \times 24 \times 10^{-16}}} = 1 \times 10^{-11} m = 10 \text{Å}$$

Ex. 3: Calculate the velocity and kinetic energy of a neutron having de-Broglie wavelength 1 Å.

Solution:

According to the de-Broglie's concept of matter wave,

$$\lambda = \frac{h}{mv} \text{ or } v = \frac{h}{m\lambda}$$

Here, $h = 6.62 \times 10^{-34}$ J-sec, $m = 1.67 \times 10^{-27}$ Kg, and $\lambda = 1\text{Å} = 10^{-10}$ m

$$\therefore v = \frac{6.62 \times 10^{-34}}{1.67 \times 10^{-27} \times 10^{-10}} = 3.96 \times 10^3 \text{ m/sec.}$$

The kinetic energy of the neutron is

$$E = \frac{1}{2}mv^2 = \frac{1}{2} \times 1.67 \times 10^{-27} \times (3.96 \times 10^3)^2 = 1.309 \times 10^{-20} \text{ J}$$

or $E = \dfrac{1.309 \times 10^{-20}}{1.6 \times 10^{-19}} = 0.082$ eV.

Ex. 4: A proton is moving with a speed of 2×10^8 m/s. Find the wavelength of the matter wave associated with it.

Solution:

The wavelength of the matter wave associated with a particle is

$$\lambda = \frac{h}{p}.$$

The momentum $p = mv = \dfrac{m_0 v}{\sqrt{1 - \dfrac{v^2}{c^2}}}$.

Here, the rest mass of proton $m_0 = 1.67 \times 10^{-27}$ kg, $v = 2 \times 10^8$ m/s,

$$p = \frac{1.67 \times 10^{-27} \times 2 \times 10^8}{\sqrt{1 - \dfrac{(2 \times 10^8)^2}{(3 \times 10^8)^2}}} = \frac{1.67 \times 10^{-27} \times 2 \times 10^8}{\sqrt{5}}$$

$$\lambda = \frac{6.63 \times 10^{-34} \times \sqrt{5}}{1.67 \times 10^{-27} \times 2 \times 10^8} = 1.47 \times 10^{-15} m = 1.47 \times 10^{-5} \text{ Å}$$

Ex. 5: A particle of rest mass m_0 has a kinetic energy K; show that its de-Broglie wavelength is given by

$$\lambda = \frac{hc}{\sqrt{K(K + 2m_0 c^2)}}.$$

Hence, calculate the wavelength of an electron of kinetic energy 1 MeV. What will be the value of λ if $K \ll m_0 c^2$?

or

A particle of rest mass m_0 has a kinetic energy K. What will be the value of λ if $K \ll m_0 c^2$?

Solution:

According to the de-Broglie's concept of matter wave,

$$\lambda = \frac{h}{mv} \tag{2.53}$$

$$m = \frac{m_0}{\sqrt{1 - \frac{v^2}{c^2}}} \Rightarrow m^2 v^2 = c^2 \left(m^2 - m_0^2 \right)$$

$$mv = c\sqrt{\left(m^2 - m_0^2 \right)}.$$

Substituting this value of mv in equation (2.53), we get

$$\lambda = \frac{h}{c\sqrt{(m^2 - m_0^2)}} = \frac{hc}{c^2 \sqrt{(m^2 - m_0^2)}} = \frac{hc}{\sqrt{c^4 (m - m_0)(m + m_0)}}$$

$$\lambda = \frac{hc}{\sqrt{(m - m_0)c^2 \{(m + m_0)c^2\}}} = \frac{hc}{\sqrt{(m - m_0)c^2 \{(m - m_0)c^2 + 2m_0 c^2\}}}$$

$$\lambda = \frac{hc}{\sqrt{K(K + 2m_0 c^2)}} \quad \left[\because (m - m_0)c^2 = K \right], \tag{2.54}$$

which is a required relation.

For an electron,

$$m_0 c^2 = \frac{9.1 \times 10^{-31} \times (3 \times 10^8)^2}{1.6 \times 10^{-19}} = 0.51 \times 10^6 \text{ eV} = 0.51 \text{ MeV}.$$

For $K = 1$ MeV, then equation (2.51) becomes

$$\lambda = \frac{hc}{\sqrt{1(1 + 2 \times 0.51)}} = \frac{hc}{\sqrt{2.02}}$$

$$\lambda = \frac{6.63 \times 10^{-34} \times 3 \times 10^8}{\sqrt{2.02} \times 1.6 \times 10^{-19} \times 10^6} = 8.78 \times 10^{-13} \text{ m} = 8.78 \times 10^{-3} \text{ Å}.$$

If $K \ll m_0 c^2$, then $K + 2m_0 c^2 = 2m_0 c^2$,

$$\lambda = \frac{hc}{\sqrt{K(K + 2m_0 c^2)}} = \frac{hc}{\sqrt{K 2 m_0 c^2}}$$

$$\lambda = \frac{h}{\sqrt{2 m_0 K}}.$$

Thus, the relativistic de-Broglie wavelength is reduced to non-relativistic de-Broglie wavelength at $K \ll m_0 c^2$.

Quantum Mechanics

Ex. 6: Calculate the wavelength associated with (i) 1 MeV electron, (ii) 1 MeV proton, and (iii) 1 MeV photon.

Solution:

(i) The rest mass of an electron is

$$m_0 c^2 = \frac{9.1 \times 10^{-31} \times (3 \times 10^8)^2}{1.6 \times 10^{-19}} = 0.51 \times 10^6 \, eV = 0.51 \, MeV$$

Since the given kinetic energy (1 MeV) of an electron is greater than its rest energy (0.51 MeV), the relativistic de-Broglie formula is applicable, which is given as

$$\lambda = \frac{hc}{\sqrt{K(K+2m_0 c^2)}} = \frac{6.63 \times 10^{-34} \times 3 \times 10^8}{1.42 \times 10^6 \times 1.6 \times 10^{-19}} = 8.75 \times 10^{-13} \, m$$

$$\lambda = 8.75 \times 10^{-3} \, \text{Å}.$$

(ii) The rest mass of proton is

$$m_0 c^2 = \frac{1.67 \times 10^{-27} \times (3 \times 10^8)^2}{1.6 \times 10^{-19}} = 937 \times 10^6 \, eV = 937 \, MeV.$$

Since the given kinetic energy (1 MeV) of proton is less than its rest energy (937 MeV), the non-relativistic de-Broglie formula is applicable, which is given as

$$\lambda = \frac{h}{\sqrt{2m_0 K}} = \frac{6.63 \times 10^{-34}}{\sqrt{2 \times 1.67 \times 10^{-27} \times 1 \times 10^6 \times 1.6 \times 10^{-19}}}$$

$$\lambda = 2.87 \times 10^{-4} \, \text{Å}.$$

(iii) The rest mass of the photon is zero, and hence the rest mass–energy is also zero. Thus, the energy of a photon is entirely kinetic and is given by

$$E = h\nu = \frac{hc}{\lambda} \text{ or } \lambda = \frac{hc}{E}.$$

Here, $E = 1$ MeV,

$$\lambda = \frac{6.63 \times 10^{-34} \times 3 \times 10^8}{1 \times 10^6 \times 1.6 \times 10^{-19}} = 1.24 \times 10^{-12} \, m = 1.24 \times 10^{-2} \, \text{Å}$$

Ex. 7: An electron has de-Broglie wavelength 2.0×10^{-12} m. Find its kinetic energy. Also find the phase and group velocity of its de-Broglie waves.

Solution:
Since the electron has very small wavelength, relativistic correction should be applied. The relativistic mass m of an electron is

$$m = \frac{m_0}{\sqrt{1-\frac{v^2}{c^2}}} \Rightarrow v = c\sqrt{1-\frac{m_0^2 \times c^2}{m^2 \times c^2}} \Rightarrow v = c\sqrt{1-\frac{E_0^2}{E^2}},$$

where E is the total energy, and E_0 is the rest energy of the moving electron.

Total energy E of the electron is given by

$$E = \sqrt{(pc)^2 + (m_0 c^2)^2}$$

$$pc = \frac{hc}{\lambda} = \frac{6.63 \times 10^{-34} \times 3 \times 10^8}{2 \times 10^{-12} \times 1.6 \times 10^{-19}} = 621.56 \text{ keV}$$

$$m_0 c^2 = \frac{9.1 \times 10^{-31} \times (3 \times 10^8)^2}{1.6 \times 10^{-19}} = 511 \text{ keV}.$$

Thus, the kinetic energy of the electron $K = E - m_0 c^2 = 804.64 - 511 = 293.64$ keV,

$$v = c\sqrt{1 - \frac{E_0^2}{E^2}} = c\sqrt{1 - \frac{(511)^2}{(804.64)^2}} = 0.7724c.$$

Thus, the group velocity of the de-Broglie waves $v_g = v = 0.7724c$.

The phase velocity of the de-Broglie waves $v_p = \frac{c^2}{v_g} = 1.29c$.

Ex. 8: Calculate the uncertainty in the velocity of an electron, which is confined in a 10 Å box.

Solution:

According to the Heisenberg uncertainty principle,

$$\Delta x \Delta p \geq \hbar \text{ or } \Delta p \geq \frac{\hbar}{\Delta x}$$

$$\Delta p \approx \frac{6.63 \times 10^{-34}}{2 \times 3.14 \times 10 \times 10^{-10}} \approx 1.054 \times 10^{-25} \text{ kgm/s}$$

Uncertainty in velocity, $m\Delta v = \Delta p$ or $\Delta v = \frac{\Delta p}{m}$,

Here, mass of the electron, $m = 9.1 \times 10^{-31}$ kg

$$\Delta v \approx \frac{1.054 \times 10^{-25}}{9.1 \times 10^{-31}} \approx 0.116 \times 10^6 \text{ m/s}.$$

Ex. 9: An electron has a speed of 5×10^3 m/s within the accuracy of 0.003%. Calculate the uncertainty in the position of the electron.

Solution:

According to the Heisenberg uncertainty principle,

$$\Delta x \Delta p \geq \hbar$$

The momentum of the electron $p = mv = 9.0 \times 10^{-31} \times 5 \times 10^3 = 4.5 \times 10^{-27}$ kgm/s.

The uncertainty in the value of p is 0.003% of this value, that is,

$$\Delta p = \frac{0.003}{100} \times 4.5 \times 10^{-27} = 1.35 \times 10^{-31} \text{ kgm/s}.$$

Therefore, the uncertainty in the position of this electron is

$$\Delta x = \frac{h}{2\pi \Delta p} = \frac{6.63 \times 10^{-34}}{2 \times 3.14 \times 1.35 \times 10^{-31}} = 7.82 \times 10^{-4} \text{ m}$$

Ex. 10: A hydrogen atom, say, has a radius of 0.5 Å. Calculate the kinetic energy needed by an electron to be confined to the atom.

Quantum Mechanics

Solution:

The radius of the atom is 0.5 Å; therefore, the uncertainty in the position is

$\Delta x = 0.5 \text{Å} = 0.5 \times 10^{-10}$ m.

According to the uncertainty principle, $\Delta x \Delta p_x \geq \hbar$ or $\Delta p_x \geq \dfrac{h}{2\pi \Delta x}$,

$p \approx \Delta p_x = \dfrac{6.63 \times 10^{-34}}{2 \times 3.14 \times 5.0 \times 10^{-11}} \geq 2.1 \times 10^{-24}$ kgm/s

The kinetic energy needed by an electron to be confined to the atom is

$\text{K.E.} = \dfrac{p^2}{2m} = \dfrac{(2.1 \times 10^{-24})^2}{2 \times 9.1 \times 10^{-31}} = \dfrac{4.1 \times 10^{-17}}{18.2 \times 1.6 \times 10^{-19}} = 15.1$ eV.

Ex. 11: Establish the relation $v_g \cdot v_p = c^2$.

Solution:

According to de-Broglie wavelength,

$$\lambda = \dfrac{h}{mv}. \qquad (2.55)$$

According to the Plank's theory of radiation, the energy of a photon is given by

$$E = h\nu \Rightarrow \nu = \dfrac{E}{h}. \qquad (2.56)$$

The velocity of de-Broglie waves is given by

$v_p = \nu \lambda$.

From equations (2.55) and (2.56),

$v_p = \dfrac{E}{h} \times \dfrac{h}{mv}$

$v_p = \dfrac{E}{mv} \qquad [\because E = mc^2]$

$$v_p = \dfrac{mc^2}{mv} \Rightarrow v_p = \dfrac{c^2}{v}. \qquad (2.57)$$

Since $v_g = v$, put the value in equation (2.57).

Therefore, $v_p = \dfrac{c^2}{v_g} \Rightarrow v_g \cdot v_p = c^2$.

Ex. 12: Calculate the energy difference between the ground state and first excited state for an electron if the length of the box is 10^{-8} cm.

Solution:

We have $E_n = \dfrac{n^2 h^2}{8mL^2}$.

The ground energy state corresponds to $n = 1$, and given $L = 10^{-8}$ cm $= 10^{-10}$ Å, then

$E_1 = \dfrac{1^2 h^2}{8mL^2} = \dfrac{(6.63 \times 10^{-34})^2}{8 \times 9.1 \times 10^{-31} \times (10^{-10})^2} = 0.603 \times 10^{-17}$ J.

In terms of eV,

$$E_1 = \frac{0.603 \times 10^{-17}}{1.6 \times 10^{-19}} = 37.7 \text{ eV}.$$

The first excited energy state corresponds to $n = 2$ and is

$$E_2 = \frac{2^2 h^2}{8mL^2} = 4E_1 = 4 \times 37.7 = 150.8 \text{ eV}.$$

Now the difference is

$$\Delta E = E_2 - E_1 = 150.8 - 37.7 = 113.1 \text{ eV}.$$

Previous Year Questions (University Examination)

1. What is the wave–particle duality?
2. What are de-Broglie's matter waves?
3. Determine the de-Broglie's wavelength of electron.
4. Determine the de-Broglie's wavelength of photon.
5. Discuss few important properties of matter waves.
6. What is the difference between matter waves and electromagnetic waves?
7. Determine the de-Broglie wavelength of Helium atom.
8. What do you mean by phase and group velocities?
9. What is the relation between group velocity and phase velocity?
10. What is uncertainty principle?
11. What is the physical significance of uncertainty principle?
12. What is time-independent Schrödinger's equation?
13. What is time-dependent Schrödinger's equation?
14. What is the physical significance of wave function?
15. Discuss the result of a one-dimensional box.
16. What do you mean by eigen function and eigen values?
17. Establish the relation $v_g \cdot v_p = c^2$.
18. Write short notes on wave–particle duality with examples of suitable experiments.
19. What are matter waves? Show that the wavelength λ associated with a particle of mass m and kinetic energy E is given by $\lambda = \dfrac{h}{\sqrt{2mE}}$.
20. Describe Davisson and Germer experiment to demonstrate the wave nature of particles.
21. Explain the difference between wave velocity and group velocity in the wave motion. Obtain an expression for the group velocity in a dispersive medium.
22. Show that the phase velocity of the de-Broglie wave is greater than the velocity of light, but the group velocity is equal to the velocity of the particle with which the waves are associated.

Quantum Mechanics

23. Distinguish between the phase velocity (v_p) and group velocity (v_g) of a wave packet. Prove that $v_p v_g = c^2$.

24. State and explain Heisenberg's uncertainty principle. Using this principle, show that the electron cannot reside in an atomic nucleus.

25. State and explain Heisenberg's uncertainty principle. Using this principle, find the binding energy of an electron in an atom.

26. State and explain Heisenberg's uncertainty principle. Apply this to find the radius of the Bohr's first orbit.

27. Derive Schrödinger's time-independent and time-dependent wave equation. What is the physical significance of state function ψ used in this equation? What conditions must it fulfil?

28. Derive an expression for the wave function and energy state of a particle confined in a one-dimensional potential box using Schrödinger's wave equation.

29. A particle of rest mass m_0 has a kinetic energy K. Show that its de-Broglie wavelength is given by
$$\lambda = \frac{hc}{\sqrt{K(K+2m_0 c^2)}}.$$
What will be the value of λ if $K \ll m_0 c^2$.

30. Can a photon and an electron of the same momentum have the same wavelength? Compare their wavelengths if the two have the same energy.

31. A particle is in motion along a line between $x = 0$ and $x = L$ with zero potential energy. At points for which $x < 0$ and $x < L$, the potential energy is infinite. The wave function for the particle in nth state is given by $\psi_n = A \sin(n\pi x/L)$. Find the expression for the normalized wave function.

32. Find the wave function and the energy eigen values for the particle in a one-dimensional potential box. Find the energy of an electron moving in one dimension in an infinitely high potential box of width 1 Å.

33. Derive Schrödinger's time-independent and time-dependent equation. What is the physical significance of state function ψ used in this equation? What conditions must it fulfil?

34. Find the de-Broglie wavelength for an electron of energy V eV.

35. A particle charge q and mass m are accelerated from rest through a potential difference V. Calculate its de-Broglie wavelength, if the particle is an electron and potential difference $V = 50$ V.

36. What is de-Broglie wavelength of an electron accelerated from rest through a potential difference of 100 V.

37. Find the de-Broglie wavelength of a 15 keV electron.

38. Calculate the velocity and kinetic energy of a neutron having de-Broglie wavelength 1 Å.

39. Calculate the wavelength associated with an electron accelerated to a potential difference of 1.25 keV.

40. A proton is moving with a speed of 2×10^8 m/s. Find the wavelength of the matter wave associated with it.

41. Calculate the de-Broglie wavelength associated with a proton moving with a velocity of equal to 1/20th of the velocity of light.

42. An electron and a photon each have a wavelength of 2.0 Å. Compare their (a) momentum, (b) total energies, and (c) ratio of kinetic energies.
43. Calculate the de-Broglie wavelength of neutron of energy of 28.8 eV.
44. Calculate the de-Broglie wavelength of an α-particle accelerated through a potential difference of 200 V.
45. Find the de-Broglie wavelength of (a) a ball of mass 1.0 kg moving with a speed of 1.0 m/s, (b) an electron traveling with a speed of 10^6 m/s.
46. Calculate the de-Broglie wavelength of a neutron having kinetic energy of 1eV.
47. Calculate the de-Broglie wavelength of a neutron of energy 12.8 MeV.
48. What voltage must be applied to an electron microscope to produce electrons of wavelength 0.40 Å?
49. The energy of a particle at absolute temperature T is of the order of kT. Calculate the wavelength of thermal neutrons at 27°C.
50. Calculate the de-Broglie wavelength associated with nitrogen atom at 3.0 atmospheric pressure and at temperature 27°C. Given mass of N_2 atom is 4.65×10^{-26} kg.
51. An enclosure filled with helium is heated to a temperature of 400 K. A beam of helium atoms emerges out of the enclosure. Calculate the de-Broglie wavelength corresponding to He atoms. Mass of He atom = 6.7×10^{-27} kg.
52. A beam of electrons of kinetic energy 100 eV passes through a thin metal foil. On a screen at a distance of 20 cm, the most intense ring observed has a diameter 2.44 cm. Calculate the spacing of the related lattice planes in the metal.
53. The radius of the first Bohr orbit in hydrogen atom is 0.53×10^{-10} m. Find the velocity of the electron in that orbit using de-Broglie theory.
54. What will be the kinetic energy of an electron if its de-Broglie wavelength equals the wavelength of the yellow line of sodium (5896 Å).
55. Calculate the wavelength associated with a (i) 1 MeV electron, (ii) 1 MeV proton, and (iii) 1 MeV photon.
56. An electron has a de-Broglie wavelength 2.0×10^{-12} m. Find its kinetic energy. Also find the phase and group velocity of its de-Broglie waves.
57. If the uncertainty in the location of the particle is equal to its de-Broglie wavelength, what is the uncertainty with velocity?
58. Calculate the smallest possible uncertainty in the position of electron moving with velocity $v = 3 \times 10^7$ m/sec.
59. An electron is confined to a box of length 1.1×10^{-8} m. Calculate the minimum uncertainty in its velocity.
60. Calculate the uncertainty in the velocity of an electron that is confined in a 10 Å box.
61. An electron has a speed of 1.05×10^4 m/s within the accuracy of 0.01%. Calculate the uncertainty in the position of the electron.
62. Hydrogen atom, say, has a radius of 0.5 Å. Calculate the kinetic energy needed by an electron to be confined to the atom.

Quantum Mechanics

63. Calculate the lowest energy of a neutron confined to the nucleus where the nucleus is considered a box with a size of 10^{-14} m.

64. An electron has a speed of 5×10^3 m/s within the accuracy of 0.003%. Calculate the uncertainty in the position of the electron.

Multiple Choice Questions

1. Light shows
   (a) Particle nature
   (b) Wave nature
   (c) Both wave and particle nature
   (d) None of the above

2. Matter waves are not associated with
   (a) Microscopic particles with less mass
   (b) Microscopic particles
   (c) Very fast-moving microscopic particles
   (d) Stationary particles

3. Matter waves are
   (a) Associated with stationary microscopic particles
   (b) Associated with moving microscopic particles
   (c) Associated with large-sized stationary objects
   (d) None of the above

4. de-Broglie wavelength is
   (a) $\lambda = \dfrac{mv}{h}$
   (b) $\lambda = \dfrac{h}{mv}$
   (c) Both (a) and (b)
   (d) None of the above

5. A moving microscopic particle is equivalent to
   (a) Single wave
   (b) Group of waves
   (c) Group of particles
   (d) Both (b) and (c)

6. Group velocity is defined as
   (a) The individual average velocity of waves
   (b) The velocity with which wave-packet moves
   (c) The velocity of light
   (d) None of the above

7. Phase velocity is defined as
   (a) The individual average velocity of waves with which the wave-packet is constructed
   (b) The velocity with which the wave-packet moves
   (c) The velocity of light
   (d) None of the above

8. The relation between group velocity and phase velocity in dispersive medium is
   (a) $v_g = v_p - \lambda \cdot \dfrac{dv_p}{d\lambda}$
   (b) $v_g = v_p$
   (c) $v_g = v_p - \dfrac{dv_p}{d\lambda}$
   (d) $v_g = v_p - \mu \cdot \dfrac{dv_p}{d\lambda}$

9. The relation between group velocity and phase velocity in non-dispersive medium is
   (a) $v_g = v_p - \lambda \cdot \dfrac{dv_p}{d\lambda}$
   (b) $v_g = v_p$
   (c) $v_g = v_p - \dfrac{dv_p}{d\lambda}$
   (d) $v_g = v_p - \mu \cdot \dfrac{dv_p}{d\lambda}$

10. The relation between group velocity and particle velocity is
    (a) $v_g = v - \lambda \cdot \dfrac{dv}{d\lambda}$
    (b) $v_g = v$
    (c) $v_g = v - \mu \cdot \dfrac{dv}{d\lambda}$
    (d) $v_g = v - \dfrac{dv}{d\lambda}$

11. The Davisson and Germer experiment confirms
    (a) The existence of stationary ether
    (b) The existence of wave associated with electron
    (c) Ether drag hypothesis
    (d) Both (a) and (c)

12. In Davisson and Germer experiment, the Nickel target acts as
    (a) Source
    (b) Observer
    (c) Three-dimensional diffraction grating
    (d) Filament

13. In Davisson and Germer experiment, the diffraction of electrons is obtained at
    (a) 50° with accelerating potential 45 V
    (b) 50° with accelerating potential 54 V
    (c) 54° with accelerating potential 50 V
    (d) 45° with accelerating potential 50 V

14. The corresponding angle in Davisson and Germer experiment is
    (a) 50°
    (b) 45°
    (c) 65°
    (d) 180°

15. Bragg's equation for the maxima is
    (a) $2n\cos\theta = d\lambda$
    (b) $2d\sin\theta = n\lambda$
    (c) $2d\tan\theta = n\lambda$
    (d) None of these

16. The Davisson and Germer experiment verifies
    (a) Ether hypothesis
    (b) de-Broglie hypothesis
    (c) Heisenberg's uncertainty principle
    (d) Bragg's equation

17. Heisenberg's uncertainty principle is mathematically expressed as
    (a) $\Delta x \times \Delta p \geq \dfrac{4\pi}{h}$
    (b) $\Delta x \times \Delta p \geq \dfrac{h}{4\pi}$
    (c) $\Delta x \times \Delta p \geq \dfrac{16h}{4\pi}$
    (d) None of the above

Quantum Mechanics

18. Which of the following is the application of Heisenberg's uncertainty principle?
    (a) Non-existence of electron inside atomic nucleus
    (b) Binding energy of an electron in an atom
    (c) Radius of Bohr's first orbit
    (d) All of the above

19. Wave function ψ is
    (a) The quantity whose variations makes a de-Broglie wave
    (b) Ether medium
    (c) Wave-packet
    (d) Particle probability density

20. The physical significance of wave function ψ is
    (a) Nothing
    (b) $|\psi|^2$
    (c) ψ
    (d) None of the above

21. $|\psi|^2$ gives
    (a) Group velocity
    (b) Particle probability density
    (c) Phase velocity
    (d) All of the above

22. The wave function ψ must fulfil the condition
    (a) It must be finite everywhere
    (b) It must be single valued
    (c) It must be continuous
    (d) All of the above

23. The characteristics of wave function ψ are
    (a) Real function, finite, and discontinuous
    (b) Complex function, single valued, finite, and continuous
    (c) Complex function, infinite, and discontinuous
    (d) Complex function, single valued, and infinite

24. The normalization of wave function is always possible if
    (a) $\int_{-\infty}^{+\infty} |\psi|^2 dv = \text{Infinite}$
    (b) $\int_{-\infty}^{+\infty} |\psi|^2 dv = \text{Finite}$
    (c) $\int_{-\infty}^{+\infty} |\psi|^2 dv = 0$
    (d) All of these

25. Schrödinger's time-independent wave equation is applicable for the particles with
    (a) Constant energy
    (b) Variable energy
    (c) Only constant potential energy
    (d) All of these

26. Schrödinger's time-independent wave equation is applicable for the particles with
    (a) Stationary particles
    (b) Relativistic particles
    (c) Non-relativistic particles
    (d) All of the above

27. Schrödinger's time-independent wave equation is
    (a) $\nabla^2 \psi + \dfrac{2mE\psi}{\hbar^2} = 0$
    (b) $i\hbar \dfrac{\partial \psi}{\partial t} = \left(-\dfrac{\hbar^2}{2m} \nabla^2 + V \right) \psi$
    (c) $E\psi = H\psi$
    (d) All of the above

28. Schrödinger's time-dependent wave equation is
    (a) $\nabla^2 \psi + \dfrac{2mE\psi}{\hbar^2} = 0$
    (b) $i\hbar \dfrac{\partial \psi}{\partial t} = \left(-\dfrac{\hbar^2}{2m}\nabla^2 + V\right)\psi$
    (c) $E\psi = H\psi$
    (d) Both (b) and (c)

29. The total energy operator is represented by
    (a) $i\hbar \dfrac{\partial}{\partial t}$
    (b) $-\dfrac{\hbar^2}{2m}\nabla^2$
    (c) ∇^2
    (d) None of the above

30. The Hamiltonian operator is represented by
    (a) $i\hbar \dfrac{\partial}{\partial t}$
    (b) $-\dfrac{\hbar^2}{2m}\nabla^2$
    (c) $-\dfrac{\hbar^2}{2m}\nabla^2 + V$
    (d) None of the above

31. The eigen value and eigen function of a particle in a one-dimensional box is given as
    (a) $E_n = \dfrac{n^2 h^2}{5mL^2}$ and $\psi_n = \sqrt{\dfrac{2}{L}} \sin\left(\dfrac{2n\pi}{L}x\right)$
    (b) $E_n = \dfrac{n^2 h^2}{8mL^2}$ and $\psi_n = \sqrt{\dfrac{2}{L}} \sin\left(\dfrac{n\pi}{L}x\right)$
    (c) $E_n = \dfrac{3n^2 h^2}{8mL^2}$ and $\psi_n = \sqrt{\dfrac{2}{L}} \sin\left(\dfrac{3n\pi}{L}x\right)$
    (d) None of the above

32. The ground (zero-point) energy level for a harmonic oscillator is
    (a) $E = \dfrac{3}{2}h\upsilon$
    (b) $E = \dfrac{5}{2}h\upsilon$
    (c) $E = \dfrac{7}{2}h\upsilon$
    (d) $E = \dfrac{1}{2}h\upsilon$

33. The Eigen function for a harmonic oscillator is given by
    (a) $\psi(x) = A f_n(x) e^{-bx^2}$
    (b) $\psi_n = \sqrt{\dfrac{2}{L}} \sin\left(\dfrac{3n\pi}{L}x\right)$
    (c) $\psi_n = \sqrt{\dfrac{2}{L}} \sin\left(\dfrac{2n\pi}{L}x\right)$
    (d) None of these

34. Which one of the following is the basic postulate of quantum mechanics?
    (a) Each dynamical variable of a moving particle can be represented by a linear operator.
    (b) A linear eigenvalue equation can be always linked with each operator.
    (c) The total energy operator coincides with the Hamiltonian operator.
    (d) Both (a) and (b)

35. The velocity and kinetic energy of a neutron having de-Broglie wavelength 1 Å are
    (a) 3.96×10^3 m/sec and 1.309×10^{-20} J
    (b) 2.96×10^3 m/sec and 2.309×10^{-20} J
    (c) 1.96×10^3 m/sec and 3.309×10^{-20} J
    (d) 0 m/sec and ∞ J

Quantum Mechanics

36. A proton is moving with a speed of 2×10^8 m/s. What is the wavelength of the matter wave associated with it?
    (a) 3.25×10^{-5} Å
    (b) 2.22×10^{-5} Å
    (c) 1.47×10^{-5} Å
    (d) 0 Å

37. The de-Broglie wavelength associated with 1 MeV electron is
    (a) 8.75×10^{-3} Å
    (b) 7.81×10^{-3} Å
    (c) 6.32×10^{-3} Å
    (d) 5.41×10^{-3} Å

38. What is the uncertainty in the velocity of an electron which is confined in a 10 Å box?
    (a) 0.021×10^5 m/s
    (b) 0.116×10^6 m/s
    (c) 1.116×10^7 m/s
    (d) 2.116×10^8 m/s

39. An electron has a speed of 5×10^3 m/s within the accuracy of 0.003%. What is the uncertainty in the position of the electron?
    (a) 4.56×10^{-4} m
    (b) 3.12×10^{-4} m
    (c) 7.82×10^{-4} m
    (d) 5.21×10^{-4} m

40. What is the energy difference between the ground state and the first excited state for an electron if the length of the box is 10^{-8} cm?
    (a) 113.1 eV
    (b) 126.4 eV
    (c) 150.5 eV
    (d) 0 eV

41. Hydrogen atom, say, has a radius of 0.5 Å. What amount of kinetic energy needed by an electron is to be confined to the atom?
    (a) 10.2 eV
    (b) 11.3 eV
    (c) 13.5 eV
    (d) 15.1 eV

42. An electron and proton are accelerated through same potential. The ratio of de-Broglie wavelength $\dfrac{\lambda_e}{\lambda_p}$ is equal to
    (a) 1
    (b) $\dfrac{m_e}{m_p}$
    (c) $\dfrac{m_p}{m_e}$
    (d) $\sqrt{\dfrac{m_p}{m_e}}$

43. For a particle in an infinite potential well in its first excited state, the probability of finding the particle at the center of the box is
    (a) 0
    (b) 0.5
    (c) 0.25
    (d) 1

44. For a particle in an infinite potential well in its ground state, the probability of finding the particle at the center of the box is
    (a) 0
    (b) 0.5
    (c) 0.25
    (d) 1

45. For a particle in an infinite potential well in its first excited state, the probability of finding the particle is maximum at
    (a) $X = 0$
    (b) $X = L$
    (c) $X = L/2$
    (d) $X = L/4$ and $X = 3L/4$

46. For a particle in an infinite potential well in its ground state, the probability of finding the particle is maximum at
    (a) $X = 0$
    (b) $X = L$
    (c) $X = L/2$
    (d) $X = L/4$ and $X = 3L/4$

47. If the wave group is narrow, then there is
    (a) Large uncertainty in momentum
    (b) Small uncertainty in momentum
    (c) No uncertainty in momentum
    (d) None of the above

48. An electron, a proton, and an alpha particle are enclosed in three rigid boxes of same width. The energy levels will be closer together for
    (a) Electron
    (b) Proton
    (c) Alpha particle
    (d) None of the above

49. The de-Broglie wavelength of an electron at rest is
    (a) 0
    (b) ∞
    (c) 100 m
    (d) 1 Km

50. If the group velocity of a particle is 3×10^6 m/s, its phase velocity is
    (a) 3×10^6 m/s
    (b) 3×10^{10} m/s
    (c) 3×10^8 m/s
    (d) 3×10^{20} m/s

BIBLIOGRAPHY

1. Feynman, R., Hibbs, A., and Styer, D. (2010). *Quantum Mechanics and Path Integrals*. Dover Publications.
2. Dirac, P. (1981). *The Principles of Quantum Mechanics* (4th ed.). Oxford Science Publications.
3. Sakurai, J. J. and Jim Napolitano, J. (2017). *Modern Quantum Mechanics* (2nd ed.). Cambridge University Press.
4. Shankar, R. (2011). *Principles of Quantum Mechanics* (2nd ed.). Plenum Press.
5. Landau, L. D. and Lifshitz, E. M. (1977). *Course of Theoretical Physics Volume 3—Quantum Mechanics: Non-Relativistic Theory* (3rd ed.). Pergamon Press.
6. von Neumann, J. (2018). *Mathematical Foundations of Quantum Mechanics*. Princeton University Press.
7. McIntyre, H. D. (2012). *Quantum Mechanics: A Paradigms Approach* (1st ed.). Pearson Addison-Wesley.
8. Townsend, J. (2012). *A Modern Approach to Quantum Mechanics* (2nd ed.). University Science Books.
9. Gasiorowicz, S. (2003). *Quantum Physics* (3rd ed.). Wiley.
10. Stone, J. (2020). *The Quantum Menagerie: A Tutorial Introduction to the Mathematics of Quantum Mechanics* (1st ed.). Sebtel Press.

Keys

1. (c)	2. (d)	3. (b)	4. (b)	5. (b)	6. (b)	7. (a)	8. (a)	9. (b)	10. (b)
11. (b)	12. (c)	13. (b)	14. (c)	15. (b)	16. (b)	17. (b)	18. (d)	19. (a)	20. (a)
21. (b)	22. (d)	23. (b)	24. (b)	25. (c)	26. (c)	27. (a)	28. (d)	29. (a)	30. (c)
31. (b)	32. (d)	33. (a)	34. (d)	35. (a)	36. (c)	37. (a)	38. (b)	39. (c)	40. (a)
41. (d)	42. (d)	43. (a)	44. (d)	45. (d)	46. (c)	47. (a)	48. (c)	49. (b)	

50. (b) [Hint: Use this relation for calculation $v_g \cdot v_p = c^2$.] If a non-dispersive medium is given in the question, then use the $v_g = v_p$ relation.

CHAPTER 3

Electromagnetic Theory

3.1 INTRODUCTION

Humans have had a lengthy history of understanding electricity and magnetism. The tangible characteristics of light have also been studied. But in contrast to optics, electricity and magnetism—now known as electromagnetics—have been believed to be governed by different physical laws. This makes sense because optical physics as it was previously understood by humans differs significantly from the physics of electricity and magnetism. For instance, the ancient Greeks and Asians were aware of lode stone between 600 and 400 BC. Since 200 BC, China has been using the compass. The Greeks described static energy as early as 400 BC. But these oddities had no real effect until the invention of telegraphy. The voltaic cell or galvanic cell was created by Luigi Galvani and Alesandro Volta in the late 1700s, which led to the development of telegraphy. It quickly became clear that information could be transmitted using just two wires attached to a voltaic cell. The development of telegraphy was therefore prompted by this potential by the early 1800s. To learn more about the characteristics of electricity and magnetism, Andre-Marie Ampere (1823) and Michael Faraday (1838) conducted tests. Ampere's law and Faraday's law are consequently called after them. In order to comprehend telegraphy better, Kirchhoff voltage and current rules were also established in 1845. The data transmission mechanism was not well comprehended despite these laws. The cause of the data transmission signal's distortion was unknown. The ideal signal would alternate between ones and zeros, but the digital signal quickly lost its shape along a data transmission line. The mathematical theory of electricity and magnetism was not complete until James Clerk Maxwell added the final term to Ampere's law, the term involving displacement current, in 1865. The term "generalized Ampere's law" now refers to Ampere's law. In recognition of James Clerk Maxwell, the entire collection of equations is now known as Maxwell's equations.

Maxwell's theory was a resounding triumph because it foresaw wave phenomena that have been seen along telegraph lines. Heinrich Hertz conducted an experiment in 1888 to demonstrate that electromagnetic (EM) fields can travel through space and travel across a room. The permittivity and permeability of matter were measured experimentally, and it was also determined that EM waves travel very quickly. However, astronomical measurements have long since made it possible to determine the speed of light (Roemer, 1676). It is also known that interference events in light can be observed. It was determined, after putting these bits of knowledge together, that electricity, magnetism, and optics are all subject to the same physical rule, or the Maxwell's equations, and that optics and electromagnetics are combined into one.

When charges are in motion, the electric and magnetic fields are associated with this motion. These associated electric and magnetic fields have space and time variation, that is, time varying fields known as EM waves shown in Figure 3.1. This phenomenon is called electromagnetism.

3.2 Scalar and Vector Fields

A function that describes a physical quantity at every location in space is called a field. A scalar field is one that has a unique numerical representation at every point in space, such as temperature $T(x, y, z, t)$ in Kelvin, density $\rho(x, y, z, t)$ in kg m^{-3}, pressure $P(x, y, z, t)$ in N m^{-2}, and so on. They are steady scalar fields if they change over time. However, they are time-varying vector fields if they do change over time.

A vector field is a function with a vector value that has a vector attached to it at every location in its domain. A field of flow at every point in space, for instance, can be used to indicate the flow velocity (m s^{-1}) in a fluid. The fluid's speed at a particular location is indicated by the velocity vector $v(x, y, z, t)$. In a steady flow, the only factor affecting the velocity vector v is distance rather than time. The gradient, $V(x, y, z)$, of a scalar-valued function V is another illustration of a vector field (x, y, z). The magnetic field $B(x, y, z, t)$, measured in tesla, and the electric field $E(x, y, z, t)$, measured in N C^{-1} or V m^{-1}, are also vector fields.

3.3 Electric and Magnetic Fields

The forces that an electrical charge encounters are used to describe the electric field and magnetic field. For instance, if we introduce a charge q into an electric field E, the charge encounters a force qE. We also know that a charge moving with velocity v in a magnetic field B experiences a magnetic force $q \cdot (v \times B)$. This implies that even after we eliminate the charge, there is still something present (a field).

The force $F = q(E + v B)$ acts on a charge if it is at the position (x, y, z) at time t. Each location (x, y, z) in space can be connected to an E and B vector. Therefore, the x, y, z, and t vector functions of the electric and magnetic forces are considered. We refer to it as

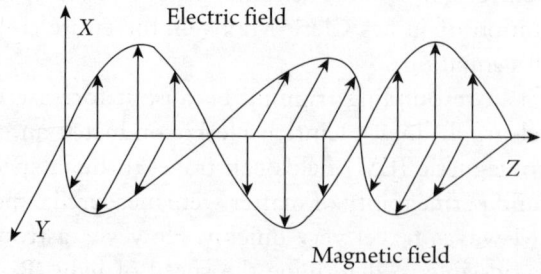

Fig. 3.1 Electromagnetic waves.

Electromagnetic Theory

an "electromagnetic field" because E or B can be determined at every location in space. Charges generate EM forces. Differential equations can be used to completely explain the connections between the field values at one location and the values at a location nearby. Therefore, knowing vector calculus is necessary in order to comprehend EM theory.

3.4 THE CONCEPT OF GRADIENT

When ascending a hill, we notice that the steepest ascent is in a path that is perpendicular to the contour lines of constant gravitational potential. We rise more slowly by a factor of cos if we ascend at an angle to this direction. Similarly, surfaces of constant V can be used to characterize a scalar function of position $V(x, y, z)$. As shown in Figure 3.2, there is a direction in which V grows most quickly at a point P. This path is parallel to the constant V through P surface. The gradient of V (or V) is a vector with magnitude equivalent to the speed at which V increases in this direction. The rate of growth of the vector, or its component, in any direction, is given by the symbol ∇V. This means that the rate of growth of V in this direction is cos times lower than in the direction of V because, at an angle to the direction of V, the distance traversed for a given small increment of V is increased in the ratio 1: cos, as shown in Figure 1.1. The rates of growth of V in the directions of the coordinate axes are $\begin{matrix} x \\ y \\ z \end{matrix}$ in Cartesian coordinates. Hence, the Cartesian components of the vector are

$$\nabla V = \left(\frac{\partial V}{\partial x}, \frac{\partial V}{\partial y}, \frac{\partial V}{\partial z} \right).$$

By using an electrostatics field as an illustration, let's demonstrate the idea of gradient. The effort required to move a unit test charge from point x to point $x + \Delta x$ is equal to the potential difference between the two points, or

$$\Delta W = V(x + \Delta x, y, z) - V(x, y, z) = \frac{\partial V}{\partial x} \Delta x. \tag{3.1}$$

But the work done against the field for the similar path is given by

$$\Delta W = -\oint E.dl = -E_x \Delta x. \tag{3.2}$$

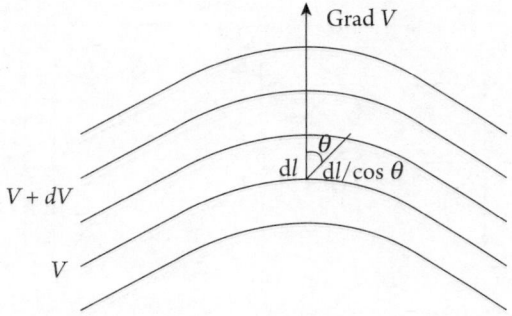

Fig. 3.2 Concept of gradient.

On comparing equations (3.1) and (3.2), we get

$$E_x = -\frac{\partial V}{\partial x}.$$

Similarly,

$$E_x = -\frac{\partial V}{\partial y} \text{ and } E_x = -\frac{\partial V}{\partial z}.$$

Hence,

$$E = -\nabla V.$$

The lines of electric force cross the equipotential surfaces at right angles, and the electric field is determined by the negative gradient of the electric potential in terms of both amplitude and direction.

3.5 THE CONCEPT OF DIVERGENCE

The amount by which a vector field D diverges or converges from a given location P is measured as its divergence. Figure 3.3 illustrates the vector field's divergence at a point P as having (a) positive divergence since the vector field expands or spreads out (source), (b) negative divergence because of vector field constricts (sink), and (c) zero divergence since the vector field neither expands nor constricts (no source, no sink). Let's discuss this in more depth now.

Consider a rectangle volume element with dimensions (Δx, Δy, Δz) and one corner at (x, y, z), as shown in Figure 3.4. Let the elements of the vector D be (Dx, Dy, as well as Dz). The flow of the vector D through a face of area $y\,z$ of the volume element has the x-component $Dx\,\Delta y\,\Delta z$. The difference between this flow at the $x + \Delta x$ face and the x-face will be

$$\Delta x \frac{\partial}{\partial x}\left(D_x \Delta y \Delta z\right) = \Delta x \Delta y dz \frac{\partial D_x}{\partial x},$$

The same justification holds true for the other two sets of faces that make up the volume part. As a result, the volume element's overall D flux is provided by

$$\int D.ds = \Delta x\, \Delta y\, \Delta z \left(\frac{\partial D_x}{\partial x} + \frac{\partial D_y}{\partial y} + \frac{\partial D_z}{\partial z}\right)$$

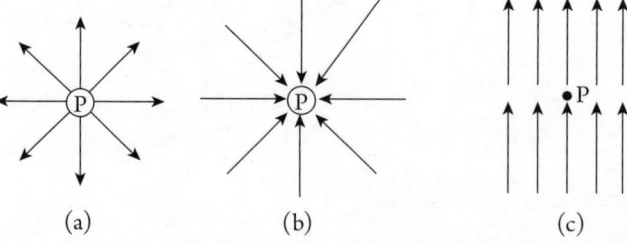

Fig. 3.3 Illustration of the divergence of a vector field at P (a) positive divergence, (b) negative divergence, and (c) zero divergence.

Electromagnetic Theory

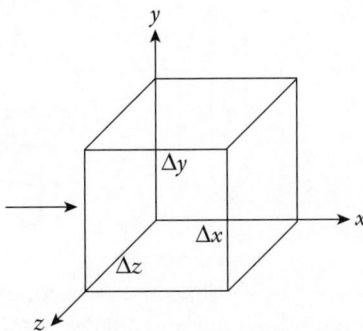

Fig. 3.4 Calculation of flux out of a volume element.

or,

$$\frac{\int D.ds}{\Delta v} = \left(\frac{\partial D_x}{\partial x} + \frac{\partial D_y}{\partial y} + \frac{\partial D_z}{\partial z} \right) = \nabla.D.$$

As a result, the outward flow from the surface of a volume element is determined by the vector's divergence and the volume of the element. Now that we understand how a vector diverges, it makes physical sense. It refers to the vector's outflow or influx per unit volume.

The flow of D out of a volume element and divergence of D are related. Therefore, the aggregate of the fluxes out of all parts of any finite volume represents the volume's overall flux. In order to integrate the divergence over the full volume, that is,

$$\int_s \vec{D}.\vec{ds} = \int_v \vec{\nabla}.\vec{D}dv,$$

where V is the amount contained within it and s is any closed surface. The divergence theorem, also known as Gauss' theorem or Gauss theorem, connects the surface integral and the volume integral. It is called after Gauss.

3.6 The Concept of Curl

A vector field's curl at a given location P is interpreted as a measurement of the field's circulation or how much it curls there. Examples include Figure 3.5(a), which depicts a vector field with a curl pointed away from the page, and Figure 3.5(b), which depicts a vector field with zero curl. We will now go into more depth about it.

Let's determine the vector field D's movement. Our loop is divided into a lot of little circles, making each little loop roughly a square. We can learn the circulation patterns around each individual rectangle. The circulation around the circle is then obtained by adding up their sums. Now, let's assume that the small cube is located in the xy plane, as shown in Figure 3.5(c).

As shown in Figure 3.6, we circle around starting at the spot (x, y) in the direction indicated by the arrows. Then the entire line sum will change to

$$\oint D.dl = D_x(1)\Delta x + D_y(2)\Delta y - D_x(3)\Delta x - D_y(4)\Delta y.$$

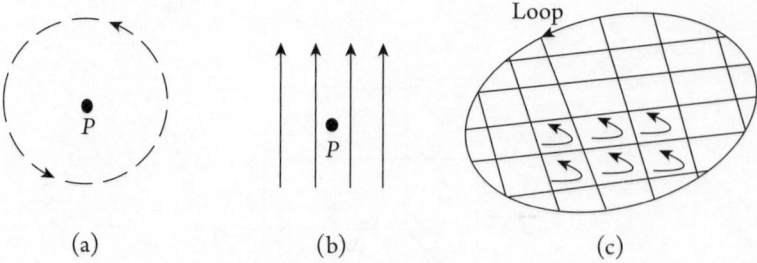

Fig. 3.5 (a) Curl at P points out of the page; (b) curl at P is zero. (c) Some surface bounded by the loop is chosen. The surface is divided into a large number of small loops so that each loop becomes approximately a square. The circulation around the loop is the sum of the circulations around the little loops.

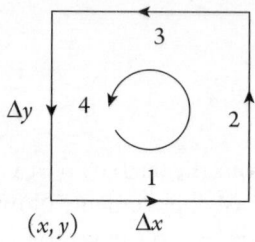

Fig. 3.6 Circulation around an element square.

Now, $D_x(1)\Delta x - D_x(3)\Delta x$ will now be, roughly speaking, negative. However, in order to be more precise, we must take the rate of change of D_x with regard to y into account.

$$D_x(1)\Delta x - \left(D_x(1) + \frac{\partial D_x}{\partial y}\Delta y\right)\Delta x = -\frac{\partial D_x}{\partial y}\Delta x \Delta y.$$

In a similar manner, the other two words in use can be written as

$$D_y(2)\Delta y - D_y(4)\Delta y = \left(D_y + \frac{\partial D_y}{\partial x}\right)\Delta y - D_y\Delta y = \frac{\partial D_y}{\partial x}\Delta x \Delta y.$$

As a result, the area around the element cube circulates

$$\left(\frac{\partial D_y}{\partial x} - \frac{\partial D_x}{\partial y}\right)\Delta x \Delta y,$$

which is normal to the surface element and a z-component of $\nabla \times D$:

$$\nabla \times D = \begin{vmatrix} i & j & k \\ \frac{\partial}{\partial x} & \frac{\partial}{\partial y} & \frac{\partial}{\partial z} \\ D_x & D_y & D_z \end{vmatrix}.$$

Thus, the movement around a differential square can be expressed as

$$\oint D \cdot dl = (\nabla \times D)_n \, ds = (\nabla \times D) \cdot ds.$$

Electromagnetic Theory

Any suitable surface can be used to fill in the loop, and we can then add the circulations around a collection of infinitesimal squares on this surface, that is,

$$\int_l \vec{D}.d\vec{l} = \int_s (\vec{\nabla} \times \vec{D}).d\vec{s}.$$

This Stokes' theorem establishes a link between the line integral and surface integral.

3.7 LAPLACIAN OPERATOR

Observe that (i) a scalar field's gradient is a vector field, (ii) a scalar field's divergence is a vector field, and (iii) a vector field's curl is a vector field. A natural differential operator that generates a scalar field from a scalar field can be built from the del operator. A scalar field is produced by applying the gradient operator to a scalar field to create a vector field, and then applying the divergence operator to this product. This is the "div grad" of a vector field, and it is provided by

$$\nabla \cdot \nabla = \frac{\partial^2}{\partial x^2} + \frac{\partial^2}{\partial y^2} + \frac{\partial^2}{\partial z^2}.$$

We refer to this operator ∇^2 as the Laplacian operator, named after Laplace, who investigated the physical uses of scalar fields (such as the potential of an inverse-square force law), which fulfil the equation $\nabla^2 V = 0$,

$$\nabla^2 V = \frac{\partial^2 V}{\partial x^2} + \frac{\partial^2 V}{\partial y^2} + \frac{\partial^2 V}{\partial z^2} = 0.$$

In the context of potential fields (such as the electrostatic potential in an electric field and the velocity potential in an ideal, frictionless fluid), such as those described by Poisson's equation $\nabla^2 V = -\rho/\varepsilon_0$, this formula has many uses in science and technology.

The Laplacian of a vector has been discussed above. We can define a vector's Laplacian because the Laplacian operator ∇^2 is a scalar operator. As it makes no sense in this situation, $\nabla^2 A$ shouldn't be thought of as the gradient of A. As opposed to this, $\nabla^2 A$ is defined as the gradient of the divergence of A minus the curl of A's curl, or

$$\nabla^2 A = \nabla(\nabla \cdot A) - \nabla \times \nabla \times A.$$

3.8 EQUATION OF CONTINUITY

Statement: It states that the total current flowing out of some volume must be equal to the rate of decrease of charge within the volume. Assuming that, charge neither created nor destroyed.

$$\vec{I} = -\frac{dq}{dt}. \tag{3.3}$$

But $\vec{I} = \int_s \vec{J}.d\vec{s}.$ (3.4)

From equations (3.3) and (3.4),

$$\int_s \vec{J}.\vec{ds} = -\int_v \frac{\partial \rho}{\partial t} dv. \qquad (3.5)$$

From Gauss' divergence theorem,

$$\int_s \vec{A}.\vec{ds} = \int_v (\vec{\nabla}.\vec{A}) dv.$$

Therefore, equation (3.5) becomes

$$\int_v (\vec{\nabla}.\vec{J}) dv = -\int_v \frac{\partial \rho}{\partial t} dv$$

$$\int_v (\vec{\nabla}.\vec{J} + \frac{\partial \rho}{\partial t}) dv = 0.$$

Since the volume is arbitrary, so integrands must vanish.

$$\boxed{\vec{\nabla}.\vec{J} + \frac{\partial \rho}{\partial t} = 0}$$

This is an equation of continuity.

For stationary currents, that is, $\frac{\partial \rho}{\partial t} = 0$.

$$\boxed{div\vec{J} = 0}$$

3.9 Ampere's Circuital Law

The line integral of magnetic field intensity around a closed path is equal to the current enclosed by the path.

$$\oint_l \vec{B}.d\vec{l} = \mu_0 \vec{I} \text{ or, } \oint_l \vec{H}.d\vec{l} = \vec{I}. \qquad (3.6)$$

Since $\vec{I} = \int_s \vec{J} \cdot d\vec{s}.$ $\qquad (3.7)$

Therefore, from equations (3.6) and (3.7),

$$\therefore \oint_l \vec{H}.d\vec{l} = \int_s \vec{J}.d\vec{s}.$$

By Stokes' theorem,

$$\oint_l \vec{A}.d\vec{l} = \int_s (\vec{\nabla} \times \vec{A}).d\vec{s}$$

$$\therefore \int_s (\vec{\nabla} \times \vec{H}).d\vec{s} = \int_s \vec{J}.d\vec{s}$$

or, $\int_s (\vec{\nabla} \times \vec{H} - \vec{J}).d\vec{s} = 0$

Electromagnetic Theory

or, $(\vec{\nabla} \times \vec{H} - \vec{J}) = 0$

$$\boxed{Curl\,\vec{H} = \vec{J}}$$

Let us examine the validity of the above equation for time varying field. On taking divergence of the above equation on both sides, we get

$$div(Curl\vec{H}) = div\vec{J}$$

$$Since,\, div(Curl\vec{A}) = 0$$

$$\therefore div\vec{J} = 0.$$

From continuity equation, $div\,\vec{J} = 0$ is for stationary currents, that is, charge density is static. Thus, Ampere's circuital law is realistic in steady state conditions but deficient for time-varying fields.

Ampere's circuital law must be modified to accommodate for time-varying fields as a result.

3.9.1 Displacement Current

Maxwell believed that the description of current density J was lacking, so something— let's say J_d —had to be added. Consequently, the overall current density is now

$$\vec{C} = \vec{J} + \vec{J_d} \text{ or } Curl\vec{H} = \vec{J} + \vec{J_d}. \tag{3.8}$$

On taking divergence of the above equation on both sides to check validity for time varying fields, we get

$$div\left(Curl\vec{H}\right) = div\left(\vec{J} + \vec{J_d}\right) \qquad [\because div\left(Curl\vec{H}\right) = 0]$$

$$\therefore div\vec{J_d} = -div\vec{J}.$$

From continuity equation we know that

$$\boxed{div\,\vec{J} = -\frac{\partial \rho}{\partial t}}$$

$$\therefore div\vec{J_d} = -\left(-\frac{\partial \rho}{\partial t}\right)$$

$$div\vec{J_d} = \frac{\partial \rho}{\partial t}.$$

Gauss' law of electrostatics in differential form is given by

$$div\vec{D} = \rho$$

$$\therefore div\vec{J_d} = \frac{\partial(div\vec{D})}{\partial t}$$

$$div\,\vec{J_d} = div\left(\frac{\partial \vec{D}}{\partial t}\right)$$

$$\vec{J}_d = \frac{\partial \vec{D}}{\partial t} \text{ or, } \vec{C} = \vec{J} + \frac{\partial \vec{D}}{\partial t}$$

$$\boxed{Curl\vec{H} = \vec{J} + \frac{\partial \vec{D}}{\partial t}}$$

The above equation is Maxwell's modified equation for time varying fields. The term that Maxwell added to Ampere's law is known as displacement current because it arises when electric displacement vector D changes with time.

3.10 Maxwell's Equations

The electric and magnetic forces are correlated with the motion of charges. EM waves are time-varying fields that have spatial and time variation in the associated electric and magnetic fields. This phenomenon is known as electromagnetism, and the fundamental rule of electromagnetism is represented by a set of equations known as Maxwell's equation.

3.10.1 Maxwell's Equations in Differential Form

The Maxwell's equations in differential form are given as follows:

$$\vec{\nabla} \cdot \vec{D} = \rho \quad \text{(Gauss' law of electrostatics)} \tag{3.9}$$

$$\vec{\nabla} \cdot \vec{B} = 0 \quad \text{(Gauss' law of magnetostatics)} \tag{3.10}$$

$$\vec{\nabla} \times \vec{E} = -\frac{\partial \vec{B}}{\partial t} \quad \text{(Faraday's law of EM induction)} \tag{3.11}$$

$$\vec{\nabla} \times \vec{H} = \vec{J} + \frac{\partial \vec{D}}{\partial t} \quad \text{(Modified Ampere's circuital law)} \tag{3.12}$$

3.10.2 Derivation of Maxwell's First Equation

According to Gauss' law of electrostatics,

$$\int_s \vec{E} \cdot d\vec{s} = \frac{\Sigma q}{\varepsilon_0}$$

$$\text{or,} \int_s \vec{E} \cdot d\vec{s} = \frac{1}{\varepsilon_0} \int_v \rho dv.$$

From Gauss' divergence theorem, changing surface integral to volume integral, we get

$$\int_v (Div\vec{E}) dv = \frac{1}{\varepsilon_0} \int_v \rho dv$$

$$\int_v Div(\varepsilon_0 \vec{E}) dv = \int_v \rho dv$$

$$Div(\varepsilon_0 \vec{E}) = \rho$$

$$Div\vec{D} = \rho$$

Electromagnetic Theory

$$\boxed{\vec{\nabla} \cdot \vec{D} = \rho}$$

The first Maxwell equation to be expressed in differential form is the above one.

3.10.3 Derivation of Maxwell's Second Equation

The flux of magnetic induction B across any closed surface is always zero because the magnetic line of force entering any arbitrary surface is exactly same as leaving it.

$$\oint_s \vec{B} \cdot \vec{ds} = 0.$$

From Gauss' divergence theorem, changing surface integral to volume integral, we get

$$\oint (Div\vec{B}) dv = 0.$$

Since the volume is arbitrary, integrand must vanish,

$$Div\vec{B} = 0$$

$$\boxed{\vec{\nabla} \cdot \vec{B} = 0}$$

The second Maxwell equation to be expressed in differential form is the above one.

3.10.4 Derivation of Maxwell's Third Equation

Faraday's law of EM induction is given by

$$e = -\frac{d\varphi}{dt} = -\frac{d}{dt} \oint \vec{B} \cdot \vec{ds} \quad \left[\because \varphi = \oint \vec{B} \cdot \vec{ds} \right] \tag{3.13}$$

We know that $e = \oint_l \vec{E} \cdot \vec{dl}$ \hfill (3.14)

From equations (3.9) and (3.10),

$$\oint_l \vec{E} \cdot \vec{dl} = -\frac{d}{dt} \int_s \vec{B} \cdot \vec{ds}$$

$$\oint_l \vec{E} \cdot \vec{dl} = -\int_s \frac{\partial \vec{B}}{\partial t} \cdot \vec{ds}$$

$$\oint_s (Curl\vec{E}) \vec{ds} = -\int_s \frac{\partial \vec{B}}{\partial t} \cdot \vec{ds}$$

$$\oint_s \left(Curl\vec{E} + \frac{\partial \vec{B}}{\partial t} \right) \vec{ds} = 0 \qquad \text{[On applying Stokes' theorem]}$$

$$\left(Curl\vec{E} + \frac{\partial \vec{B}}{\partial t} \right) = 0$$

$$Curl\vec{E} = -\frac{\partial \vec{B}}{\partial t}$$

$$\boxed{\vec{\nabla} \times \vec{E} = -\frac{\partial \vec{B}}{\partial t}}$$

The third Maxwell equation to be expressed in differential form is the above one.

3.10.5 Derivation of Maxwell's Fourth Equation

The line integral of magnetic field intensity around a closed path is equal to the current enclosed by the path.

$$\oint_l \vec{B} \cdot d\vec{l} = \mu_0 \vec{I} \text{ or, } \oint_l \vec{H} \cdot d\vec{l} = \vec{I}. \tag{3.15}$$

Since $\vec{I} = \int_s \vec{J} \cdot d\vec{s}$ \hfill (3.16)

From equations (3.11) and (3.12),

$$\therefore \oint_l \vec{H} \cdot d\vec{l} = \int_s \vec{J} \cdot d\vec{s}$$

By Stokes' theorem,

$$\oint_l \vec{A} \cdot d\vec{l} = \int_s (Curl\ \vec{A}) d\vec{s}$$

$$\therefore \int_s (Curl\ \vec{H}) d\vec{s} = \int_s \vec{J} \cdot d\vec{s}$$

$$\text{or,} \int_s (Curl\ \vec{H} - \vec{J}) d\vec{s} = 0$$

$$\text{or, } (Curl\ \vec{H} - \vec{J}) = 0$$

$$\boxed{Curl\ \vec{H} = \vec{J}}$$

Maxwell believed that the description of current density J was lacking, so something—let's say J_d—had to be added. Consequently, the overall current density is now

$$\vec{C} = \vec{J} + \vec{J_d} \text{ or } Curl\vec{H} = \vec{J} + \vec{J_d}. \tag{3.17}$$

On taking divergence of the above equation on both sides to check validity for time varying fields, we get

$$div(Curl\vec{H}) = div(\vec{J} + \vec{J_d}) \qquad \left[\because div(Curl\vec{H}) = 0\right]$$

$$\therefore div\vec{J_d} = -div\vec{J}.$$

From continuity equation we know that

$$\boxed{div\vec{J} = -\frac{\partial \rho}{\partial t}}$$

The differential form of Gauss' law of electrostatics is given by

$$div\vec{D} = \rho$$

Electromagnetic Theory

$$\therefore div\vec{J}_d = -\left(-\frac{\partial \rho}{\partial t}\right)$$

$$div\vec{J}_d = \frac{\partial \rho}{\partial t}$$

$$\therefore div\vec{J}_d = \frac{\partial(div\vec{D})}{\partial t}$$

$$div\vec{J}_d = div\left(\frac{\partial \vec{D}}{\partial t}\right)$$

$$\vec{J}_d = \frac{\partial \vec{D}}{\partial t} \text{ or, } \vec{C} = \vec{J} + \frac{\partial \vec{D}}{\partial t}$$

$$\boxed{Curl\,\vec{H} = \vec{J} + \frac{\partial \vec{D}}{\partial t}}$$

The fourth Maxwell equation to be expressed in differential form is the above one.

3.11 Maxwell's First Equation in Integral Form

Maxwell's first equation in differential form is

$$\vec{\nabla}\cdot\vec{D} = \rho.$$

Integrating it w.r.t. volume V on both sides, we get

$$\int_v (\vec{\nabla}\cdot\vec{D})\,dv = \int_v \rho\,dv.$$

Applying Gauss' divergence theorem,

$$\int_s \vec{D}\cdot d\vec{s} = \int_v \rho\,dv$$

$$\int_s \vec{E}\cdot d\vec{s} = \frac{1}{\varepsilon_0}\int_v \rho\,dv.$$

Physical Significance: It stands in for the electrostatics principle known as Gauss' law, which says that the electric flux through any closed hypothetical surface is equal to $\frac{1}{\varepsilon_0}$ times the total charge enclosed by the surface.

3.12 Maxwell's Second Equation in Integral Form

Maxwell's second equation in differential form is

$$\vec{\nabla}\cdot\vec{B} = 0.$$

Integrating it w.r.t. volume V on both sides, we get

$$\int_v (\vec{\nabla}\cdot\vec{B})\,dv = 0.$$

Applying Gauss' divergence theorem,

$$\boxed{\int_s \vec{B} \cdot d\vec{s} = 0}$$

Physical Significance: In magnetostatics, it symbolizes Gauss' law, which asserts that the net magnetic flux through any closed surface is zero. It also explains the absence of isolated poles or magnetic monopoles.

3.13 Maxwell's Third Equation in Integral Form

Maxwell's third equation in differential form is

$$\vec{\nabla} \times \vec{E} = -\frac{\partial \vec{B}}{\partial t}.$$

Integrating it w.r.t. surface S on both sides, we get

$$\int_s (\vec{\nabla} \times \vec{E}) \cdot d\vec{s} = -\int_s \left(\frac{\partial \vec{B}}{\partial t}\right) \cdot d\vec{s}.$$

Applying Stokes' theorem,

$$\int_l \vec{E} \cdot d\vec{l} = -\int_s \left(\frac{\partial \vec{B}}{\partial t}\right) \cdot d\vec{s} \Rightarrow \int_l \vec{E} \cdot d\vec{l} = -\frac{\partial \phi}{dt}.$$

Physical Significance: It represents the electric field is produced with changing magnetic flux through an open surface, according to Faraday's rule of EM induction.

3.14 Maxwell's Fourth Equation in Integral Form

Maxwell's third equation in differential form is

$$\vec{\nabla} \times \vec{H} = \vec{J} + \frac{\partial \vec{D}}{\partial t}.$$

Integrating it w.r.t. surface S on both sides, we get

$$\int_s (\vec{\nabla} \times \vec{E}) \cdot d\vec{s} = \int_s \left(\vec{J} + \frac{\partial \vec{D}}{\partial t}\right) \cdot d\vec{s}.$$

Applying Stokes' theorem,

$$\int_l \vec{H} \cdot d\vec{l} = \int_s \vec{J} \cdot d\vec{s} + \int_s \left[\frac{\partial \vec{D}}{\partial t}\right] \cdot d\vec{s}.$$

Physical Significance: It represents a modified version of Ampere's circuital law, which says that the generation of a magnetic field by displacement current results from a changing electric field.

Electromagnetic Theory

3.15 Poynting Vector

The main feature of EM waves is that it can transport energy from one point to another point. The rate of energy flow per unit area in plane EM waves is defined by a vector $\vec{s}$ called Poynting vector.

$$\vec{S} = \vec{E} \times \vec{H} \tag{3.18}$$

or, $\vec{S} = \dfrac{1}{\mu_0}(\vec{E} \times \vec{B}).$ \hfill (3.19)

From equation (3.14), it is clear that the direction of Poynting vector $\vec{s}$ is perpendicular to both $\vec{E}$ and $\vec{H}$ while $\vec{s}$ must be along $\hat{k}$.

3.16 Poynting Theorem

Statement: The sum of the time rate of change of EM energy within a certain volume and the time rate of energy flowing out through the boundary surface is equal to the power transferred into the EM field.

Proof: The Maxwell's equations in differential form are given as

$$\vec{\nabla}\cdot\vec{D} = \rho \quad \text{(Gauss' law of electrostatics)} \tag{3.20}$$

$$\vec{\nabla}\cdot\vec{B} = 0 \quad \text{(Gauss' law of magnetostatics)} \tag{3.21}$$

$$\vec{\nabla}\times\vec{E} = -\frac{\partial \vec{B}}{\partial t} \quad \text{(Faraday's law of EM induction)} \tag{3.22}$$

$$\vec{\nabla}\times\vec{H} = \vec{J} + \frac{\partial \vec{D}}{\partial t} \quad \text{(Modified Ampere's circuital law)} \tag{3.23}$$

Multiplying equation (3.22) by $\vec{H}$ and equation (3.23) by $\vec{E}$, we get

$$\vec{H}\left(\vec{\nabla}\times\vec{E}\right) = -\vec{H}\cdot\frac{\partial \vec{B}}{\partial t}. \tag{3.24}$$

$$\vec{E}\left(\vec{\nabla}\times\vec{H}\right) = \vec{E}\cdot\left(\vec{J} + \frac{\partial \vec{D}}{\partial t}\right). \tag{3.25}$$

Subtracting equation (3.25) from equation (3.24), we get

$$\vec{H}\left(\vec{\nabla}\times\vec{E}\right) - \vec{E}\left(\vec{\nabla}\times\vec{H}\right) = -\vec{H}\cdot\frac{\partial \vec{B}}{\partial t} - \vec{E}\cdot\left(\vec{J} + \frac{\partial \vec{D}}{\partial t}\right).$$

Since $\vec{\nabla}\cdot\left(\vec{A}\times\vec{B}\right) = \vec{B}\left(\vec{\nabla}\times\vec{A}\right) - \vec{A}\left(\vec{\nabla}\times\vec{B}\right)$

$$\therefore \vec{\nabla}\cdot\left(\vec{E}\times\vec{H}\right) = -\left(\vec{H}\cdot\frac{\partial \vec{B}}{\partial t} + \vec{E}\cdot\frac{\partial \vec{D}}{\partial t}\right) - \vec{E}\cdot\vec{J}. \tag{3.26}$$

On evaluating the value of term $\left(\vec{H}\cdot\dfrac{\partial \vec{B}}{\partial t} + \vec{E}\cdot\dfrac{\partial \vec{D}}{\partial t}\right)$, we get

Since $\vec{B} = \mu\vec{H}$ and $\vec{D} = \varepsilon\vec{E}$, we get

$$\vec{H} \cdot \frac{\partial \vec{B}}{\partial t} = \mu \vec{H} \cdot \frac{\partial \vec{H}}{\partial t} = \frac{1}{2}\mu \frac{\partial \vec{H}.\vec{H}}{\partial t} = \frac{1}{2} \frac{\partial \mu \vec{H}.\vec{H}}{\partial t} = \frac{1}{2} \frac{\partial \vec{B}.\vec{H}}{\partial t}$$

$$\Rightarrow \vec{H} \cdot \frac{\partial \vec{B}}{\partial t} = \frac{\partial}{\partial t}\left(\frac{1}{2}\vec{B}.\vec{H}\right).$$

Similarly, $\vec{E} \cdot \frac{\partial \vec{D}}{\partial t} = \frac{\partial}{\partial t}\left(\frac{1}{2}\vec{E}.\vec{D}\right).$

Equation (3.26) becomes

$$\vec{\nabla} \cdot (\vec{E} \times \vec{H}) = -\frac{\partial}{\partial t}\left[\frac{1}{2}(\vec{B}.\vec{H} + \vec{E}.\vec{D})\right] - \vec{E}.\vec{J}.$$

Integrating the above equation w.r.t. volume, we get

$$\int_v \vec{\nabla} \cdot (\vec{E} \times \vec{H}) dv = -\int_v \frac{\partial}{\partial t}\left[\frac{1}{2}(\vec{B}.\vec{H} + \vec{E}.\vec{D})\right] dv - \int_v \vec{E}.\vec{J} dv.$$

Applying Gauss' divergence theorem, we get

$$\oint_v (\vec{E} \times \vec{H}) \cdot d\vec{s} = -\int_v \frac{\partial}{\partial t}\left[\frac{1}{2}(\vec{B}.\vec{H} + \vec{E}.\vec{D})\right] dv - \int_v \vec{E}.\vec{J} dv$$

$$\oint_s (\vec{E} \times \vec{H}) \cdot d\vec{s} = -\frac{\partial}{\partial t}\int_v \left[\frac{1}{2}(\vec{B}.\vec{H} + \vec{E}.\vec{D})\right] dv - \int_v \vec{E}.\vec{J} dv.$$

Rearranging the terms in the above equation, we get

$$\boxed{-\int_v \vec{E}.\vec{J} dv = \frac{d}{dt}\int_v \left[\frac{1}{2}(\vec{B}.\vec{H} + \vec{E}.\vec{D})\right] dv + \oint_s (\vec{E} \times \vec{H}) \cdot d\vec{s}}$$

The above equation is expressed as Poynting theorem. The energy conservation rule in electromagnetism and the work–energy theorem are other names for this theorem.

3.16.1 The Physical Significance of Terms in Poynting Theorem

The physical significance of the terms in Poynting theorem is given as follows:

1. The term $\int_v \vec{E}.\vec{J} dv; \int_v \frac{\vec{E} \cdot \vec{I}}{A} dv = \int_v \frac{\vec{P}}{A} dv.$

   This term represents the power transferred into EM field through the motion of free charges in a volume.

2. The term $\frac{d}{dt}\int_v \left[\frac{1}{2}(\vec{B} \cdot \vec{H} + \vec{E} \cdot \vec{D})\right] dv.$

Electromagnetic Theory

The term $\frac{1}{2}\vec{B}.\vec{H}$ and $\frac{1}{2}\vec{E}.\vec{D}$ indicates, for magnetic and electric fields, the stored energy. The entire energy contained in an EM field within a specific volume is equal to their sum.

3. The term $\oint_s (\vec{E} \times \vec{H}) \cdot d\vec{s}$.

This term represents the amount of EM energy crossing the closed surface per second or the rate of flow of outward energy through the surface s enclosing volume v.

3.17 Maxwell's Equation in Free Space

The Maxwell's equations in differential form are given by

$$\vec{\nabla}.\vec{D} = \rho \text{ (Gauss' law of electrostatics)} \quad (3.27)$$

$$\vec{\nabla}.\vec{B} = 0 \text{ (Gauss' law of magnetostatics)} \quad (3.28)$$

$$\vec{\nabla} \times \vec{E} = -\frac{\partial \vec{B}}{\partial t} \text{ (Faraday's law of EM induction)} \quad (3.29)$$

$$\vec{\nabla} \times \vec{H} = \vec{J} + \frac{\partial \vec{D}}{\partial t} \text{ (Modified Ampere's circuital law)} \quad (3.30)$$

For free space, we know that

$$\rho = 0, \vec{J} = 0, \vec{D} = \varepsilon_0 \vec{E}, \vec{B} = \mu_0 \vec{H}.$$

Put these values in equations (3.27–3.30), we get

$$\vec{\nabla}.\vec{E} = 0 \quad (3.31)$$

$$\vec{\nabla}.\vec{H} = 0 \quad (3.32)$$

$$\vec{\nabla} \times \vec{E} = -\mu_0 \frac{\partial \vec{H}}{\partial t} \quad (3.33)$$

$$\vec{\nabla} \times \vec{H} = \varepsilon_0 \frac{\partial \vec{E}}{\partial t} \quad (3.34)$$

Taking curl of equation (3.33) both sides, we get

$$\vec{\nabla} \times (\vec{\nabla} \times \vec{E}) = -\mu_0 \frac{\partial (\vec{\nabla} \times \vec{H})}{\partial t}. \quad (3.35)$$

On substituting the value of $(\vec{\nabla} \times \vec{H})$ from equation (3.34),

$$\vec{\nabla} \times (\vec{\nabla} \times \vec{E}) = -\mu_0 \varepsilon_0 \frac{\partial^2 \vec{E}}{\partial t^2}. \quad (3.36)$$

Since we know that,

$$\vec{\nabla} \times (\vec{\nabla} \times \vec{A}) = Grad(div\vec{A}) - \nabla^2 \vec{A}.$$

Equation (3.36) becomes

$$Grad(div\vec{E}) - \nabla^2 \vec{E} = -\mu_0 \varepsilon_0 \frac{\partial^2 \vec{E}}{\partial t^2}.$$

$Grad(div\vec{E}) = 0$ from equation (3.31), therefore,

$$\nabla^2 \vec{E} = \mu_0 \varepsilon_0 \frac{\partial^2 \vec{E}}{\partial t^2}. \tag{3.37}$$

Similarly, taking curl of equation (3.34) and solving, we get

$$\nabla^2 \vec{H} = \mu_0 \varepsilon_0 \frac{\partial^2 \vec{H}}{\partial t^2}. \tag{3.38}$$

Equations (3.37) and (3.38) represent wave equation for $\vec{E}$ and $\vec{H}$ in free space.

The general wave equation can be written as $\nabla^2 u = \mu_0 \varepsilon_0 \frac{\partial^2 u}{\partial t^2},$ \hfill (3.39)

where u is a scalar quantity and can stand for one of the component of $\vec{E}$ and $\vec{H}$.

But, from wave mechanics, we know that the general wave equation for progressive wave is

$$\nabla^2 \Psi = \frac{1}{v^2} \frac{\partial^2 \Psi}{\partial t^2}, \tag{3.40}$$

where v is the velocity of wave.

On comparing equation (3.39) and equation (3.40),

$$\frac{1}{v^2} = \mu_0 \varepsilon_0 \Rightarrow v = \frac{1}{\sqrt{\mu_0 \varepsilon_0}} = 3 \times 10^8 \text{ m/s}.$$

Thus, the EM wave travels with speed of light in free space.

3.17.1 Electromagnetic Waves Are Transverse in Nature

From the above discussion, we obtained that

$$\mu_0 \varepsilon_0 = \frac{1}{c^2}.$$

Therefore, the EM wave equations (3.37) and (3.38) become

$$\nabla^2 \vec{E} - \frac{1}{c^2} \frac{\partial^2 \vec{E}}{\partial t^2} = 0 \tag{3.41}$$

$$\nabla^2 \vec{H} - \frac{1}{c^2} \frac{\partial^2 \vec{H}}{\partial t^2} = 0. \tag{3.42}$$

The solutions of the above equations are

$$\vec{E} = E_0 e^{i\vec{k}\cdot\vec{r} - i\omega t} \tag{3.43}$$

$$\vec{H} = E_0 e^{i\vec{k}\cdot\vec{r} - i\omega t} \tag{3.44}$$

where $\vec{k}$ is the propagation vector and $\vec{r}$ is the position vector.

Electromagnetic Theory

From Maxwell's equation (3.31) and equation (3.32) in free space, we have
$\vec{\nabla}.\vec{E} = 0$ and $\vec{\nabla}.\vec{H} = 0$.

Let us evaluate the value of $\vec{\nabla}.\vec{E}$,

$$\vec{\nabla}.\vec{E} = \left(\hat{i}\frac{\partial}{\partial x} + \hat{j}\frac{\partial}{\partial y} + \hat{k}\frac{\partial}{\partial z}\right) \cdot \left[\left(\hat{i}E_{0x} + \hat{j}E_{0y} + \hat{k}E_{0z}\right)e^{i\vec{k}.\vec{r} - i\omega t}\right].$$

Since $\vec{k}.\vec{r} = \left(\hat{i}k_x + \hat{j}k_y + \hat{k}k_z\right) \cdot \left(\hat{i}x + \hat{j}y + \hat{k}z\right)$

or, $\vec{k}.\vec{r} = \left(xk_x + yk_y + zk_z\right)$

$$\therefore \vec{\nabla}.\vec{E} = \left(\hat{i}\frac{\partial}{\partial x} + \hat{j}\frac{\partial}{\partial y} + \hat{k}\frac{\partial}{\partial z}\right) \cdot \left[\left(\hat{i}E_{0x} + \hat{j}E_{0y} + \hat{k}E_{0z}\right)e^{i\left(xk_x + yk_y + zk_z\right) - i\omega t}\right]$$

$$\vec{\nabla}.\vec{E} = \left(ik_x E_{0x} + ik_y E_{0y} + ik_z E_{0z}\right)e^{i\vec{k}.\vec{r} - i\omega t}$$

$$\vec{\nabla}.\vec{E} = i\left(\hat{i}k_x + \hat{j}k_y + \hat{k}k_z\right) \cdot \left(\hat{i}E_{0x} + \hat{j}E_{0y} + \hat{k}E_{0z}\right)e^{i\vec{k}.\vec{r} - i\omega t}$$

$$\vec{\nabla}.\vec{E} = i\vec{k}.\vec{E}. \tag{3.45}$$

Similarly, $\vec{\nabla}.\vec{H} = i\vec{k}.\vec{H}.$ (3.46)

Since $\vec{\nabla}.\vec{E} = 0 \Rightarrow i\vec{k}.\vec{E} = 0 \Rightarrow \vec{k}.\vec{E} = 0.$ (3.47)

Thus, propagation vector k and electric field vector $\vec{E}$ is perpendicular to each other.

Since $\vec{\nabla}.\vec{H} = 0 \Rightarrow i\vec{k}.\vec{H} = 0 \Rightarrow \vec{k}.\vec{H} = 0.$ (3.48)

Thus, propagation vector $\vec{k}$ and magnetic field vector $\vec{H}$ is perpendicular to each other. Thus, propagation vector k is perpendicular to both electric field vector $\vec{E}$ and magnetic field vector $\vec{H}$ as shown in Figure 3.7. Hence, EM waves are transverse in nature.

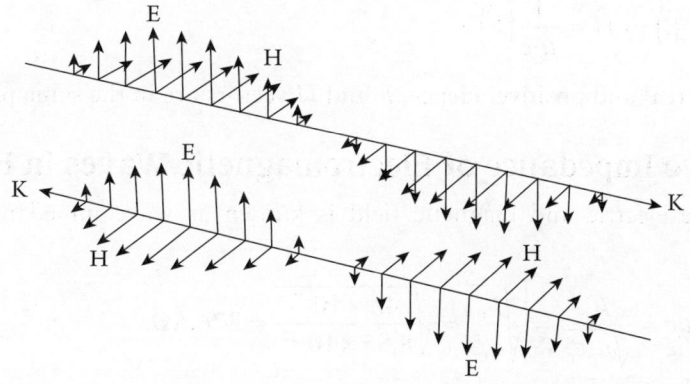

Fig. 3.7 A schematic diagram of transverse nature of electromagnetic waves.

3.17.2 $\vec{E}$ and $\vec{H}$ Vectors Are Perpendicular to Each Other in Electromagnetic Waves

The Maxwell's third equation in free space is given by

$$\vec{\nabla} \times \vec{E} = -\mu_0 \frac{\partial \vec{H}}{\partial t}.$$

Thus, $i(\vec{k} \times \vec{E}) = -\mu_0 \dfrac{\partial \left(H_0 e^{i\vec{k}\cdot\vec{r} - i\omega t}\right)}{\partial t}$

$$i(\vec{k} \times \vec{E}) = \mu_0 (i\omega) \vec{H}$$

or, $(\vec{k} \times \vec{E}) = \mu_0 \omega \vec{H}.$ \hfill (3.49)

Equation (3.49) shows that $\vec{k}$, $\vec{E}$, and $\vec{H}$ are mutually perpendicular. Hence, it is proved that $\vec{E}$ and $\vec{H}$ are perpendicular to each other.

3.17.3 $\vec{E}$ and $\vec{H}$ Vectors Are in Same Phase in Electromagnetic Waves

From equation (3.49), we know that

$$(\vec{k} \times \vec{E}) = \mu_0 \omega \vec{H}$$

$$\vec{H} = \frac{1}{\mu_0 \omega}(\vec{k} \times \vec{E}).$$

Since $\vec{k} = \hat{n} k; k = \dfrac{2\pi}{\lambda} = \dfrac{\omega}{c}$

$$\vec{H} = \frac{k}{\mu_0 \omega}(\hat{n} \times \vec{E})$$

$$\vec{H} = \frac{1}{\mu_0 c}(\hat{n} \times \vec{E}) \Rightarrow \vec{H} = \frac{1}{\mu_0 c}(\vec{E}). \hfill (3.50)$$

Thus, $\mu_0 c$ is real and positive. Hence, E and H vectors are in the same phase.

3.17.4 Wave Impedance of Electromagnetic Waves in Free Space

The ratio of the electric and magnetic field is known as wave impedance and denoted by Z_0.

$$Z_0 = \left|\frac{\vec{E}}{\vec{H}}\right| = \mu_0 c = \frac{\mu_0}{\sqrt{\mu_0 \varepsilon_0}} = \sqrt{\frac{\mu_0}{\varepsilon_0}} = \sqrt{\frac{4\pi \times 10^{-7}}{8.85 \times 10^{-12}}} = 376.6\,\Omega.$$

Thus, wave impedance of EM waves in free space is $376.6\,\Omega$.

Electromagnetic Theory

3.18 Maxwell's Equation in Nonconducting Medium

The Maxwell's equations in differential form are given by

$$\vec{\nabla}.\vec{D} = \rho \quad \text{(Gauss' law of electrostatics)} \tag{3.51}$$

$$\vec{\nabla}.\vec{B} = 0 \quad \text{(Gauss' law of magnetostatics)} \tag{3.52}$$

$$\vec{\nabla} \times \vec{E} = -\frac{\partial \vec{B}}{\partial t} \quad \text{(Faraday's law of EM induction)} \tag{3.53}$$

$$\vec{\nabla} \times \vec{H} = \vec{J} + \frac{\partial \vec{D}}{\partial t} \quad \text{(Modified Ampere's circuital law)} \tag{3.54}$$

For free space, we know that

$$\rho = 0, J = 0, \vec{D} = \varepsilon \vec{E}, \vec{B} = \mu \vec{H}$$

Put these values in equation (3.51–3.54), we get

$$\vec{\nabla}.\vec{E} = 0 \tag{3.55}$$

$$\vec{\nabla}.\vec{H} = 0 \tag{3.56}$$

$$\vec{\nabla} \times \vec{E} = -\mu \frac{\partial \vec{H}}{\partial t} \tag{3.57}$$

$$\vec{\nabla} \times \vec{H} = \varepsilon \frac{\partial \vec{E}}{\partial t} \tag{3.58}$$

Taking curl of equation (3.57) both sides, we get

$$\vec{\nabla} \times \left(\vec{\nabla} \times \vec{E}\right) = -\mu \frac{\partial \left(\vec{\nabla} \times \vec{H}\right)}{\partial t} \tag{3.59}$$

On substituting the value of $\left(\vec{\nabla} \times H\right)$ from equation (3.58),

$$\vec{\nabla} \times \left(\vec{\nabla} \times \vec{E}\right) = -\mu\varepsilon \frac{\partial^2 \vec{E}}{\partial t^2}. \tag{3.60}$$

Since we know that

$$\vec{\nabla} \times \left(\vec{\nabla} \times \vec{A}\right) = Grad\left(div\vec{A}\right) - \nabla^2 \vec{A}.$$

Equation (3.60) becomes

$$Grad\left(div\vec{E}\right) - \nabla^2 \vec{E} = -\mu\varepsilon \frac{\partial^2 \vec{E}}{\partial t^2}.$$

$Grad\left(div\vec{E}\right) = 0$, from equation (3.55), therefore,

$$\nabla^2 \vec{E} = \mu\varepsilon \frac{\partial^2 \vec{E}}{\partial t^2}. \tag{3.61}$$

Similarly, taking curl of equation (3.58) and solving, we get

$$\nabla^2 \vec{H} = \mu\varepsilon \frac{\partial^2 \vec{H}}{\partial t^2}. \tag{3.62}$$

Equations (3.61) and (3.62) represent wave equation for $\vec{E}$ and $\vec{H}$ in free space.
The general wave equation can be written as

$$\nabla^2 u = \mu\varepsilon \frac{\partial^2 u}{\partial t^2}, \tag{3.63}$$

where u is a scalar quantity and can stand for one of the components of $\vec{E}$ and $\vec{H}$.

But, from wave mechanics, we know that the general wave equation for progressive wave is

$$\nabla^2 \Psi = \frac{1}{v^2} \frac{\partial^2 \Psi}{\partial t^2}, \tag{3.64}$$

where v is the velocity of wave.
On comparing equation (3.63) and equation (3.64),

$$\frac{1}{v^2} = \mu\varepsilon \Rightarrow v = \frac{1}{\sqrt{\mu\varepsilon}} < 3 \times 10^8 \text{ m/s}.$$

Thus, the EM wave travels with the velocity less than speed of light in nonconducting medium.

3.19 Maxwell's Equation in Conducting Media

The Maxwell's equations in differential form are given by

$$\vec{\nabla}\cdot\vec{D} = \rho \quad \text{(Gauss' law of electrostatics)} \tag{3.65}$$

$$\vec{\nabla}\cdot\vec{B} = 0 \quad \text{(Gauss' law of magnetostatics)} \tag{3.66}$$

$$\vec{\nabla}\times\vec{E} = -\frac{\partial \vec{B}}{\partial t} \quad \text{(Faraday's law of EM induction)} \tag{3.67}$$

$$\vec{\nabla}\times\vec{H} = \vec{J} + \frac{\partial \vec{D}}{\partial t} \quad \text{(Modified Ampere's circuital law)} \tag{3.68}$$

In conducting medium, we know that

$$\rho = 0, J = \sigma E, \vec{D} = \varepsilon\vec{E}, \vec{B} = \mu\vec{H}$$

Put these values in equation (3.65–3.68), we get

$$\vec{\nabla}\cdot\vec{E} = 0 \tag{3.69}$$

$$\vec{\nabla}\cdot\vec{H} = 0 \tag{3.70}$$

$$\vec{\nabla}\times\vec{E} = -\mu\frac{\partial \vec{H}}{\partial t} \tag{3.71}$$

$$\vec{\nabla}\times\vec{H} = \sigma\vec{E} + \varepsilon\frac{\partial \vec{E}}{\partial t} \tag{3.72}$$

Electromagnetic Theory

Taking curl of equation (3.71) on both sides, we get

$$\vec{\nabla} \times (\vec{\nabla} \times \vec{E}) = -\mu \frac{\partial (\vec{\nabla} \times \vec{H})}{\partial t}. \tag{3.73}$$

On substituting the value of $(\vec{\nabla} \times \vec{H})$ from equation (3.72),

$$\vec{\nabla} \times (\vec{\nabla} \times \vec{E}) = -\mu \frac{\partial \left(\sigma \vec{E} + \varepsilon \frac{\partial \vec{E}}{\partial t} \right)}{\partial t}. \tag{3.74}$$

Since we know that

$$\vec{\nabla} \times (\vec{\nabla} \times \vec{A}) = Grad(div\vec{A}) - \nabla^2 \vec{A}$$

Equation (3.74) becomes

$$Grad(div\vec{E}) - \nabla^2 \vec{E} = -\mu \frac{\partial \left(\sigma \vec{E} + \varepsilon \frac{\partial \vec{E}}{\partial t} \right)}{\partial t}.$$

$Grad(div\vec{E}) = 0$, from equation (3.69), therefore,

$$\nabla^2 \vec{E} = \mu\sigma \frac{\partial \vec{E}}{\partial t} + \mu\varepsilon \frac{\partial^2 \vec{E}}{\partial t^2}$$

$$\nabla^2 \vec{E} - \mu\sigma \frac{\partial \vec{E}}{\partial t} - \mu\varepsilon \frac{\partial^2 \vec{E}}{\partial t^2} = 0. \tag{3.75}$$

Similarly, taking curl of equation (3.72) and solving, we get

$$\nabla^2 \vec{H} - \mu\sigma \frac{\partial \vec{H}}{\partial t} - \mu\varepsilon \frac{\partial^2 \vec{H}}{\partial t^2} = 0. \tag{3.76}$$

Equations (3.75) and (3.76) represent wave equation for $\vec{E}$ and $\vec{H}$ in conducting medium. The general wave equation can be written as

$$\nabla^2 \Psi - \mu\sigma \frac{\partial \Psi}{\partial t} - \mu\varepsilon \frac{\partial^2 \Psi}{\partial t^2} = 0, \tag{3.77}$$

where Ψ is a scalar quantity and can stand for one of the components of $\vec{E}$ and $\vec{H}$. The solutions of equations (3.75), (3.76), and (3.77) are

$$\vec{E} = E_0 e^{i\vec{k}\cdot\vec{r} - i\omega t} \tag{3.78}$$

$$\vec{H} = H_0 e^{i\vec{k}\cdot\vec{r} - i\omega t} \tag{3.79}$$

$$\Psi = \Psi_0 e^{i\vec{k}\cdot\vec{r} - i\omega t}, \tag{3.80}$$

where $\vec{k}$ is the propagation vector and $\vec{r}$ is the position vector.

On substituting the value of Ψ in equation (3.77), we get

$$-k^2\Psi + i\omega\mu\sigma\Psi + \omega^2\mu\varepsilon\Psi = 0.$$

On rearranging the terms,

$$k^2 = \mu\omega^2\varepsilon\left(1 + \frac{i\sigma}{\omega\varepsilon}\right). \tag{3.81}$$

As k is complex, we can write

$$k = \alpha + i\beta$$

$$k^2 = \alpha^2 - \beta^2 + 2i\alpha\beta. \tag{3.82}$$

Now comparing equations (3.81) and (3.82), we get

$$\alpha^2 - \beta^2 = \mu\varepsilon\omega^2 \tag{3.83}$$

$$2\alpha\beta = \mu\omega\sigma. \tag{3.84}$$

Solving equations (3.83) and (3.84), we get

$$\alpha = \omega\sqrt{\mu\varepsilon}\left[\frac{\sqrt{1 + \left(\frac{\sigma}{\omega\varepsilon}\right)^2} + 1}{2}\right]^{\frac{1}{2}} \tag{3.85}$$

Real Part

$$\beta = \omega\sqrt{\mu\varepsilon}\left[\frac{\sqrt{1 + \left(\frac{\sigma}{\omega\varepsilon}\right)^2} - 1}{2}\right]^{\frac{1}{2}} \tag{3.86}$$

Imaginary Part

Now in terms of α and β, the field vectors E and H take the form,

$$\vec{E} = E_0 e^{i(\alpha+i\beta)\hat{n}\cdot r - i\omega t} = E_0 e^{-\beta\hat{n}\cdot r} e^{i(\alpha\hat{n}\cdot r - \omega t)}$$

$$\vec{H} = H_0 e^{i(\alpha+i\beta)\hat{n}\cdot r - i\omega t} = H_0 e^{-\beta\hat{n}\cdot r} e^{i(\alpha\hat{n}\cdot r - \omega t)}.$$

From the above equations, it is obvious that amplitudes are exponentially decaying due to the term $e^{-\beta\hat{n}\cdot r}$. This quantity is a measure of attenuation known as attenuation coefficient.

Here, the velocity of the waves can be determined as follows:

$$v = \frac{\omega}{k} = \frac{\omega}{\alpha} = \frac{1}{\sqrt{\mu\varepsilon}}\left[\frac{\sqrt{1 + \left(\frac{\sigma}{\omega\varepsilon}\right)^2} + 1}{2}\right]^{-\frac{1}{2}}. \tag{3.87}$$

Electromagnetic Theory

Case-1: For a poor conductor $\left(\dfrac{\sigma}{\omega\varepsilon}\right) \ll 1$:

Then, $\alpha = \omega\sqrt{\mu\varepsilon}$ and $\beta = \omega\sqrt{\mu\varepsilon}\left[\dfrac{1+\dfrac{1}{2}\left(\dfrac{\sigma}{\omega\varepsilon}\right)-1}{2}\right]^{\frac{1}{2}} = \dfrac{\sigma}{2}\sqrt{\dfrac{\mu}{\varepsilon}}.$

The propagation vector is given by

$$\vec{k} = \omega\sqrt{\mu\varepsilon} + i\dfrac{\sigma}{2}\sqrt{\dfrac{\mu}{\varepsilon}}.$$

Case-2: For a good conductor $\left(\dfrac{\sigma}{\omega\varepsilon}\right) \gg 1$:

$$\alpha = \beta = \sqrt{\dfrac{\mu\sigma\omega}{2}}.$$

The propagation vector is given by

$$\vec{k} = \sqrt{\dfrac{\mu\sigma\omega}{2}}(1+i).$$

3.20 SKIN DEPTH OR PENETRATION DEPTH

The penetration depth of EM waves in conducting medium in which the strength of electric field is reduced to $\dfrac{1}{e}$ times of its original value as shown in Figure 3.8.

The term $\dfrac{1}{\beta}$ measures the depth at which EM wave entering a conductor is damped to $\dfrac{1}{e}$ times of its initial amplitude is known as skin depth or penetration depth.

Therefore, the skin depth is given by

$$\delta = \dfrac{1}{\beta} = \dfrac{1}{\omega\sqrt{\mu\varepsilon}}\left[\dfrac{\sqrt{1+\left(\dfrac{\sigma}{\omega\varepsilon}\right)^2}-1}{2}\right]^{-\frac{1}{2}}.$$

Case-1: For a good conductor $\left(\dfrac{\sigma}{\omega\varepsilon}\right) \gg 1$:

$$\delta = \sqrt{\dfrac{2}{\mu\sigma\omega}} = \sqrt{\dfrac{1}{\mu\sigma\pi f}} = \dfrac{1}{\sqrt{\mu\sigma\pi f}}.$$

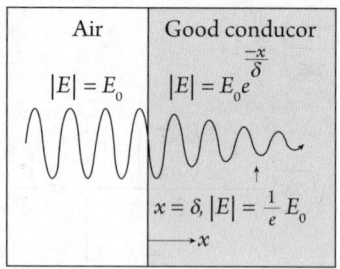

Fig. 3.8 A schematic diagram of penetration depth in good conductor.

<u>**Case-2:**</u> For a poor conductor $\left(\dfrac{\sigma}{\omega\varepsilon}\right) \ll 1$:

$$\delta = \frac{2}{\sigma}\sqrt{\frac{\varepsilon}{\mu}}.$$

3.21 Applications of Electromagnetism

1. Numerous domestic appliances use electromagnetism as their fundamental operating principle. Lighting, kitchen appliances, air conditioning systems, and so on are a few of these uses.
2. Lighting systems account for the majority of energy use in both residential and business structures. There were lots of fluorescent lighting devices used in these lighting systems. Fluorescent bulb ballasts operate on the electromagnetism principle, which results in a high voltage being produced when the light is turned on.
3. Electric motors are used in cooling devices like fans, blowers, and air conditioners. These engines operate according to the electromagnetism branch of EM induction. According to the Lorenz force theory, the electric motor is propelled by the magnetic field created by the electric current in any electrical appliance. Depending on the purpose, these motors range in size, power, and price.
4. Electromagnetism is used in kitchen equipment like induction cooktops, microwaves, electric mixers and grinders, toasters, and so on.
5. Alarm devices employ electromagnetically operated electrical bells. EM coils that move the striker against the bell in these bells create the sound. A bell is rung as long as the coil is charged because the iron striker is drawn to it.
6. When the striker makes contact with the bell, the electromagnets become demagnetized. Under the strain of the spring, the striker returns to its initial position, re-establishing electrical contact. The switch is opened after this procedure has been repeated.
7. Doors are locked using securing systems for security systems, which are typically magnetic locking systems. These systems either require a security code or the use of a magnetic card to open them.
8. The number of keys recorded on the magnetic tape of the card is read by a magnetic card reader on the doors. The door opens when the key that is stored in memory fits the information on the card.

Electromagnetic Theory

9. Loudspeakers are used by entertainment devices such as televisions, radios, and stereos. This device is made up of an electromagnet that is fastened to a membrane or cone that is encircled by the magnetic flow generated by a permanent magnet.

10. The speaker's electromagnet and membrane travel back and forth when the current flowing through the electromagnet is changed. When the current is changed at the same wavelengths as the sound waves, the speaker vibrates, further amplifying the sound waves.

11. Electromagnetism is the basis for almost all tools and equipment used in industries. Iron, cobalt, nickel, and other materials that naturally react to magnetic fields are used in the construction of such devices. Starting from small control instruments to the large power equipments, the electromagnetism is used at least at one stage of their working. Most industries rely on generators and motors as their main sources of power and drive systems, respectively. Generators convert the mechanical to electrical energy, whereas the motors convert electrical energy to mechanical energy.

12. In the majority of instances, IC engines power the generators, which are used to provide electricity when the mains power is interrupted. There are various motor classifications that are used in industries. These are used for things like transportation systems, hoists, lifts, and cranes.

13. A number of instruments and actuating devices rely on electromagnetism. Magnetoresistive sensors, fluxgate sensors, Hall-effect sensors, and other devices are examples of EM sensors. These sensors turn the electrical indication from a physical quantity, such as flow, pressure, level, proximity, and so on, into the physical quantity.

14. Actuators are the final control components that drive the load under particular circumstances. All of these actuator devices, such as solenoid valves, relays, motors, and so on, operate on the electromagnetism concept. This is the modern technology of transportation systems that use the concept of electromagnetism. These are called high speed trains, which use powerful electromagnets to develop the speed.

15. Using the fundamental magnetism concepts of electromagnetic suspension (EMS) and electrodynamic suspension (EDS), these trains will float over a guideway. Electromagnets used in EMS are drawn to the iron tracks by the train's body.

16. The train is raised by an attractive force created by these magnets wrapping around the guided tracks. In EDS, generated currents lift the train in the conductive guideways against the repellent force.

17. A guideway is nothing more than a pattern of tracks and magnetic coils that have been specifically designed, spaced at regular intervals. A maglev railway traveling along this guideway is suspended by the magnetic levitation phenomenon with only magnetic fields acting as supports. It is the process of transmitting information from a source to a receiver. This transmission of energy over long distances is carried out through EM waves at high frequencies. These waves are also called microwaves or high frequency radio waves.

18. Assume that EM energy is created when sound energy is transformed into mobile phone energy. This EM energy is sent from the transmitter to the recipient by

radio waves. When they reach the listener, these EM waves are once more converted into sound energy.

19. Communication networks can be either analogue or digital, depending on the type of baseband signal. This system can be classified as a baseband or carrier transmission system depending on the type of transmitting signal.

20. Through a waveguide or transmission line, EM fields generated by time-varying sources are transmitted. When these EM fields spread without any link or conducting medium to the sources, they produce an EM wave radiation.

21. The entire range of EM radiation's potential wavelengths and frequencies is represented by the EM spectrum. Low-frequency, ultra-low frequency, medium frequency, high-frequency, ultra-high frequency, super high frequency, and so on are some of these frequency categories.

22. Today's high-tech medical devices, including implants, magnetic resonance imaging, and thermal therapy for cancer, heavily rely on EM fields (MRI).

23. Most medical applications use RF range frequencies. Advanced technology that uses electromagnetism to examine the human body's minute details is used in MRI scans.

24. An alternative medical practice known as EM therapy claims to be able to cure illness by exposing the body to EM radiation or pulsed EM fields. This kind of medicine is used to address a variety of illnesses, including nervous disorders, diabetes, spinal cord injuries, ulcers, and asthma.

25. The electromagnetism concept underlies the operation of many medical devices, including scanners, x-ray machines, and other devices.

Solved Problems

Ex. 1: Earth receives 2 calories of solar energy per cm² sec as an average over a year for whole surface. What are the amplitudes of average electric and magnetic field radiation?

Solution:

Poynting vector is given by

$$\vec{S} = \vec{E} \times \vec{H} = EH \sin 90^0 = EH$$

$$\vec{S} = EH = \frac{2 \times 4.2 \times 10^4}{60} = 1400 \text{ joule}/m^2 s. \quad (3.88)$$

Since $\left|\frac{\vec{E}}{\vec{H}}\right| = 377 \Omega.$ (3.89)

Multiplying equation (3.88) and equation (3.89), we get

$$E^2 = 527240 \Rightarrow E_{avg} = 726.1 V/m$$

$$H_{avg} = \frac{1400}{E_{avg}} = 1.928 A - \text{turns}/m$$

Electromagnetic Theory

The amplitudes are: $E_0 = E_{avg}\sqrt{2} = 1026.7 \, V/m$

$H_0 = H_{avg}\sqrt{2} = 2.726 \, A-turns/m.$

Ex. 2: A lamp radiates 500 W power uniformly in all directions. Calculate the electric and magnetic field intensity at 1 m distance from the lamp.

Solution:

Poynting vector is given by

$$\vec{S} = \vec{E} \times \vec{H} = EH\sin 90^0 = EH$$

$$\vec{S} = \frac{P}{Area} = \frac{500}{4\pi r^2} = \frac{500}{4\pi(1)^2} = \frac{500}{4\pi} = EH. \tag{3.90}$$

Since $\left|\frac{\vec{E}}{\vec{H}}\right| = 377\Omega.$ \hfill (3.91)

Multiplying equation (3.90) and equation (3.91), we get

$H^2 = 0.105 \Rightarrow H_{avg} = 0.33 \, A-turns/m$

$\therefore E_{avg} = 120.63 \, V/m.$

Ex. 3: Calculate the skin depth for frequency of 10^{20} Hz for silver if $\mu_0 = 4\pi \times 10^{-7}$ weber/A-m, $\sigma = 3 \times 10^7$ S/m.

Solution:

We know that

$$\delta = \frac{1}{\sqrt{\mu\sigma\pi f}} = \frac{1}{\sqrt{4\pi \times 10^{-7} \times 3 \times 10^7 \times \pi \times 10^{20}}}$$

$\delta = 0.091 \times 10^{-10}$ m.

Ex. 4: If the magnitude of $\vec{H}$ in a plane wave is 1 amp/m, find the magnitude for $\vec{E}$ for plane wave in free space.

Solution:

Given that $\vec{H} = 1$ amp/m.

In free space, $\frac{\vec{E}}{\vec{H}} = 377\Omega \Rightarrow \vec{E} = 377 \times \vec{H} = 377 \, v/m.$

Previous Year Questions (University Examination)

1. Explain the concept of gradient, divergence, and curl of any vector and give its physical significance.
2. Derive Maxwell's equations in differential form. Explain the physical significance of each equation.
3. Derive Maxwell's equations in integral form. Explain the physical significance of each equation.

4. Explain the concept of Maxwell's displacement current and show how it led to the modification of the Ampere's law.
5. Deduce Maxwell's equation for free space and prove that the EM waves are transverse. Also prove that the velocity of plane EM wave in the vacuum is given by $c = \dfrac{1}{\sqrt{\mu_0 \varepsilon_0}}$.
6. Deduce an expression for intrinsic impedance of EM waves. Show that for free space its value is equal to 120π.
7. Write down Maxwell's equations for free space and show that E, H, and direction of wave propagation form a set of orthogonal vectors.
8. What is Poynting vector? Deduce Poynting theorem for the flow of energy in an EM field. Give its physical significance also.
9. Deduce Maxwell's equation for nonconducting medium. Prove that the velocity of plane EM wave in nonconducting medium is less than the velocity of light.
10. What is skin depth? Show that for poor conductors, skin depth is independent of frequency of the wave.
11. Deduce the wave equation for EM waves in conducting medium.
12. If the earth receives 5 cal min^{-1} cm^{-2} solar energy, what are the amplitudes of electric and magnetic fields of radiation?
13. Assuming that all the energy from a 1000 watt lamp is radiated uniformly; calculate the average values of the intensities of electric and magnetic fields of radiation at a distance of 2 m from the lamp.
14. For silver, $\mu = \mu_0$ and $\sigma = 3 \times 10^7$ mhos/m. Calculate the skin depth at 10 Hz frequency.
15. If the upper atmosphere of earth receives 1360 W/m² energy from the sun. What will be the peak values of electric and magnetic fields at the layer?
16. Determine the conduction current and displacement current densities in a material having conductivity of 10^{-4} S/m and relative permittivity 2.25. The electric field in the material is $E = 5 \times 10^{-6} \sin(9 \times 10^9) V/m$.
17. Calculate the displacement current through a parallel air-filled capacitor having plates of area 10 cm² separated by a distance 2 mm and connected to a 360 V, 1 MHz source.
18. A copper wire carries a conduction current of 1 ampere. Determine the displacement current in the wire at 100 MH.
19. The relative permittivity of distilled water is 81. Calculate refractive index and velocity of light in it.
20. The electric field in an EM wave is given by $E = E_0 \sin\omega(t - x/c)$, where $E_0 = 1000 \, N/c$. Find the energy contained in a cylinder of cross-section 10^{-3} m² and length 100 cm along the X-axis.
21. A plane monochromatic linearly polarized wave is traveling eastward. The wave is polarized with E directed vertically up and down. Write expressions for E, H, and B provided that $E_0 = 0.1 \, V/m$.
22. The plane EM wave propagating in the positive X-direction has a wavelength 7.0 mm. The electric field is in the Y-direction and its maximum magnitude is 42 V/m. Write suitable equations for the electric and magnetic fields as a function of x and t.

Electromagnetic Theory

Multiple Choice Questions

1. The equation of continuity is
   (a) $div\vec{j} + \frac{\partial \rho}{\partial t} = 0$
   (b) $div\vec{j} - \frac{\partial \rho}{\partial t} = 0$
   (c) $div\vec{D} = \rho$
   (d) $Curl\vec{H} = \vec{j}$

2. The equation of continuity represents the conservation law of
   (a) Energy
   (b) Momentum
   (c) Charge
   (d) None of these

3. Maxwell's first equation is the differential form of
   (a) Gauss' law of electrostatics
   (b) Gauss' law of magneto-statics
   (c) Faraday's law of EM-induction
   (d) Ampere's law

4. Maxwell's second equation is the differential form of
   (a) Gauss' law of electrostatics
   (b) Gauss' law of magneto-statics
   (c) Faraday's law of EM-induction
   (d) Ampere's law

5. Maxwell's third equation is the differential form of
   (a) Gauss' law of electrostatics
   (b) Gauss' law of magneto-statics
   (c) Faraday's law of EM-induction
   (d) Ampere's law

6. Maxwell's fourth equation is the differential form of
   (a) Gauss' law of electrostatics
   (b) Gauss' law of magneto-statics
   (c) Faraday's law of EM-induction
   (d) Modified Ampere's law

7. Poynting theorem represents
   (a) Conservation of energy
   (b) Conservation of momentum
   (c) Conservation of charge
   (d) None of these

8. In conducting medium, the EM waves are
   (a) amplified
   (b) attenuated
   (c) Both (a) and (b)
   (d) None of the above

9. The characteristic impedance of free space is
   (a) $0 \, \Omega$
   (b) $1 \, \Omega$
   (c) $377 \, \Omega$
   (d) None of the above

10. The skin depth is given by
    (a) $\frac{1}{\beta}$
    (b) $\frac{1}{\sigma}$
    (c) Both (a) and (b)
    (d) None of the above

11. Displacement current is due to
    (a) Displacement of electric charges
    (b) Time varying magnetic field
    (c) Time varying electric field
    (d) Both (b) and (c)

12. Which of the following equations shows the nonexistence of magnetic monopoles in nature
    (a) $Div \vec{D} = \rho$
    (b) $Div \vec{B} = 0$
    (c) $Div \vec{E} = 0$
    (d) None of the above

13. The displacement current in a good conductor is negligible compared to the conduction current at any frequency
    (a) More than the optical frequencies
    (b) Less than the optical frequencies
    (c) Equal to the optical frequencies
    (d) None of the above

14. The displacement current density can be expressed as time rate of
    (a) $\dfrac{\partial \vec{B}}{\partial t}$
    (b) $\dfrac{\partial \vec{H}}{\partial t}$
    (c) $\dfrac{\partial \vec{K}}{\partial t}$
    (d) $\dfrac{\partial \vec{D}}{\partial t}$

15. The speed of EM waves in free space (vacuum) is
    (a) More than speed of light
    (b) Less than speed of light
    (c) Equal to the speed of light
    (d) None of these

16. The speed of EM waves in free space (vacuum) can be expressed as
    (a) $v = \dfrac{1}{\sqrt{\mu_0 \varepsilon_0}}$
    (b) $v = \sqrt{\mu_0 \varepsilon_0}$
    (c) $v = \sqrt{\dfrac{\mu_0}{\varepsilon_0}}$
    (d) $v = \sqrt{\dfrac{\varepsilon_0}{\mu_0}}$

17. For the propagation of EM waves in free space, the phase difference between its electric and magnetic field vectors $\vec{E}$ and $\vec{H}$ is
    (a) 0
    (b) $\dfrac{\pi}{2}$
    (c) π
    (d) $\dfrac{\pi}{4}$

18. The speed of EM waves in a dielectric medium is
    (a) More than speed of light
    (b) Less than speed of light
    (c) Equal to the speed of light
    (d) None of these

19. The dimension of the quantity $\dfrac{1}{\sqrt{\mu_0 \varepsilon_0}}$ is
    (a) Momentum
    (b) Energy
    (c) Force
    (d) Velocity

20. Light is
    (a) EM waves
    (b) de-Broglie waves
    (c) Mechanical waves
    (d) None of these

21. The E, K, and H vectors for EM waves in free space are mutually
    (a) Co-planar
    (b) Co-linear
    (c) Orthogonal
    (d) None of these

Electromagnetic Theory

22. The electric field vector $\vec{E}$ of an EM wave in free space satisfies the equation
    (a) $\nabla^2 \vec{E} = \mu_0 \varepsilon_0 \dfrac{\partial^2 \vec{E}}{\partial t^2}$
    (b) $\nabla^2 \vec{E} = \mu_0 \varepsilon_0 \dfrac{\partial \vec{E}}{\partial t}$
    (c) $\nabla^2 \vec{E} = \dfrac{1}{\sqrt{\mu_0 \varepsilon_0}} \dfrac{\partial^2 \vec{E}}{\partial t^2}$
    (d) $\nabla^2 \vec{E} = \sqrt{\dfrac{\varepsilon_0}{\mu_0}} \dfrac{\partial^2 \vec{E}}{\partial t^2}$

23. In an EM wave, the direction of magnetic field $\vec{H}$ is
    (a) Parallel to the electric field
    (b) Antiparallel to the electric field
    (c) Perpendicular to the electric field
    (d) None of these

24. The intrinsic impedance of EM waves is
    (a) 177 Ω
    (b) 277 Ω
    (c) 377 Ω
    (d) 477 Ω

25. The ratio of electric field $\vec{E}$ and magnetic field vector $\vec{H}$ has the dimension of
    (a) Impedance
    (b) Capacitance
    (c) Inductance
    (d) None of the above

26. The rate of energy flow per unit area in plane EM waves is represented by
    (a) Propagation vector
    (b) Electric displacement vector
    (c) Displacement current density vector
    (d) Poynting vector

27. Poynting vector can be expressed as
    (a) $\vec{S} = \vec{E} \times \vec{H}$
    (b) $\vec{S} = \dfrac{1}{\mu_0}(\vec{E} \times \vec{B})$
    (c) $2d \tan\theta = n\lambda$
    (d) Both (a) and (b)

28. The unit of Poynting vector is
    (a) Watt-meter^2
    (b) Watt-meter
    (c) Watt/meter^2
    (d) Watt/meter

29. The magnitude of Poynting vector represents the total energy transfer by EM waves in
    (a) Per unit time per unit length
    (b) Per unit time per unit area
    (c) Per unit time per unit volume
    (d) Per unit time

30. Poynting theorem represents the conservation law of
    (a) Energy
    (b) Linear momentum
    (c) Angular momentum
    (d) None of the above

31. An electric force (F_e) and a magnetic force (F_m) were experienced by a charged particle (q) traveling at a speed of v through an area of electromagnetic radiation, then the ratio of magnetic force to electric force is
    (a) $\dfrac{F_m}{F_e} = \dfrac{c}{v}$
    (b) $\dfrac{F_m}{F_e} = \dfrac{v}{c}$
    (c) $\dfrac{F_m}{F_e} = \left(\dfrac{c}{v}\right)^2$
    (d) $\dfrac{F_m}{F_e} = \left(\dfrac{v}{c}\right)^2$

32. The energy of EM wave in vacuum is given by the expression
    (a) $\dfrac{E^2}{2\varepsilon_0} + \dfrac{B^2}{2\mu_0}$
    (b) $\dfrac{\varepsilon_0 E^2}{2} + \dfrac{B^2}{2\mu_0}$
    (c) $\dfrac{\varepsilon_0 E^2}{2} + \dfrac{\mu_0 B^2}{2}$
    (d) $\dfrac{E^2}{2\varepsilon_0} + \dfrac{\mu_0 B^2}{2}$

33. The intrinsic impedance of EM waves is
    (a) $\sqrt{\dfrac{\mu_0}{\varepsilon_0}}$
    (b) $\dfrac{1}{\sqrt{\mu_0 \varepsilon_0}}$
    (c) $\sqrt{\dfrac{\varepsilon_0}{\mu_0}}$
    (d) All of the above

34. The propagation constant $\vec{k}$ of EM waves in a conducting medium is
    (a) A real number
    (b) An imaginary number
    (c) A complex number
    (d) None of the above

35. The magnetic field vector of EM waves in a conducting medium
    (a) Lags behind the electric field vector in phase
    (b) Leads the electric field vector in phase
    (c) Is in phase with electric field vector
    (d) None of the above

36. In conducting medium, the EM waves are
    (a) Amplified
    (b) Attenuated
    (c) Both (a) and (b)
    (d) None of the above

37. The EM wave velocity in nonconducting medium is
    (a) $\dfrac{1}{\sqrt{\mu\varepsilon}}$
    (b) $\dfrac{1}{\sqrt{\mu_0 \varepsilon_0}}$
    (c) $\sqrt{\dfrac{\mu_0}{\varepsilon_0}}$
    (d) $\sqrt{\dfrac{\varepsilon_0}{\mu_0}}$

38. The energy density in the electric and magnetic fields of an EM wave is
    (a) Different
    (b) 1.5
    (c) L/C
    (d) Same

39. The depth at which EM wave entering a conductor is damped to $\dfrac{1}{e}$ times of its initial amplitude is known as
    (a) Measured depth
    (b) Variable depth
    (c) Constant depth
    (d) Skin depth

40. For a good conductor, the skin depth is given as
    (a) $\dfrac{1}{\sqrt{\mu\sigma\pi f}}$
    (b) $\dfrac{2}{\sigma}\sqrt{\dfrac{\varepsilon}{\mu}}$
    (c) $\dfrac{1}{\sqrt{\mu\sigma\pi f v k}}$
    (d) All of the above

Electromagnetic Theory

41. For a poor conductor, the skin depth is given as
    (a) $\dfrac{1}{\sqrt{\mu\sigma\pi f}}$
    (b) $\dfrac{2}{\sigma}\sqrt{\dfrac{\varepsilon}{\mu}}$
    (c) $\dfrac{1}{\sqrt{\mu\sigma\pi fvk}}$
    (d) All of the above

42. For a poor conductor, the skin depth is given as
    (a) $\dfrac{1}{\sqrt{\mu\sigma\pi f}}$
    (b) $\dfrac{2}{\sigma}\sqrt{\dfrac{\varepsilon}{\mu}}$
    (c) $\dfrac{1}{\sqrt{\mu\sigma\pi fvk}}$
    (d) All of the above

43. The EM waves are transverse in nature, if they propagate in
    (a) Free space
    (b) Isotropic dielectric
    (c) Conducting medium
    (d) All of the above

44. The field vectors $\vec{E}$ and $\vec{H}$ are not in the same phase if EM wave propagates in
    (a) Free space
    (b) Isotropic medium
    (c) Conducting medium
    (d) Anisotropic dielectric

45. The amplitudes of electric and magnetic field vectors of an EM wave propagating in free space are related to each other by the relation
    (a) $E_0 B_0 = c$
    (b) $E_0 = B_0 c$
    (c) $B_0 = cE_0$
    (d) $E_0 B_0 = c^2$

46. The electric and magnetic fields share the energy of EM wave in the ratio
    (a) 1:2
    (b) 2:1
    (c) 1:1
    (d) 1:4

47. The direction of propagation of EM wave is given by
    (a) $\vec{E}\cdot\vec{B}$
    (b) $\vec{E}$
    (c) $\vec{B}$
    (d) $\vec{E}\times\vec{B}$

48. The effects of displacement current can be measured in a conductor carrying alternating currents only at
    (a) Optical frequencies
    (b) Acoustical frequencies
    (c) X-ray frequencies
    (d) Gamma ray frequencies

49. A lamp radiates 500 W power uniformly in all directions, then electric intensity at 1 m distance from the lamp is
    (a) 120.63 V/m
    (b) 0.311 V/m
    (c) 0.684 V/m
    (d) 0 V/m

50. The Poynting vector $\vec{S}$ of an EM wave is
    (a) $\vec{S}=\vec{E}\times\vec{B}$
    (b) $\vec{S}=\vec{E}\times\vec{H}$
    (c) $\vec{S}=\vec{E}/\vec{B}$
    (d) $\vec{S}=\vec{E}/\vec{H}$

BIBLIOGRAPHY

1. Zangwill, A. (2013). *Modern Electrodynamics*. Cambridge University Press.
2. Purcell, E. M. and Morin, D. J. (2013). *Electricity and Magnetism*. Cambridge University Press.
3. Schwartz, M. (1987). *Principles of Electrodynamics*. Dover Publications.
4. Grant I. S. and Phillips, W. R. (2008). *Electromagnetism*. John Wiley & Sons.
5. Reitz, R. J., Milford, F. J., and Christy, R. W. (2008). *Foundations of Electromagnetic Theory*. Pearson.
6. Hammond, P. (1997). *Electromagnetism for Engineers: An Introductory Course*. Oxford University Press.
7. Stratton. J. A. (2007). *Electromagnetic Theory*. Wiley.
8. Jackson, J. D. (1999). *Classical Electrodynamics*. Wiley.
9. Griffiths, D. (2012). *Introduction to Electrodynamics*. Addison Wesley.
10. Pollack, G. L. and Stump, D. R. (2002). *Electromagnetism*. Addison Wesley.

Keys

1. (a)	2. (c)	3. (a)	4. (b)	5. (c)	6. (d)	7. (a)	8. (b)	9. (c)	10. (a)
11. (c)	12. (c)	13. (b)	14. (d)	15. (c)	16. (a)	17. (a)	18. (b)	19. (d)	20. (a)
21. (c)	22. (a)	23. (a)	24. (c)	25. (a)	26. (d)	27. (d)	28. (c)	29. (b)	30. (a)
31. (b)	32. (b)	33. (a)	34. (c)	35. (a)	36. (b)	37. (a)	38. (d)	39. (d)	40. (a)
41. (a)	42. (a)	43. (d)	44. (c)	45. (b)	46. (c)	47. (d)	48. (a)	49. (a)	50. (b)

Statistical Mechanics

4.1 Introduction

Statistical mechanics bridges the gaps between the laws of thermodynamics and the internal structure of the matter. Some examples are as follows:

1. Assembly of atoms in gaseous or liquid helium.
2. Assembly of water molecules in solid, liquid, or vapor state.
3. Assembly of free electrons in metal.

The behavior of all these abovementioned assemblies is totally different in different phases. Therefore, it is most significant to relate the macroscopic behavior of the system to its microscopic structure.

In this mechanics, most probable behavior of assembly are studied instead of individual particle interactions or behavior.

The behavior of assembly that is repeated a maximum time is known as most probable behavior.

4.2 Phase Space

Six coordinates can fully characterize the state of any system:

1. Three for describing the position x, y, z and three for momentum P_x, P_y, P_z.
2. The combined position and momentum space (x, y, z, P_x, P_y, P_z) is called phase space.
3. The momentum space represents the energy of state,
$$E = \frac{P_x^2 + P_y^2 + P_z^2}{2m} = \frac{P^2}{2m}.$$

For a system of N particles, there exists $3N$ position coordinates and $3N$ momentum coordinates. A single particle in phase space is known as a phase point, and the space occupied by it is known as μ-space.

4.3 Volume Element of μ-Space

Consider a particle having the position and momentum coordinates in the range $x, x + dx$; $y, y + dy$; $z, z + dz$ and $P_x, P_x + dP_x$; $P_y, P_y + dP_y$; $P_z, P_z + dP_z$.

Then, at any instant, the phase point will lie in the volume element,

$$d\Gamma_{min} = (dx\ dy\ dz\ dP_x\ dP_y\ dP_z)_{min}$$

$$d\Gamma_{min} = (dx\ dP_x)_{min} (dy\ dP_y)_{min} (dz\ dP_z)_{min}.$$

Using uncertainty relation,

$$d\Gamma_{min} = \hbar \times \hbar \times \hbar = \hbar^3 \cong h^3 \quad [\because \hbar \cong h].$$

A point in phase space is actually a cell whose minimum volume is of the order of $\hbar^3$ or h^3.

4.4 Number of Accessible Microstates or Phase Cells in the Energy Range E and E+dE

Consider the particles having the energy in range E and $E + dE$. The volume of the assembly of the particles is V.

Volume in the phase space available to the particles = $\iiint dx\ dy\ dz \iiint dP_x\ dP_y\ dP_z$.

Since the minimum volume of unit cell = $\hbar^3 \cong h^3$.

The number of unit cells in the above volume,

$$g = \frac{\iiint dx\ dy\ dz \iiint dP_x\ dP_y\ dP_z}{h^3}. \tag{4.1}$$

Where $\iiint dx\ dy\ dz = V \rightarrow$ Physical volume occupied by the particle.

$\iiint dP_x\ dP_y\ dP_z = V_p \rightarrow$ Volume element of momentum space.

4.5 Evaluation of $\iiint dV_p$

A surface with constant energy E in the momentum space will be a sphere.

The radius of sphere $P = \sqrt{2mE}$.

The equation of concentric sphere in energy range E and $E + dE$.

Therefore, $dV_p = 4\pi P^2 dP$

$$\iiint dP_x\ dP_y\ dP_z = 4\pi (2mE) \cdot d \cdot (2mE)^{\frac{1}{2}}$$

$$\iiint dP_x\ dP_y\ dP_z = 4\pi (2m)^{\frac{3}{2}} E \times \frac{1}{2} E^{-\frac{1}{2}} = 2\pi (2m)^{\frac{3}{2}} E^{\frac{1}{2}}.$$

Put the above value in equation (4.1), $g(E) = \frac{V}{h^3} 2\pi (2m)^{\frac{3}{2}} E^{\frac{1}{2}}$

$$g(E)dE = 2\pi V \left[\frac{2m}{h^2}\right]^{\frac{3}{2}} E^{\frac{1}{2}} dE.$$

This is an expression of the number of cells for particles in the energy range E and $E + dE$. The number of cells doubles for electrons because electrons have two allowed values of the

Statistical Mechanics

spin quantum numbers. Therefore, the expression of the number of cells for electrons in the energy range E and $E + dE$ is given by,

$$g(E)dE = 4\pi V \left[\frac{2m}{h^2}\right]^{\frac{3}{2}} E^{\frac{1}{2}} dE. \tag{4.2}$$

4.6 Density of Microstates

It is defined as number of quantum states per unit energy difference given as

$$g(E) = 4\pi V \left[\frac{2m}{h^2}\right]^{\frac{3}{2}} E^{\frac{1}{2}}.$$

In terms of momentum,

$$g(P) = \frac{4\pi V P^2}{h^3}.$$

4.7 Ensemble

A system is defined as a collection of number of particles. An ensemble is defined as a collection of a large number of microscopically identical but essentially independent systems. The term microscopically identical means that the system constituting an ensemble satisfies the same macroscopic conditions, such as volume, energy, pressure, total number of particles and so on. In an ensemble, the system plays the same role as the non-interacting molecules do in a gas. The microscopic identity of the systems constituting an example may be achieved by using the same values of some set of macroscopic parameters that uniquely determine the equilibrium state of the system.

4.7.1 Classification of Ensemble

There may be many types of ensembles; out of them, the most widely used are the micro canonical, canonical, and grand canonical ensembles.

4.7.2 Micro-canonical Ensemble

It is a grouping of several vitally important independent systems that have the same energy E, volume V, and number of particles N. We make the straightforward assumption that each particle is identical in nature. The solid, impermeable, and well-insulated walls that separate the different systems in a microcanonical ensemble ensure that the values of E, V, and N for any given system are unaffected by the presence of other systems, as shown in Figure 4.1.

4.7.3 Canonical Ensemble

It is a vast number of fundamentally unrelated systems that have the same temperature T, volume V, and number of identical particles N. By placing each system in thermal contact with a sizable heat reservoir that is kept at a constant temperature T and putting all of the systems into thermal contact with one another, the temperature of all the systems may be equalized. A canonical ensemble's distinct systems are divided by stiff, impermeable,

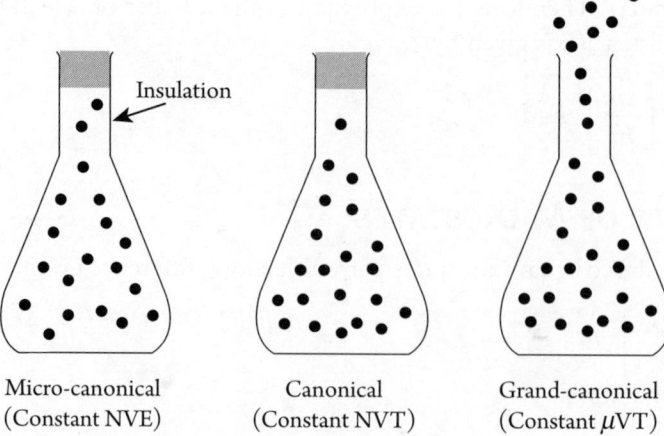

Fig. 4.1 Statistical ensembles.

but conducting walls. The conductivity of the partition walls allows for the transfer of heat between the systems. As a result, T will become the common temperature for all systems.

4.7.4 Grand Canonical Ensemble

It consists of a huge number of fundamentally unrelated systems with the same temperature (T), volume (V), and chemical potential (μ) that are all combined into one system. A vast canonical ensemble has hard, porous, and conducting barriers separating the various components. The interchange of heat energy and particles between the systems occurs because the dividing walls are conducting and permeable, bringing all the systems to the same temperature T and chemical potential μ.

4.8 CLASSIFICATION OF STATISTICS

Generally, there are three types of statistics, which are as follows:

1. Maxwell–Boltzmann (MB) statistics
2. Bose–Einstein (BE) statistics
3. Fermi–Dirac (FD) statistics

4.8.1 Maxwell–Boltzmann (MB) Statistics

The basic postulates of MB statistics are as follows:
  a. The associated particles are distinguishable.
  b. Each energy state can contain any number of particles.
  c. Total number of particles in the entire system is constant.
  d. Total energy of all the particles in the entire system is constant.
  e. Particles are spinless.

Examples: gas molecules at high temperature and low pressure.

Statistical Mechanics

4.8.2 Bose–Einstein Statistics

The basic postulates of BE statistics are as follows:
a. The associated particles are identical and *indistinguishable*.
b. Each energy state can contain any number of particles.
c. Total energy and total number of particles of the entire system is constant.
d. The particles have zero or integral spin.
e. The wave function of the system is symmetric under the positional exchange of any two particles.

Examples: photon, phonon, all mesons; these are known as bosons.

4.8.3 Fermi–Dirac Statistics

The basic postulates of FD statistics are as follows:
a. Particles are identical and indistinguishable.
b. Total energy and total number of particles of the entire system are constant.
c. Particles have half-integral spin.
d. Particles obey Pauli's exclusion principle, that is, no two particles in a single system can have the same value for each of the four quantum numbers. In other words, a single energy state can contain at best a single particle with appropriate spin.
e. The wave function of the system is antisymmetric under the positional exchange of any two particles.

Examples: electron, proton, neutron, all hyperons ($\Lambda, \Sigma, \Xi, \Omega$) and so on; these are known as fermions.

Now, here we are going to discuss Maxwell–Boltzmann (MB) statistics, Bose–Einstein (BE) statistics and Fermi–Dirac (FD) statistics in detail one by one.

4.9 Maxwell–Boltzmann Statistics

Maxwell–Boltzmann statistics is classical statistics, which is given for the classical particles.
Following are the basic postulates of MB statistics:
a. The associated particles are distinguishable.
b. Each energy state can contain any number of particles.
c. Total number of particles in the entire system is constant.
d. Total energy of all the particles in the entire system is constant.
e. Particles are spinless.

Example: gas molecules at high temperature and low pressure.

Classical Particles: Classical particles are identical but far enough to be distinguishable. The wave functions of the classical particles do not overlap with each other.

Distinguishable: Two particles are said to be distinguishable if their separation is large in compared to their de-Broglie wavelength. For distinguishable particles, it can be determined if two particles have changed their places or not.

4.10 Maxwell–Boltzmann Distribution Law

Let E be the total energy of the entire system, which is constant, N be the total number of identical, distinguishable particles (also constant), and V be the total volume of the system.

We now focus on the number of particles siting in given energy levels $E_1, E_2, E_3, E_4, \ldots E_n$, which are available within the system. The energy levels are fixed for the system.

The number of particles in each energy levels are variable and given by $n_1, n_2, n_3, n_4, \ldots n_r$, as shown in Figure 4.2.

Then, the total probability of a distribution of N particles among K-cells is given by

$$W = \frac{N!(g_1)^{n_1} \cdot (g_2)^{n_2} \cdot (g_3)^{n_3} \cdots (g_k)^{n_k}}{n_1! n_2! n_3! \ldots n_k!}$$

$$W = N! \prod_{r=1}^{K} \frac{(g_r)^{n_r}}{n_r!}. \tag{4.3}$$

Taking natural logarithm of equation (4.3), we get

$$\log W = \log N! + \log \left[\prod_{r=1}^{K} \frac{(g_r)^{n_r}}{n_r!} \right], \tag{4.4}$$

Since the total number of particles are fixed, that is,

$$n_1 + n_2 + n_3 + n_4 + \ldots + n_r + \ldots + n_k = N$$

or, $\sum_{r=1}^{k} n_r = N.$

Since force of interaction between the particles are negligible, that is,

$$E_1 n_1 + E_2 n_2 + E_3 n_3 + E_4 n_4 + \ldots + E_r n_r + \ldots + E_k n_k = U$$

$$\sum_{r=1}^{k} E_r n_r = U.$$

Therefore,

$$\sum_{r=1}^{k} dn_r = dN = 0. \tag{4.5}$$

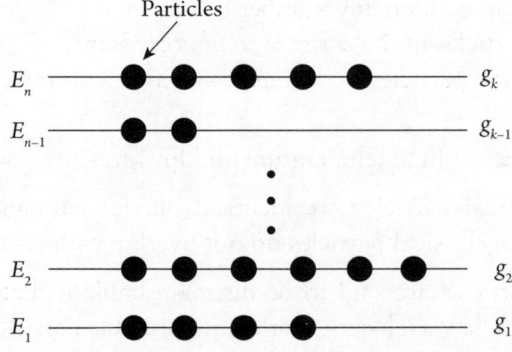

Fig. 4.2 Distribution of particles among different energy levels.

Statistical Mechanics

$$\sum_{r=1}^{k} E_r dn_r = dU = 0. \tag{4.6}$$

For most probable distribution, we should have maximum value W_{max}.

In order to obtain maximum value of W_{max}, apply the mathematical concept of maxima, that is, $d \log W = 0$ or $\frac{1}{W} \cdot dW = 0$, implies that $W \neq 0$, Since W is the function of $n_1, n_2, n_3, n_4, \ldots n_k$, therefore $dW = 0$.

Hence, $d \log W = \frac{d \log W}{dn_1} \cdot dn_1 + \frac{d \log W}{dn_2} \cdot dn_2 + \frac{d \log W}{dn_3} \cdot dn_3 \ldots + \frac{d \log W}{dn_k} \cdot dn_k$

$$d \log W = \sum_{r=1}^{k} \frac{d \log W}{dn_r} \cdot dn_r. \tag{4.7}$$

Now combining equations (4.5), (4.6), and (4.7) by means of Lagrange's method of undetermined multipliers. Let α and β be these multipliers independent of n_r. On multiplying equation (4.5) by $-\alpha$ and equation (4.6) by $-\beta$ and adding these equations to equation (4.7), we get

$$\sum_{r=1}^{K} \left[\frac{\partial}{\partial n_r} (\log W) - \alpha - \beta E_r \right] dn_r = 0. \tag{4.8}$$

This is the condition for most probable distribution.

On applying Stirling's formula, which is given as
$\log n! = n \log n - n$

Using Stirling's formula, equation (4.4) can be written as

$$\log W = N \log N - N + \sum_{r=1}^{K} \left(n_r \log g_r - n_r \log n_r + n_r \right).$$

Differentiating the above equation w.r.t. n_r and equating to zero, we get

$$\sum \frac{\partial}{\partial n_r} (\log W) = \log g_r - \left[\log n_r + \frac{n_r}{n_r} \right] + 1 = \log g_r - \log n_r$$

$$\sum \frac{\partial}{\partial n_r} (\log W) = - \log \frac{n_r}{g_r}. \tag{4.9}$$

On substituting the above term value from equation (4.6) to equation (4.8), we get

$$\sum_{r=1}^{K} \left[-\log \frac{n_r}{g_r} - \alpha - \beta E_r \right] dn_r = 0$$

$$\log \frac{n_r}{g_r} = -\alpha - \beta E_r.$$

Taking antilog both sides, we get

$$\frac{n_r}{g_r} = e^{-\alpha} e^{-\beta E_r},$$

$$n_r = g_r e^{-\alpha} e^{-\beta E_r} \Rightarrow n_r = \frac{g_r}{e^{\alpha+\beta E_r}}. \qquad (4.10)$$

The above relation gives the number of particles lying in r^{th} energy level that has quantum states (microstates). This is known as Maxwell–Boltzmann Distribution Law.

4.11 Evaluation of β

The number of the particles lying in r^{th} energy level is given by

$$n_r = g_r e^{-\alpha} e^{-\beta E_r}.$$

The average energy $\bar{U}$ of the molecules is given by

$$\bar{U} = \frac{U}{N} = \frac{\sum_r E_r n_r}{\sum_r n_r} = \frac{\sum_r E_r g_r e^{-\alpha} e^{-\beta E_r}}{\sum_r g_r e^{-\alpha} e^{-\beta E_r}}.$$

For continuous variation of energy, the above equation can be replaced by the integral so that it becomes

$$\bar{U} = \frac{\int_0^\infty E g(E) \cdot e^{-\beta E}}{\int_0^\infty g(E) \cdot e^{-\beta E} dE}.$$

Here, $g(E) = 2\pi V \left\{ \frac{2m}{h^2} \right\}^{\frac{3}{2}} E^{\frac{1}{2}}$. By substituting the value of $g(E)$ and solving, we get

$$\bar{U} = \frac{3}{2\beta}.$$

According to kinetic theory,

$$\frac{3}{2\beta} = \frac{3}{2} KT \text{ or } \beta = \frac{1}{KT}.$$

4.12 Determination of $e^{-\alpha}$

The total number of particles in the system is given by

$$N = \sum_r n_r = \sum_r g_r e^{-\alpha} e^{-E_r/KT}$$

$$N = e^{-\alpha} \sum_r g_r e^{-E_r/KT}. \qquad (4.11)$$

Therefore, for continuous variation of energy,

$$e^{-\alpha} = \frac{N}{\int_0^\infty g(E) dE e^{-E/KT}}$$

$$e^{-\alpha} = \frac{N}{2\pi V \left[\frac{2m}{h^2} \right]^{\frac{3}{2}} \int_0^\infty E^{\frac{1}{2}} \cdot e^{-E/KT} dE}. \qquad (4.12)$$

Statistical Mechanics

4.13 Evaluation of Integral

$$I = 2\pi V \left[\frac{2m}{h^2}\right]^{\frac{3}{2}} \int_0^\infty E^{\frac{1}{2}} \cdot e^{-E/KT} dE.$$

Let $\dfrac{E}{KT} = x$; then, $E = KTx \Rightarrow dE = KTdx$;

$$I = 2\pi V \left[\frac{2m}{h^2}\right]^{\frac{3}{2}} \int_0^\infty (KTx)^{\frac{1}{2}} \cdot e^{-x} KT dx$$

$$I = 2\pi V \left[\frac{2mKT}{h^2}\right]^{\frac{3}{2}} \int_0^\infty (x)^{\frac{3}{2}-1} \cdot e^{-x} dx.$$

Since the Gamma function is given by

$$\int_0^\infty (x)^{n-1} \cdot e^{-x} dx = n.$$

Therefore,

$$\int_0^\infty (x)^{\frac{3}{2}-1} \cdot e^{-x} dx = \frac{3}{2} = \frac{1}{2}\sqrt{\pi}.$$

Hence, $I = 2\pi V \left[\dfrac{2mKT}{h^2}\right]^{\frac{3}{2}} \times \dfrac{1}{2}\sqrt{\pi} = V \left[\dfrac{2\pi mKT}{h^2}\right]^{\frac{3}{2}}.$ \hfill (4.13)

On substituting the integral value in equation (4.12), we get

$$e^{-\alpha} = \frac{N}{V}\left[\frac{h^2}{2\pi mKT}\right]^{\frac{3}{2}}.$$

Let $e^{-\alpha} = A$; putting this value in equation (4.11), we get

$$N = A \sum_r g_r e^{-E_r/KT}$$

$$\frac{N}{A} = Z = \sum_r g_r e^{-E_r/KT}.$$

This quantity Z is called the Boltzmann partition function or simply the partition function.

4.14 Maxwell–Boltzmann Energy Distribution Function

The average number of particles per quantum state (E_r) in the energy level is represented by the energy distribution function $f(E_r)$, which is given by

$$f(E_r) = \frac{n_r}{g_r}.$$

Put the value of n_r from equation (4.10), we get

$$f(E_r) = \frac{g_r e^{-\alpha} e^{-\beta E_r}}{g_r} = e^{-\alpha} e^{-\beta E_r}. \tag{4.14}$$

On Substituting the value of $e^{-\alpha}$ and β, we get

$$f(E) = \frac{N}{V}\left[\frac{h^2}{2\pi m KT}\right]^{\frac{3}{2}} e^{-E/KT}. \tag{4.15}$$

This is an expression of Maxwell–Boltzmann energy distribution function.

4.15 Maxwell–Boltzmann Energy Distribution Law

If energy levels are very close together, then number of particles $n(E)\,dE$, whose energies lie between E and $E + dE$ is given by

$$n(E)dE = f(E) \cdot g(E)dE$$

$$n(E)dE = \frac{N}{V}\left[\frac{h^2}{2\pi m KT}\right]^{\frac{3}{2}} e^{-E/KT} \cdot 2\pi V \left[\frac{2m}{h^2}\right]^{\frac{3}{2}} E^{\frac{1}{2}} dE$$

$$n(E)dE = \frac{2\pi N}{[\pi KT]^{\frac{3}{2}}} E^{\frac{1}{2}} \cdot e^{-E/KT} dE. \tag{4.16}$$

4.16 Maxwell–Boltzmann Speed or Velocity Distribution Law

The number of particles $n(E)\,dE$, whose energies lie between E and $E + dE$ is given by

$$n(E)dE = \frac{2\pi N}{[\pi KT]^{\frac{3}{2}}} E^{\frac{1}{2}} \cdot e^{-E/KT} dE. \tag{4.17}$$

The energy of an ideal gas is purely kinetic,

$$E = \frac{1}{2}mv^2 \Rightarrow dE = \frac{1}{2}m \cdot 2v dv = mv dv.$$

Put the value of E and dE in equation (4.18), we get

$$n(v)dv = \frac{2\pi N}{[\pi KT]^{\frac{3}{2}}} \left[\frac{1}{2}mv^2\right]^{\frac{1}{2}} \cdot e^{-\frac{1}{2}mv^2/KT} mv dv$$

$$n(v)dv = 4\pi N \left[\frac{m}{2\pi KT}\right]^{\frac{3}{2}} v^2 \cdot e^{-mv^2/2KT} dv. \tag{4.18}$$

The above expression is known as Maxwell–Boltzmann speed or velocity distribution law. The speed or velocity distribution plot for a number of molecules at different temperatures is shown in Figure 4.3.

Statistical Mechanics

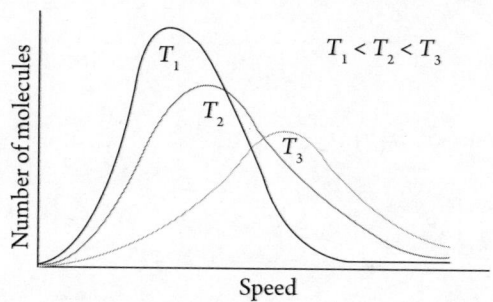

Fig. 4.3 The number of molecules [$n(v)$] versus speed at different temperatures.

4.17 Most Probable Speed

The number of molecules lying in velocity range v and $v + dv$ is given by

$$n(v) = 4\pi N \left[\frac{m}{2\pi KT}\right]^{\frac{3}{2}} v^2 \cdot e^{-mv^2/2KT}.$$

Taking log of both sides, we get

$$\log n(v) = \log\left\{4\pi N \left[\frac{m}{2\pi KT}\right]^{\frac{3}{2}}\right\} + 2\log v - \frac{mv^2}{2KT}.$$

Differentiating w.r.t. v, we get

$$\frac{1}{n(v)} \cdot \frac{dn(v)}{dv} = 0 + \frac{2}{v} - \frac{2mv}{2KT}$$

$$\frac{dn(v)}{dv} = \left[\frac{2}{v} - \frac{mv}{KT}\right] n(v).$$

At most probable speed $v = v_p$; $n(v)$ should be maximum,

i.e., $\left[\dfrac{dn(v)}{dv}\right]_{v=v_p} = 0$

$$\left[\frac{2}{v_p} - \frac{mv_p}{KT}\right] = 0$$

$$v_p = \sqrt{\frac{2KT}{m}} = 1.414\sqrt{\frac{KT}{m}}.$$

This is an expression of most probable speed of molecules in Maxwell–Boltzmann statistics.

4.18 Average Speed

The average speed of gas molecules can be given by

$$\bar{v} = \int_0^\infty v n(v) dv$$

$$\bar{v} = \frac{1}{N} \int_0^\infty v \, n(v) \, dv.$$

On substituting the value of $n(v)dv$, we get

$$\bar{v} = \frac{1}{N} \int_0^\infty v \cdot 4\pi N \left[\frac{m}{2\pi KT}\right]^{\frac{3}{2}} v^2 \cdot e^{-mv^2/2KT} \, dv$$

$$\bar{v} = 4\pi \left[\frac{m}{2\pi KT}\right]^{\frac{3}{2}} \int_0^\infty v^3 \cdot e^{-mv^2/2KT} \, dv$$

Let $\dfrac{mv^2}{2KT} = x$; $v = \left[\dfrac{2KTx}{m}\right]^{\frac{1}{2}} \Rightarrow dv = \left[\dfrac{2KT}{m}\right]^{\frac{1}{2}} \cdot \dfrac{1}{2} x^{-\frac{1}{2}} dx.$

On substituting the value, we get

$$\bar{v} = 4\pi \left[\frac{m}{2\pi KT}\right]^{\frac{3}{2}} \int_0^\infty \left[\frac{2KTx}{m}\right]^{\frac{3}{2}} \cdot e^{-x} \left[\frac{2KT}{m}\right]^{\frac{1}{2}} \cdot \frac{1}{2} x^{-\frac{1}{2}} dx$$

$$\bar{v} = \frac{4\pi}{2} \left[\frac{m}{2\pi KT}\right]^{\frac{3}{2}} \left[\frac{2KT}{m}\right]^{\frac{3}{2}} \left[\frac{2KT}{m}\right]^{\frac{1}{2}} \int_0^\infty x \cdot e^{-x} dx.$$

Using Gamma functions,

$$\int_0^\infty (x)^{n-1} \cdot e^{-x} dx = n$$

$$\int_0^\infty (x)^{2-1} \cdot e^{-x} dx = 2 = 1.$$

On substituting the integral value in the above equation, we get

$$\therefore \bar{v} = \sqrt{\frac{8KT}{\pi m}} = 1.596 \sqrt{\frac{KT}{m}}.$$

This is an expression of average speed of molecules in Maxwell–Boltzmann statistics.

4.19 ROOT-MEAN-SQUARE SPEED

The root-mean-square velocity of the gas molecules may be given by

$$\bar{v}^2 = \frac{1}{N} \int_0^\infty v^2 n(v) \, dv.$$

On substituting the value of $n(v)dv$, we get

$$\bar{v}^2 = \frac{1}{N} \int_0^\infty v^2 \cdot 4\pi N \left[\frac{m}{2\pi KT}\right]^{\frac{3}{2}} v^2 \cdot e^{-mv^2/2KT} \, dv$$

Statistical Mechanics

$$\bar{v}^2 = 4\pi \left[\frac{m}{2\pi KT}\right]^{\frac{3}{2}} \int_0^\infty v^4 \cdot e^{-mv^2/2KT} dv.$$

Let $\dfrac{mv^2}{2KT} = x$; $v = \left[\dfrac{2KTx}{m}\right]^{\frac{1}{2}} \Rightarrow dv = \left[\dfrac{2KT}{m}\right]^{\frac{1}{2}} \cdot \dfrac{1}{2} x^{-\frac{1}{2}} dx.$

On substituting the value, we get

$$\bar{v}^2 = 4\pi \left[\frac{m}{2\pi KT}\right]^{\frac{3}{2}} \int_0^\infty \left[\frac{2KTx}{m}\right]^2 \cdot e^{-x} \left[\frac{2KT}{m}\right]^{\frac{1}{2}} \cdot \frac{1}{2} x^{-\frac{1}{2}} dx$$

$$\bar{v}^2 = \frac{4}{\sqrt{\pi}} \cdot \left(\frac{KT}{m}\right) \int_0^\infty x^{3/2} \cdot e^{-x} dx.$$

Using Gamma functions,

$$\int_0^\infty (x)^{n-1} \cdot e^{-x} dx = n$$

$$\int_0^\infty (x)^{5/2-1} \cdot e^{-x} dx = \frac{5}{2} = \frac{3\sqrt{\pi}}{4}.$$

On substituting integral value in the above equation, we get

$$\bar{v}^2 = \sqrt{\frac{3KT}{m}} = 1.732 \sqrt{\frac{KT}{m}}.$$

This is an expression of RMS speed of molecules in Maxwell–Boltzmann statistics.

On comparing the most probable speed, average speed, and RMS speed in Maxwell–Boltzmann statistics, we get a relation as

$$\frac{v_p}{\sqrt{2}} = \frac{\bar{v}}{\sqrt{8/\pi}} = \frac{v_{r.m.s.}}{\sqrt{3}} = \sqrt{\frac{KT}{m}}.$$

The plot of the number of molecules versus different speeds is shown in Figure 4.4. It is concluded that rms speed > average speed > most probable speed.

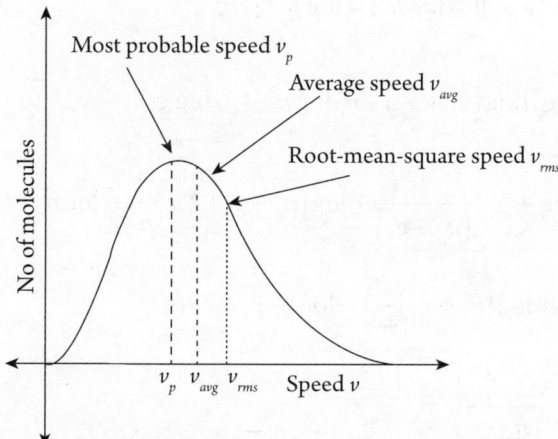

Fig. 4.4 The number of molecules [$n(v)$] versus velocity at any arbitrary temperature.

4.20 Bose–Einstein Statistics

The fundamental principles of BE statistics are as follows:

a. The related particles are identical and not able to be identified as different or distinct.
b. Any number of particles can be present in any energy state.
c. The overall system energy and particle count are both constant.
d. The particles' spin is integral or zero.
e. Under any two-particle positional interchange, the system's wave function is symmetric. Examples of bosons include the photon and the phonon, which are all mesons.

4.21 Bose–Einstein Distribution Law

Consider the assembly of Boson particles; the overall number of independent ways (W) of distributing the particles $(n_1, n_2, n_3, n_4, \ldots n_s)$ among the quantum states at different energy levels is given as

$$W = \frac{(n_1 + g_1 - 1)!}{n_1! g_1 - 1!} \cdot \frac{(n_2 + g_2 - 1)!}{n_2! g_2 - 1!} \cdots \frac{(n_s + g_s - 1)!}{n_s! g_s - 1!}$$

$$W = \prod_{r=1}^{s} \frac{(n_r + g_r - 1)!}{n_r! g_r - 1!}. \tag{4.19}$$

Neglecting 1 in comparison to n and g_r, we get

$$W = \prod_{r=1}^{s} \frac{(n_r + g_r)!}{n_r! g_r!}. \tag{4.20}$$

Condition for most probable distribution is given as

$$\frac{\partial}{\partial n_r}(\log W) - \alpha - \beta E_r = 0. \tag{4.21}$$

Taking log of both side and using Stirling's formula,

$$\log W = \sum_{r=1}^{s} \left[\log(n_r + g_r)! - \log n_r! - \log g_r! \right]$$

$$\log W = \sum_{r=1}^{s} \left[(n_r + g_r)\log(n_r + g_r) - n_r \log n_r - g_r \log g_r \right]$$

$$\frac{\partial}{\partial n_r}\log W = \sum_{r=1}^{s} \left[(n_r + g_r)\frac{1}{(n_r + g_r)} + \log(n_r + g_r) - n_r \frac{1}{n_r} - \log n_r \right]$$

$$\frac{\partial}{\partial n_r}\log W = \sum_{r=1}^{s} \left[1 + \log(n_r + g_r) - 1 - \log n_r \right]$$

$$\frac{\partial}{\partial n_r}\log W = \log\left[\frac{(n_r + g_r)}{n_r}\right].$$

On substituting the above value in equation (4.21), we get

$$\log\left[\frac{(n_r + g_r)}{n_r}\right] - \alpha - \beta E_r = 0$$

$$\log\left[\frac{(n_r + g_r)}{n_r}\right] = \alpha + \beta E_r$$

$$1 + \frac{g_r}{n_r} = e^{\alpha + \beta E_r}$$

$$\frac{g_r}{n_r} = e^{\alpha + \beta E_r} - 1$$

$$n_r = \frac{g_r}{e^\alpha e^{\beta E_r} - 1}. \tag{4.22}$$

The above relation gives the number of Boson particles lying in s^{th} energy level that has quantum states (microstates). This is known as Bose–Einstein distribution law.

4.22 BOSE–EINSTEIN ENERGY DISTRIBUTION FUNCTION

The Bose–Einstein energy distribution function $f(E_r)$ is defined by

$$f(E_r) = \frac{n_r}{g_r} = \frac{1}{e^\alpha e^{\beta E_r} - 1}$$

$$\text{or, } f(E) = \frac{n_r}{g_r} = \frac{1}{e^\alpha e^{\beta E} - 1}. \tag{4.23}$$

This is an expression for Bose–Einstein energy distribution function.

4.23 BOSE–EINSTEIN ENERGY DISTRIBUTION LAW

The number of particles lying in the range E and $E + dE$ are given by

$$n(E)dE = f(E) \cdot g(E)dE$$

$$n(E)dE = \frac{g(E)dE}{e^\alpha e^{\beta E} - 1},$$

where $\beta = \frac{1}{KT}$ and $e^\alpha = \frac{V}{N}\left[\frac{2\pi m KT}{h^2}\right]^{\frac{3}{2}}$,

On substituting the value of $g(E)dE$, we get

$$n(E)dE = 2\pi V \left[\frac{m}{h^2}\right]^{\frac{3}{2}} \frac{E^{\frac{1}{2}}}{e^\alpha e^{\beta E} - 1}.$$

The above relation is known as Bose–Einstein energy distribution Law which indicates that e^α must be greater than unity. When e^α is very small ($e^\alpha \ll 1$) compared to unity, the Maxwell–Boltzmann distribution law is not applicable but the Bose–Einstein and Fermi–Dirac distribution laws are applicable. In such cases, the collection of particles is said to be in the state of degeneracy. The e^{α} is the determining factor of degeneracy and is known as degeneracy parameter.

4.24 PLANCK'S RADIATION FORMULA

The density of microstates in terms of momentum is given by

$$g(P) = \frac{4\pi V P^2}{h^3}.$$

Therefore, the total number of eigen states between momenta p and $p + dp$ is given by

$$g(P)dp = \frac{4\pi V P^2 dp}{h^3}. \tag{4.24}$$

For a photon, $P = \dfrac{h\upsilon}{c} \Rightarrow dP = \dfrac{h\,d\upsilon}{c}$,

In order to obtain the total number of eigen states between frequencies υ and $\upsilon + d\upsilon$ substitute the values of momentum P and dP in equation (4.24), we get

$$g(\upsilon)d\upsilon = \frac{4\pi V \upsilon^2 d\upsilon}{c^3}. \tag{4.25}$$

Taking into account the doubling of states due to the polarization of photons (that is, two modes of propagation of each photon), the total number of eigen states between frequencies υ and $\upsilon + d\upsilon$ is given by

$$g(\upsilon)d\upsilon = \frac{8\pi V \upsilon^2 d\upsilon}{c^3}. \tag{4.26}$$

The Bose–Einstein distribution law is given as

$$dn = \frac{g(\upsilon)d\upsilon}{e^{\alpha}e^{\beta E} - 1}. \tag{4.27}$$

In this case, $\alpha = 0$; $\beta = \dfrac{1}{KT}$; $E = h\upsilon$. By putting the value of $g(\upsilon)d\upsilon$ from equation (4.26) in equation (4.27), we get

$$dn = \frac{8\pi V \upsilon^2}{c^3} \cdot \frac{d\upsilon}{e^{\frac{h\upsilon}{KT}} - 1}$$

$$\frac{dn}{V} = \frac{8\pi \upsilon^2}{c^3} \cdot \frac{d\upsilon}{e^{\frac{h\upsilon}{KT}} - 1}. \tag{4.28}$$

This equation represents the total number of photons per unit volume lying in frequency range υ and $\upsilon + d\upsilon$.

The energy density of radiation of frequencies between υ and $\upsilon + d\upsilon$ can be obtained by multiplying equation (4.28) by photons energy $E = h\upsilon$. Therefore,

$$E_{\upsilon} = \left[\frac{dn}{V}h\upsilon\right] = \frac{8\pi \upsilon^2 d\upsilon}{c^3} \cdot \frac{h\upsilon}{e^{\frac{h\upsilon}{KT}} - 1}$$

Statistical Mechanics

$$E_\upsilon = \frac{8\pi h \upsilon^3}{c^3} \cdot \frac{d\upsilon}{e^{\frac{h\upsilon}{KT}} - 1}. \tag{4.29}$$

This is the well know Planck's law of radiation in terms of frequency.

Since $\upsilon = \frac{c}{\lambda} \Rightarrow d\upsilon = -\frac{c\,d\lambda}{\lambda^2}$, equation (4.29) in terms of wavelength λ can be written as

$$E_\lambda d\lambda = \frac{8\pi h}{\lambda^3} \cdot \frac{\left(-\frac{c\,d\lambda}{\lambda^2}\right)}{e^{\frac{hc}{\lambda KT}} - 1}$$

$$E_\lambda d\lambda = \frac{8\pi hc}{\lambda^5} \cdot \frac{d\lambda}{e^{\frac{hc}{\lambda KT}} - 1}. \tag{4.30}$$

This is the well know Planck's law of radiation in terms of wavelength.

4.25 Derivation of Various Laws Related with Black Body

1. **Rayleigh–Jeans Law:**

   For small values of $\frac{hc}{\lambda KT} \ll 1$, that is, in the region of long wavelength, the exponential term in equation (4.30) can be expanded and retaining only the first term as

   $$e^{\frac{hc}{\lambda KT}} \approx 1 + \frac{hc}{\lambda KT}.$$

   On substituting this value in equation (4.30), we get

   $$E_\lambda d\lambda = \frac{8\pi hc}{\lambda^5} \cdot \frac{d\lambda}{\left(1 + \frac{hc}{\lambda KT} - 1\right)} = \frac{8\pi KT}{\lambda^4} d\lambda.$$

   This is Rayleigh–Jeans law.

2. **Wien's Displacement Law:**

   From Planck's radiation law, we have

   $$E_\lambda d\lambda = \frac{8\pi hc}{\lambda^5} \cdot \frac{d\lambda}{e^{\frac{hc}{\lambda KT}} - 1}. \tag{4.31}$$

   At constant temperature T of a black body, the wavelength λ_m at which the energy density is maximum is given by

   $$\left[\frac{dE_\lambda}{d\lambda}\right] = 0.$$

   Taking log of both sides of equation (4.31), we get

   $$\log E_\lambda = \log(8\pi hc) - 5\log \lambda - \log\left(e^{\frac{hc}{\lambda KT}} - 1\right).$$

Differentiating the above equation w.r.t. $d\lambda$, we get

$$\frac{1}{E_\lambda}\left[\frac{dE_\lambda}{d\lambda}\right] = 0 - \frac{5}{\lambda} - \frac{1}{e^{\frac{hc}{\lambda KT}} - 1}\left(e^{\frac{hc}{\lambda KT}}\right)\left(-\frac{hc}{\lambda^2 KT}\right)$$

$$\frac{1}{E_\lambda}\left[\frac{dE_\lambda}{d\lambda}\right] = -\frac{5}{\lambda} + \frac{hc}{\lambda^2 KT} \cdot \frac{e^{\frac{hc}{\lambda KT}}}{e^{\frac{hc}{\lambda KT}} - 1} = \frac{1}{\lambda}\left[-5 + \frac{hc}{\lambda^2 KT} \cdot \frac{e^{\frac{hc}{\lambda KT}}}{e^{\frac{hc}{\lambda KT}} - 1}\right].$$

At $\lambda = \lambda_m$, $\left[\frac{dE_\lambda}{d\lambda}\right] = 0$.

Therefore, $-5 + \frac{hc}{\lambda^2 KT} \cdot \frac{e^{\frac{hc}{\lambda KT}}}{e^{\frac{hc}{\lambda KT}} - 1} = 0$.

Let $\frac{hc}{\lambda_m KT} = x$. Then, $-5 + \frac{xe^x}{e^x - 1} = 0$,

or, $\frac{xe^x}{e^x - 1} = 5$,

or, $\left(1 - \frac{x}{5}\right)e^x = 1$.

On solving the above equation by trial-and-error method, we get

$\frac{hc}{\lambda_m KT} = 4.9651$.

Therefore, $\lambda_m T = \frac{hc}{4.9651} = constant$.

This is Wien's displacement law.

3. **Stefan's Boltzmann Law:**

The energy density of the total radiation in a black body enclosure is given by

$$E = \int_0^\infty E(\lambda)d\lambda = \int_0^\infty \frac{8\pi hc}{\lambda^5} \cdot \frac{d\lambda}{e^{\frac{hc}{\lambda KT}} - 1}.$$

Let $\frac{hc}{\lambda KT} = x \Rightarrow \frac{hc}{KTx} = \lambda;\ d\lambda = -\frac{hc}{KTx^2}dx$.

When $\lambda = 0;\ x = \infty$ and $\lambda = \infty;\ x = 0$.

Hence, $E = \int_\infty^0 \frac{8\pi hc}{1} \cdot \left(\frac{KTx}{hc}\right)^5 \cdot \frac{1}{e^x - 1}\left(-\frac{hc}{KTx^2}dx\right)$

$$E = \int_0^\infty \left(\frac{8\pi K^4 T^4}{h^3 c^3}\right) \cdot \frac{x^3}{e^x - 1}dx = \left(\frac{8\pi K^4 T^4}{h^3 c^3}\right)\int_0^\infty \frac{x^3}{e^x - 1}dx.$$

The value of integral is $\int_0^\infty \frac{x^3}{e^x - 1}dx = \frac{\pi^4}{15}$, thus

Statistical Mechanics

$$E = \left(\frac{8\pi K^4 T^4}{h^3 c^3}\right) \cdot \frac{\pi^4}{15}$$

$$E = \frac{4}{c}\left(\frac{2\pi^5 K^4}{15 h^3 c^2}\right) T^4. \tag{4.32}$$

It can be shown that the total radiated power by a black body at a given temperature is given by

$$P = \frac{cE}{4}.$$

Substituting the value of E from equation (4.32) in the above expression, we get

$$P = \left(\frac{2\pi^5 K^4}{15 h^3 c^2}\right) T^4 \tag{4.33}$$

or, $P = \sigma T^4$, \hfill (4.34)

where $\sigma = \left(\frac{2\pi^5 K^4}{15 h^3 c^2}\right).$

Equation (4.34) is known as Stefan's Boltzmann law of radiation. The law states that the total power P of all wavelengths radiated by a black body per unit area per second is proportional to the fourth power of its absolute temperature. The constant σ is called Stefan's constant. The value of Stefan's constant is calculated and given as

$$\sigma = 5.67 \times 10^{-8} \frac{J}{m^2 s K^4} = 5.67 \times 10^{-8} \frac{W}{m^2 K^4}.$$

4.26 Fermi–Dirac Statistics

The basic postulates of FD statistics are as follows:
a. The related particles are identical and not able to be identified as different or distinct.
b. The overall system energy and particle count are both constant.
c. Particles have half-integral spin.
d. Pauli's exclusion principle, which states that no two particles in the same system can have the same value for each of the four quantum numbers, is observed in particle behavior. In other words, the maximum number of particles with the proper spin that can exist in a single energy state is one.
e. Under any positional exchange of any two particles, the system's wave function is antisymmetric.

Examples include the electron, proton, neutron, and all hyperon (Λ, Σ, Ξ, Ω), which are also referred to as fermions.

4.27 Fermi–Dirac Distribution

Consider an assembly of such particles that are identical and weakly interacting and exhibiting spin angular momentum $\frac{\hbar}{2}$. The system is in thermal equilibrium at temperature T, total energy U, volume V, and total number of particles N.

Let n_r identical particles are to be distributed among g_r quantum states in r^{th} energy level such that maximum one particle can occupy one quantum state. The total number of ways of arranging n_r particles among g_r quantum states in r^{th} energy level is given by

$$W = \frac{g_r(g_r-1)(g_r-2)\ldots(g_r-n_r+1)(g_r-n_r)!}{(g_r-n_r)} = \frac{g_r!}{(g_r-n_r)!}.$$

In each of these arrangements, the number of permutations of n_r particles among themselves is $n_r!$. Since the particles are indistinguishable, these permutations do not give independent arrangements. Hence, the actual number of independent permutations of n_r particles among g_r quantum states is given by

$$W = \frac{g_r!}{n_r!(g_r-n_r)!}. \tag{4.35}$$

The total number W of independent ways of obtaining a distribution of $(n_1, n_2, n_3, \ldots n_s)$ particles among the quantum states in the various energy levels, with a maximum of one particle per quantum state, is the product of expressions given by equation (4.35) for $r = 1, 2, 3, \ldots s$.

Thus,

$$W = \frac{g_1!}{n_1!(g_1-n_1)!} \cdot \frac{g_2!}{n_2!(g_2-n_2)!} \cdot \frac{g_3!}{n_3!(g_3-n_3)!} \cdots \frac{g_s!}{n_s!(g_s-n_s)!}$$

$$W = \prod_{r=1}^{s} \frac{g_r!}{n_r!(g_r-n_r)!}. \tag{4.36}$$

The condition for most probable distribution is given by

$$\frac{\partial}{\partial n_r}(\log W) - \alpha - \beta E_r = 0. \tag{4.37}$$

Taking logarithm of both sides of equation (4.36), we get

$$\log W = \sum_{r=1}^{s} \left[\log g_r! - \log n_r! - \log(g_r-n_r)! \right]. \tag{4.38}$$

Now, applying Stirling's formula, we have

$$\log W = \sum_{r=1}^{s} \left[(g_r \log g_r - n_r \log n_r - (g_r-n_r)\log(g_r-n_r) \right]$$

Differentiating the above equation w.r.t. n_r, we get

$$\frac{\partial}{\partial n_r} \log W = 0 - \left(n_r \frac{1}{n_r} + \log n_r \right) - \left[-\frac{(g_r-n_r)}{(g_r-n_r)} - \log(g_r-n_r) \right]$$

$$\frac{\partial}{\partial n_r} \log W = -\log n_r + \log(g_r-n_r)$$

$$\frac{\partial}{\partial n_r} \log W = \log\left(\frac{g_r-n_r}{n_r}\right).$$

Statistical Mechanics

Substituting this value in equation (4.37), we get

$$\log\left(\frac{g_r - n_r}{n_r}\right) - \alpha - \beta E_r = 0$$

$$\log\left(\frac{g_r - n_r}{n_r}\right) = \alpha + \beta E_r.$$

Taking antilogarithm of both sides, we get

$$\left(\frac{g_r - n_r}{n_r}\right) = e^{\alpha + \beta E_r}$$

$$n_r = \frac{g_r}{e^{\alpha + \beta E_r} + 1}. \tag{4.39}$$

This is an expression of the Fermi–Dirac energy distribution law for a system of identical fermions.

4.28 Fermi–Dirac Energy Distribution Function or Fermi Function (Occupation Index)

Fermi function is defined as

$$f(E) = \frac{n_r}{g_r} = \text{number of fermions per quantum states}$$

$$f(E) = \frac{1}{e^{\alpha + \beta E} + 1}. \tag{4.40}$$

Since $\beta = \frac{1}{KT}$,

$$f(E) = \frac{1}{e^\alpha e^{E/KT} + 1}. \tag{4.41}$$

Since the electrons obey Pauli's exclusion principle, which states that two electrons in the same orbital or quantum state must have opposite spins. This can be possible only if $\alpha = -\frac{E_F}{KT}$, where E_F is the Fermi energy level. Thus, the Fermi function becomes

$$f(E) = \frac{1}{e^{(E-E_F)/KT} + 1}. \tag{4.42}$$

This is an expression of Fermi function or occupation index.

Fermi Energy

The Fermi energy at absolute zero temperature is denoted by E_F and this is considered as constant over a large range of temperature.

The Fermi function is given by

$$f(E) = \frac{1}{e^{(E-E_F)/KT} + 1}. \tag{4.43}$$

From equation (4.42), the value of Fermi function can be evaluated at low temperature as follows:

At $T = 0$ K,

if $E < E_F$, $f(E) = \dfrac{1}{e^{-\infty}+1} = 1$ (Filled valence band.) \hfill (4.44)

If $E > E_F$, $f(E) = \dfrac{1}{e^{\infty}+1} = 0$ (Empty conduction band.) \hfill (4.45)

At any other temperature $T > 0$, when $E = E_F$,

$$f(E) = \dfrac{1}{e^{\frac{0}{kT}}+1} = \dfrac{1}{2}$$ (Half-filled valence and conduction band.) \hfill (4.46)

Hence, all the energy states are filled below the Fermi level, while all the energy states above it are empty at $T = 0$ K, as shown by equations (4.44) and (4.45), respectively. However, at any other temperature $T > 0$, only half of the energy states are filled above and below the Fermi level, as shown by equation (4.46).

4.29 Fermi–Dirac Energy Distribution Law

The number of particles lying in the range E and $E + dE$ are given by

$$n(E)dE = f(E) \cdot g(E)dE$$

$$n(E)dE = \dfrac{g(E)dE}{e^{\alpha}e^{\beta E}+1},$$

where $\beta = \dfrac{1}{KT}$ and $e^{\alpha} = \dfrac{V}{N}\left[\dfrac{2\pi m KT}{h^2}\right]^{\frac{3}{2}}$.

On substituting the value of $g(E)dE$, we get

$$n(E)dE = 4\pi V \left[\dfrac{2m}{h^2}\right]^{\frac{3}{2}} \dfrac{E^{\frac{1}{2}}}{e^{(E-E_F)/KT}+1}.$$

The above relation is known as FD energy distribution law. The plot of the Fermi–Dirac energy distribution at different temperatures is shown in Figure 4.5.

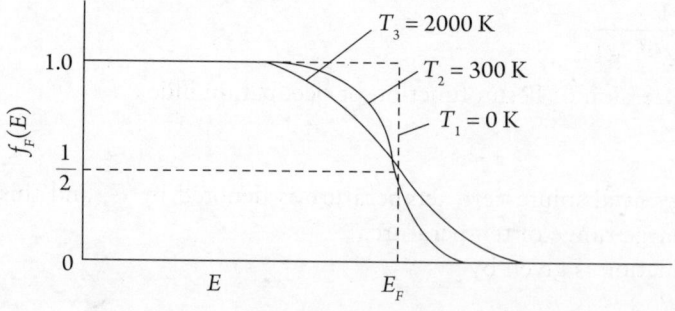

Fig. 4.5 The Fermi–Dirac distribution for several temperatures.

Statistical Mechanics

Fermi Energy for Free Electron in a Metal

The total number N of free electrons in a metal of volume V is given by

$$N = \int_0^\infty n(E)dE = \int_0^\infty f(E)g(E)dE = \int_0^{E_F} f(E)g(E)dE + \int_{E_F}^\infty f(E)g(E)dE. \tag{4.47}$$

Now, at $T = 0$ K, if $E \leq E_F$, then $f(E) = 1$ and if $E \geq E_F$, then $f(E) = 0$. Hence, in equation (4.47), the second integral is zero and in the first integral $f(E) = 1$. Therefore, equation (4.47) becomes

$$N = \int_0^{E_F} g(E)dE = 4\pi V \left(\frac{2m}{h^2}\right)^{\frac{3}{2}} \int_0^{E_F} E^{\frac{1}{2}} dE = 4\pi V \left(\frac{2m}{h^2}\right)^{\frac{3}{2}} \cdot \frac{2}{3} E_F^{3/2} = \frac{8\pi V}{3}\left(\frac{2mE_F}{h^2}\right)^{\frac{3}{2}}.$$

From the above equation, we get

$$E_F = \frac{h^2}{2m}\left(\frac{3N}{8\pi V}\right)^{\frac{2}{3}}$$

$$E_F = \frac{h^2}{2m}\left(\frac{3n}{8\pi}\right)^{\frac{2}{3}}.$$

This is an expression of Fermi energy for free electrons in metal, where $n = \dfrac{N}{V}$ is the number of free electrons per unit volume, that is, the free electron density.

Mean Internal Energy of Free Electrons in a Metal at Absolute Zero Temperature

Mean internal energy of free electrons in a metal at absolute zero temperature is given by

$$\bar{U}_0 = \frac{1}{N}\int_0^\infty E n(E)dE = \frac{1}{N}\int_0^\infty E f(E)g(E)dE$$

$$= \frac{1}{N}\int_0^{E_F} E f(E)g(E)dE + \frac{1}{N}\int_{E_F}^\infty E f(E)g(E)dE. \tag{4.48}$$

Now, at $T = 0$ K, if $E \leq E_F$, then $f(E) = 1$ and if $E \geq E_F$, then $f(E) = 0$. Hence, in equation (4.48), the second integral is zero, and in the first integral, $f(E) = 1$. Therefore, equation (4.48) becomes

$$\bar{U}_0 = \frac{1}{N}\int_0^{E_F} E g(E)dE = \frac{1}{N}\cdot 4\pi V\left(\frac{2m}{h^2}\right)^{\frac{3}{2}}\int_0^{E_F} E\cdot E^{\frac{1}{2}} dE = \frac{4\pi V}{N}\left(\frac{2m}{h^2}\right)^{\frac{3}{2}}\int_0^{E_F} E^{\frac{3}{2}} dE$$

$$\bar{U}_0 = \frac{4\pi V}{N}\left(\frac{2m}{h^2}\right)^{\frac{3}{2}} \cdot \frac{2}{5} E_F^{\frac{5}{2}} = \left[\frac{8\pi V}{5N}\left(\frac{2m}{h^2}\right)^{\frac{3}{2}} E_F^{\frac{3}{2}}\right] E_F.$$

On substituting the value of E_F inside the bracket of the above equation, we get

$$\bar{U}_0 = \left[\frac{8\pi V}{5N}\left(\frac{2m}{h^2}\right)^{\frac{3}{2}}\left(\frac{h^2}{2m}\right)^{\frac{3}{2}}\frac{3N}{8\pi V}\right] E_F$$

$$\bar{U}_0 = \frac{3}{5} E_F.$$

This is the mean internal energy of free electrons in a metal at absolute zero temperature.

Fermi Temperature

The Fermi temperature (T_F) is defined as the ratio of the Fermi energy (E_F) at absolute zero to Boltzmann's constant (k). Thus,

$$T_F = \frac{E_F}{k}.$$

Relation between the Fermi Temperature and the Density of Free Electrons in a Metal

Since the Fermi temperature (T_F) is given by

$$T_F = \frac{E_F}{k}.$$

On substituting the value of E_F in the above equation, we get

$$T_F = \frac{h^2}{2mk}\left(\frac{3n}{8\pi}\right)^{\frac{2}{3}} = \frac{h^2}{2mk}\left(\frac{3}{8\pi}\right)^{\frac{2}{3}} n^{\frac{2}{3}}.$$

On substituting the value of constant terms, we get

$$T_F = \frac{(6.63 \times 10^{-34})^2}{2 \times 9.1 \times 10^{-31} \times 1.38 \times 10^{-23}} \left(\frac{3}{8\pi}\right)^{\frac{2}{3}} n^{\frac{2}{3}}$$

$$T_F = 4.23 \times 10^{-15} n^{\frac{2}{3}},$$

where $n = \frac{N}{V}$ is the number of free electrons per unit volume, that is, the free electron density.

Relation between the Fermi Temperature and Degeneracy Parameter

The degeneracy parameter A is given by

$$A = \frac{N}{V}\left[\frac{h^2}{2\pi mKT}\right]^{\frac{3}{2}} = n\left[\frac{h^2}{2\pi mKT}\right]^{\frac{3}{2}}, \tag{4.49}$$

where $n = \frac{N}{V}$ is the number of free electrons per unit volume, that is, the free electron density.

$$n = \frac{8\pi}{3}\left(\frac{2mkT_F}{h^2}\right)^{\frac{3}{2}}. \tag{4.50}$$

On substituting the value of n in equation (4.49), we get

$$A = \frac{8\pi}{3}\left(\frac{2mkT_F}{h^2}\right)^{\frac{3}{2}} \left[\frac{h^2}{2\pi mKT}\right]^{\frac{3}{2}}$$

Statistical Mechanics

$$A = \frac{8\pi}{3\sqrt{\pi}}\left(\frac{T_F}{T}\right)^{\frac{3}{2}} = 1.5\left(\frac{T_F}{T}\right)^{\frac{3}{2}}.$$

$$\frac{T_F}{T} = \left(\frac{A}{1.5}\right)^{\frac{2}{3}}.$$

The above equation shows that the degeneracy condition $A > 1$ is equivalent to $T_F > 1$. Since the value of T_F for free electrons in a metal is very large. Therefore, the free electron gas is highly degenerate in metals.

Solved Problems

Ex. 1: Calculate the probability that the speed of oxygen molecule lies between 100 and 101 m/sec at 200 K.

Solution:

The probability that a molecule possesses speed between v and $v + dv$ is given by

$$P(v)dv = 4\pi\left[\frac{m}{2\pi KT}\right]^{\frac{3}{2}} v^2 \cdot e^{-mv^2/2KT} dv.$$

Here, $m = 3\,a.m.u. = \dfrac{32}{6\times 10^{23}} g = \dfrac{32}{6\times 10^{26}}$ Kg, $v = 100$ m/s,

$dv = 101 - 100 = 1$ m/s, $k = 1.38\times 10^{-23}$ J/K, $T = 200$ K.

$$P(v)dv = 4\times 3.14\left[\frac{\frac{32}{6\times 10^{26}}}{2\times 3.141\times 1.38\times 10^{-23}\times 200}\right]^{\frac{3}{2}} (100)^2 \cdot e^{-\frac{32}{6\times 10^{26}}\times \frac{(100)^2}{2\times 1.38\times 10^{-23}\times 100}} dv.$$

$$P(v)dv = 6.11\times 10^{-4}.$$

Ex. 2: Calculate the value of root-mean-square speed of a molecule of hydrogen at N.T.P. The Boltzmann constant is 1.38×10^{-23} J/K and Avogadro number is 6×10^{26} per Kg mol.

Solution:

We have

$$v_{r.m.s.} = \sqrt{\frac{2KT}{m}}.$$

Here, $K = 1.38\times 10^{-23}$ J/K, $T = 273$ K, and $m = \dfrac{2}{6\times 10^{-26}}$ Kg.

$$\therefore v_{r.m.s.} = \sqrt{\frac{2\times 1.38\times 10^{-23}\times 273}{\frac{2}{6\times 10^{-26}}}} = 1838 \text{ m/s}.$$

Ex. 3: Calculate the root-mean-square speed and most probable speed of a gas whose density is 1.4 gm/l at a pressure of 10^5 N/m².

Solution:

We have $P = 10^5 \, \text{N/m}^2$, $\rho = 1.4 \frac{gm}{l} = 1.4 \, \text{Kg/m}^3$

$$v_{r.m.s.} = \sqrt{\frac{3P}{\rho}} = \sqrt{\frac{3 \times 10^5}{1.4}} = 4.6 \times 10^2 \, \text{m/sec}$$

and most probable speed,

$$v_{r.m.s.} = \sqrt{2/3} \, v_{r.m.s.} = \sqrt{2/3} \times 4.6 \times 10^2 = 3.77 \times 10^2 \, \text{m/s}.$$

Ex. 4: Calculate the surface temperature of the sun and moon given that $\lambda_m = 4753 \, \text{Å}$ and 14 μm, respectively. λ_m is the wavelength of maximum intensity of emission.

Solution:

According to Wien's displacement law,

$$\lambda_m T = \text{constant} = 0.2898 \times 10^{-2}$$

$$\therefore T = \frac{0.2898 \times 10^{-2}}{\lambda_m}$$

(i) For the sun, $\lambda_m = 4753 \, \text{Å} = 4753 \times 10^{-10} \, \text{m}$

$$T_s = \frac{0.2898 \times 10^{-2}}{4753 \times 10^{-10}} = 6097 \, \text{K}.$$

(ii) For the moon,

$$T_s = \frac{0.2898 \times 10^{-2}}{14 \times 10^{-6}} = 207 \, \text{K}.$$

Ex. 5: A body at 1500 K emits maximum energy at a wavelength 20000 Å. If the sun emits maximum energy at wavelength 5500 Å, what would be the temperature of the sun?

Solution:

According to Wien's displacement law,

$$\lambda_m T = \lambda'_m T'$$

or, $T = \dfrac{\lambda_m T}{\lambda'_m} = \dfrac{20000 \, \text{Å} \times 1500 \, \text{K}}{5500 \, \text{Å}} = 5454 \, \text{K}.$

Ex. 6: Calculate the average energy of an oscillator of frequency 5.6×10^{12} per second at $T = 330$ K, treating it as (i) classical oscillator and (ii) Planck's oscillator.

Solution:

The average energy of a classical oscillator is given by

$$KT = (1.38 \times 10^{-23}) \times 330 = 4.554 \times 10^{-21} \, \text{J}.$$

The average energy of Planck's oscillator is given by

$$\frac{hc}{e^{h\nu/KT} - 1} = \frac{6.64 \times 10^{-34} \times 3 \times 10^8}{e^{\frac{6.64 \times 10^{-34} \times 5.6 \times 10^{12}}{4.554 \times 10^{-21}}} - 1} = 2.9450 \times 10^{-21} \, \text{J}$$

Statistical Mechanics

Ex. 7: A gas has two indistinguishable identical particles in the rth energy level in which three independent quantum states are available. Find the possible number of microstates of the gas according to the Bose–Einstein statistics.

Solution:

We have $n_r = 2$, and $g_r = 3$.

According to the B–E statistics, total number W of the microstates is given by the exact formula

$$W = \frac{(n_1 + g_1 - 1)!}{n_1!(g_1 - 1)!} \cdot \frac{(n_2 + g_2 - 1)!}{n_2(g_2 - 1)!} \cdot \ldots \cdot \frac{(n_r + g_r - 1)!}{n_r!(g_r - 1)!}.$$

The two particles are in the rth energy level, and the other levels are empty. Therefore,

$$W = \frac{(n_r + g_r - 1)!}{n_r!(g_r - 1)!} = \frac{(2+3-1)}{2!(3-1)!}$$

$$= \frac{4!}{2!2!} = \frac{4 \times 3 \times 2 \times 1}{(2 \times 1)(2 \times 1)} = 6 \text{ microstates}.$$

Ex. 8: Treating liquid $_2He^4 - I$ as an ideal Bose–Einstein gas, find the critical temperature T_B at which there is the transition of liquid He – I to liquid He – II. Given that the molar volume of the liquid He at the critical temperature = $27.4 \times 10^{-6} m^3$, the mass of $_2He^4$ atom = 6.65×10^{-27} kg, Avogadro's number $N = 6.63 \times 10^{23}$/mol, Planck's constant $h = 6.63 \times 10^{-34}$ Js, and Boltzmann's constant $k = 1.38 \times 10^{-23}$ J/K.

Solution:

$$T_B = \left(\frac{h^2}{2\pi mk}\right)\left(\frac{N}{2.612 \times V}\right)^{2/3}$$

$$= \frac{(6.63 \times 10^{-34})^2}{2 \times 3.14 \times 6.65 \times 10^{-27} \times 1.38 \times 10^{-23}} \left(\frac{6.023 \times 10^{23}}{2.612 \times 27.4 \times 10^{-6}}\right)^{2/3}$$

$$= \frac{6.63^2 \times 10^{-18}}{57.63} \left(\frac{6.023 \times 10 \times 10^{27}}{2.612 \times 2.74}\right)^{2/3}$$

$$= 0.763 \left(\frac{60.23}{7.16}\right)^{2/3} = 0.764 \times (8.41)^{2/3}$$

$$= 0.764 \times 4.135 = 3.15 \text{ K}.$$

Ex. 9: For an ideal Bose–Einstein gas, the condensation temperature is T_B. Find the temperature at which the number of molecules in the zero-energy state ($E = 0$) is 7/8 times the total number of molecules in the gas.

Solution:

$$n_0 = N\left[1 - \left(\frac{T}{T_B}\right)^{3/2}\right]$$

$$\frac{7}{8}N = N\left[1 - \left(\frac{T}{T_B}\right)^{3/2}\right],$$

whence $\left(\dfrac{T}{T_B}\right)^{3/2} = \dfrac{1}{8}$.

Therefore $T = \left(\dfrac{1}{8}\right)^{2/3}$

$T_B = \dfrac{1}{4}T_B$.

Ex. 10: Calculate the temperature of the moon and the surface of the sun, having the wavelength of maximum intensity of emission 14 μm and 4753 Å respectively.

Solution:

According to Wein's displacement law

$\lambda_m T = $ constant $= 0.2898 \times 10^{-2}$ mK

$\therefore T = \dfrac{0.2898 \times 10^{-2}}{\lambda_m}$

(i) For the moon, $T_m = \dfrac{0.2898 \times 10^{-2}}{14 \times 10^{-6}} = 207$ K

(ii) For the sun, $\lambda_m = 4753$ Å $= 4753 \times 10^{-10}$ m

$T_s = \dfrac{0.2898 \times 10^{-2}}{4753 \times 10^{-10}} = 6097$ K.

Ex. 11: A body at 1500 K emits maximum energy at a wavelength 20000 Å. If the sun emits maximum energy at wavelength 5500 Å, what would be the temperature of the sun?

Solution:

By Wien's displacement law

$\lambda_m T = \lambda'_m T'$

or $T = \dfrac{\lambda_m T}{\lambda'_m}$

$= \dfrac{20000 \text{ Å} \times 1500 \text{ K}}{5500 \text{ Å}} = 5454$ K.

Ex. 12: Calculate the average energy of (i) classical oscillator, (ii) Planck's oscillator. Given an oscillator frequency is 5.7×10^{12} Hz at $T = 331$ K.

Solution:

The formula for a classical oscillator's average energy is

$kT = (1.380 \times 10^{-23}) \times 330$

$= 4.554 \times 10^{-21}$ J.

The formula for Planck's oscillator's average energy is

$\dfrac{hc}{(e^{h\nu/kT} - 1)}$

Statistical Mechanics

$$= \frac{(6.626 \times 10^{-34}) \times (5.6 \times 10^{12})}{\left\{ \exp\left(\frac{6.626 \times 10^{-34}}{1.380 \times 10^{-23} \times 300}\right) - 1 \right\}}$$

$= 2.9450 \times 10^{-21}$ J.

Ex. 13: Demonstrate how Rayleigh–Jeans law for longer wavelengths and Wien's law for shorter wavelengths may be derived from Planck's law.

Solution:

Planck's law is

$$E_\lambda d\lambda = \frac{8\pi hc}{\lambda^5} \frac{d\lambda}{e^{ch/\lambda kT} - 1}.$$

For shorter wavelengths, the exponential term $e^{hc/\lambda kT}$ becomes large compared to unity so that Planck's law reduces to

$$E_\lambda d\lambda = \frac{8\pi hc}{\lambda^5} \frac{d\lambda}{e^{ch/\lambda kT}}$$

$$= 8\pi hc \cdot \lambda^{-5} \cdot e^{-ch/\lambda kT} \cdot d\lambda$$

$$= A\, \lambda^{-5} e^{-a/\lambda T} \cdot d\lambda,$$

where A and a are constants. This is known as Wien's law.

For longer wavelengths, the exponential term $e^{hc/\lambda kT}$ is small and can be expanded to

$$\left(1 + \frac{hc}{\lambda kT}\right),$$

so that Planck's law becomes

$$E_\lambda d\lambda = \frac{8\pi hc}{\lambda^5} \frac{dx}{\left(1 + \frac{hc}{\lambda kT} - 1\right)} = \frac{8\pi hc}{\lambda^5} \cdot \frac{\lambda kT}{hc} \cdot d\lambda$$

$$= \frac{8\pi kT}{\lambda^4} d\lambda, \text{ which is Rayleigh–Jeans law.}$$

Previous Year Questions (University Examination)

1. Derive Maxwell–Boltzmann (MB) law of distribution of energy among the molecules of an ideal gas and derive the law of distribution of speed among the molecules.

2. Derive expressions for the most probable, average, and root-mean-square speeds in Maxwell–Boltzmann statistics.

3. Show that the most probable, average, and root-mean-square speeds in Maxwell–Boltzmann statistics are related as given below,

$$\frac{v_p}{\sqrt{2}} = \frac{\bar{v}}{\sqrt{8/\pi}} = \frac{v_{r.m.s.}}{\sqrt{3}} = \sqrt{\frac{KT}{m}}.$$

4. Deduce the law of distribution of energy of photons in Bose–Einstein (BE) statistics.

5. Derive Planck's law of black body radiation using Bose–Einstein distribution law.

6. Employing Planck's law of radiation, derive Stefan's law of thermal radiation and Wien's displacement law.
7. Obtain the expression for Fermi function or occupation index.
8. Compare the basic postulates of the Maxwell–Boltzmann (MB), Bose–Einstein (BE), and Fermi–Dirac (FD) statistics.
9. Deduce the law of distribution of energy of particles in Fermi–Dirac (FD) statistics.
10. Derive an expression of density of microstates in terms of momentum.
11. Discuss essential requirements of Bose–Einstein statistics and explain the criteria for applicability of the Bose–Einstein statistics.
12. A gas has only three indistinguishable identical particles in the rth energy level in which four independent states are available. Find the possible number of microstates of the gas according to the Bose–Einstein statistics.
13. The transition of liquid He – I to liquid He – II takes place at the Bose temperature 3.15 K. Find the temperature at which the number of Helium atoms in the zero-energy state will be 3/4 times the total number of atoms.
14. What are Bose temperature and Bose–Einstein condensation?

Multiple Choice Questions

1. The Maxwell–Boltzmann distribution law is given by
   (a) $n_r = \dfrac{g_r}{e^{\alpha+\beta E_r}+1}$
   (b) $n_r = \dfrac{g_r}{e^{\alpha+\beta E_r}-1}$
   (c) $n_r = \dfrac{g_r}{e^{\alpha+\beta E_r}}$
   (d) None of these

2. The root-mean-square speed of a molecule varies with temperature T as
   (a) $\sqrt{T}$
   (b) $\dfrac{1}{\sqrt{T}}$
   (c) $T^{\frac{3}{2}}$
   (d) $T^{-\frac{3}{2}}$

3. The correct ratio of $v_{mp}, \bar{v},$ and v_{rms} is
   (a) $1:2:\sqrt{3}$
   (b) $\sqrt{2}:\dfrac{2\sqrt{2}}{\pi}:\sqrt{3}$
   (c) $\dfrac{\sqrt{2}}{\pi}:2:\sqrt{3}$
   (d) $\sqrt{2}:\dfrac{\sqrt{2}}{\pi}:\sqrt{3}$

4. The Maxwell–Boltzmann distribution law is given by
   (a) $n_r = \dfrac{g_r}{e^{\alpha+\beta E_r}+1}$
   (b) $n_r = \dfrac{g_r}{e^{\alpha+\beta E_r}-1}$
   (c) $n_r = \dfrac{g_r}{e^{\alpha+\beta E_r}}$
   (d) None of these

5. Wien's distribution law is a special case of
   (a) Stefan's law
   (b) Rayleigh–Jean's distribution law
   (c) Kirchhoff's law
   (d) Planck's distribution law

Statistical Mechanics

6. Pauli's exclusion principle applies to
   (a) MB statistics
   (b) BE statistics
   (c) FD statistics
   (d) All of these

7. The Fermi–Dirac distribution law is given by
   (a) $n_r = \dfrac{g_r}{e^{\alpha+\beta E_r}+1}$
   (b) $n_r = \dfrac{g_r}{e^{\alpha+\beta E_r}-1}$
   (c) $n_r = \dfrac{g_r}{e^{\alpha+\beta E_r}}$
   (d) None of these

8. If $v_{mp}, \bar{v}$, and v_{rms} are the most probable speed, average speed, and root-mean-square speed of the molecule of a gas, then
   (a) $v_{mp} > \bar{v} > v_{rms}$
   (b) $v_{rms} > \bar{v} > v_{mp}$
   (c) $\bar{v} > v_{mp} > v_{rms}$
   (d) None of these

9. Statistics appropriate to photon is
   (a) FD statistics
   (b) BE statistics
   (c) MB statistics
   (d) None of the above

10. Which one of the particles does not have a spin $\dfrac{1}{2}$?
    (a) Proton
    (b) Neutron
    (c) Photon
    (d) Neutrino

11. Statistics applicable to fermions is
    (a) FD statistics
    (b) BE statistics
    (c) MB statistics
    (d) None of the above

12. MB statistics are applicable on
    (a) Classical particles
    (b) Photons
    (c) Electrons
    (d) All of these

13. In the frequency versus energy density curve of black body radiation at different temperatures, as the temperature is increased, the peak of the individual curves shifts toward
    (a) Remains stationary
    (b) Higher frequency side
    (c) Lower frequency side
    (d) None of the above

14. Statistical methods give greater accuracy, when the number of observations is
    (a) Very large
    (b) Average
    (c) Very small
    (d) None of the above

Bibliography

1. Fowler, R. H. (1929). *Statistical Mechanics: The Theory of the Properties of Matter in Equilibrium.* Cambridge University Press.
2. Richard, C. T. (1938). *The Principles of Statistical Mechanics.* Clarendon Press.
3. Landau, L. D. and E. M. Lifshitz (1980). *Statistical Physics: Volume 5.* 3rd ed. Butterworth-Heinemann.
4. Ter Haar, D. (1995). *Elements of Statistical Mechanics.* Butterworth-Heinemann.
5. Huang, K. (1963). *Statistical Mechanics.* Wiley.

6. Ryogo, K., et al. (1965). *Statistical Mechanics*. North-Holland.
7. Penrose, O. (1970). *Foundations of Statistical Mechanics: A Deductive Treatment*. Pergamon.
8. McQuarrie, Donald A. (1975). *Statistical mechanics*. Harper & Row.
9. Reichl, Linda E. (1980). *A Modern Course in Statistical Physics*. Edward Arnold.
10. Pathria, P. K. and Beale, P. (2021). *Statistical Mechanics*, 4th edn. Elsevier/Academic Press.

Keys

1. (c)   2. (a)   3. (b)   4. (b)   5. (d)   6. (c)   7. (a)   8. (b)   9. (c)   10. (c)
11. (a)   12. (a)   13. (b)   14. (a)

CHAPTER 5

Lasers

5.1 Introduction

The word laser is an abbreviation for "light amplification by stimulated emission of radiation." Historically, the laser is the outgrowth of maser, which means "microwave amplification by stimulated emission of radiation." If the stimulated radiation lies in optical region, the device is called optical maser or laser. Laser beam is highly monochromatic, highly coherent, intense, and highly collimated with a small diversion.

Using ruby as the active material, T. H. Maiman invented the first laser system in 1960. Thus, it is known as Ruby Laser.

5.2 Absorption, Spontaneous Emission, and Stimulated Emission of Radiation

Absorption of Radiation: An atom has a number of quantized states. Initially, an atom is in ground state, that is, all its electrons possess lowest possible energy state. When energy is given to an atom, it goes to excited state, that is, its electron jumps to a higher energy state by absorbing a quantum of radiation or photon. This process is called the absorption of radiation shown in Figure 5.1.

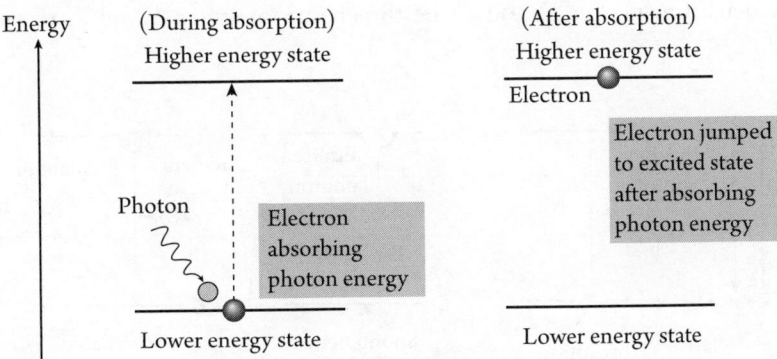

Fig. 5.1 Absorption of radiation or photons.

If E_1 and E_2 are the energies of an electron in initial and final states and v is the frequency of absorbed radiation, then

$$E_2 - E_1 = hv$$

$$v = \frac{E_2 - E_1}{h},$$

where h is Plank's constant. Thus, absorption is a stimulated process.

5.3 Spontaneous Emission

An atom in an excited state remains for only about 10^{-8} sec. After staying 10^{-8} sec, the atom jumps automatically to a lower energy state, emitting a radiation. If initially an atom is in excited state, then it spontaneously jumps to ground state, emitting a photon of frequency v, given by

$$v = \frac{E_2 - E_1}{h}.$$

This process is called spontaneous emission of radiation. If there are a large number of atoms in an excited state, the photons emitted by different atoms have a random phase, and hence they are incoherent.

5.4 Stimulated Emission

If an atom is in an excited state, then an incident photon of energy $\Delta E = hv$ causes the atom to jump to a lower energy state, emitting an additional photon of the same frequency. Thus, the two photons of the same frequency are released at the same time. This phenomenon is called stimulated emission of radiation. These two photons are coherent and travel in the same direction, as shown in Figure 5.2.

5.5 Einstein's A and B Coefficients

Consider an assembly of atoms in equilibrium at temperature T with radiation of frequency v and energy density $u(v)$. Let E_1 and E_2 be the energy of states 1 and 2, respectively, with $E_1 < E_2$.

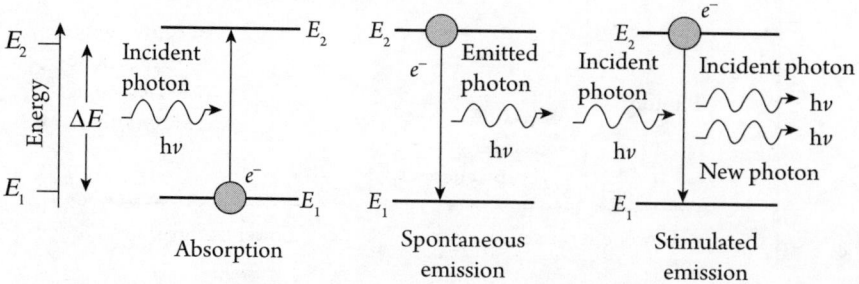

Fig. 5.2 Absorption, spontaneous emission, and stimulated emission process.

Let N_1 and N_2 be the number of atoms in energy states 1 and 2, respectively, at any instant. The probable rate of absorption transition from $1 \to 2$ depends on the properties of states 1 and 2, and is thus proportional to the energy density $u(v)$ of the radiation of frequency v incident on the atom.

Thus,

$$P_{12} = B_{12}\, u(v), \tag{5.1}$$

where B_{12}, which depends on the properties of states 1 and 2, is called Einstein's coefficient of absorption of radiation. The number of atoms in state 1 that can absorb a photon and give rise to absorption per unit time,

$$N_1 P_{12} = N_1 B_{12}\, u(v). \tag{5.2}$$

The probability of spontaneous emission $2 \to 1$ depends only on the properties of states 1 and 2. According to Einstein, the probable rate spontaneous emission is denoted by

$$(P_{21})_{\text{spontaneous}} = A_{21}. \tag{5.3}$$

The A_{21} is called Einstein's coefficient of spontaneous emission of radiation.

The probability of stimulated emission from energy state E_2 to energy state E_1 depends on the energy of incident radiation as well as on the properties of two energy states involved and is given

$$(P_{21})_{\text{stimulated}} = B_{21}\, u(v), \tag{5.4}$$

where B_{21} is Einstein's coefficient of stimulated emission of radiation.

The total probability of emission transition $2 \to 1$ is the sum of spontaneous and stimulated emission probabilities, that is,

$$P_{21} = A_{21} + B_{21}\, u(v). \tag{5.5}$$

Conversely, the number of photons in state 2 that can cause emission process (spontaneous + stimulated) per unit time is given by

$$N_2 P_{21} = N_2\, [A_{21} + B_{21}\, u(v)]. \tag{5.6}$$

At equilibrium, the absorption and emission rate must be equal, that is,

$$N_1 P_{12} = N_2 P_{21}$$

$$N_1 B_{12}\, u(v) = N_2\, [A_{21} + B_{21}\, u(v)]$$

$$N_1 B_{12}\, u(v) = N_2 A_{21} + N_2 B_{21}\, u(v)]$$

$$[N_1 B_{12} - N_2 B_{21}]\, u(v) = N_2 A_{21}$$

$$u(v) = \frac{N_2 A_{21}}{[N_1 B_{12} - N_2 B_{21}]}$$

$$u(v) = \frac{A_{21}}{B_{21}} \frac{1}{\left[\dfrac{N_1 B_{12}}{N_2 B_{21}} - 1\right]}. \tag{5.7}$$

The number of atoms N_1 and N_2 in energy states E_1 and E_2 in thermal equilibrium at temperature T is provided by the Boltzmann distribution rule, which is

$$N_1 = N_0 e^{-E_1/KT} \text{ and } N_2 = N_0 e^{-\frac{E_2}{KT}},$$

where N_0 is the total number of atoms present, and k is Boltzmann's constant.

$$\frac{N_2}{N_1} = \frac{e^{-\frac{E_2}{KT}}}{e^{-\frac{E_1}{KT}}} = e^{-\frac{E_2 - E_1}{KT}}.$$

However, $E_2 - E_1 = h\nu$ (energy of photons emitted or absorbed),

$$\frac{N_2}{N_1} = e^{-\frac{h\nu}{KT}}.$$

Therefore, $\dfrac{N_1}{N_2} = e^{\frac{h\nu}{KT}}$.

On substituting the above value in equation (5.7), we get

$$u(\nu) = \frac{A_{21}}{B_{21}} \frac{1}{\left[e^{\frac{h\nu}{KT}} \frac{B_{12}}{B_{21}} - 1 \right]}. \tag{5.8}$$

The Plank's radiation formula is given by

$$u(\nu) = \frac{8\pi h \nu^3}{c^3} \cdot \frac{1}{e^{\frac{h\nu}{KT}} - 1}. \tag{5.9}$$

On comparing equation (5.8) and equation (5.9), we get

$$\frac{A_{21}}{B_{21}} = \frac{8\pi h \nu^3}{c^3} \quad \text{and} \quad \frac{B_{12}}{B_{21}} = 1. \tag{5.10}$$

From equation (5.10), we can easily conclude that

$B_{12} = B_{21}$.

Thus, the probability of stimulated emission is the same as that of (induced) absorption.

$$\frac{A_{21}}{B_{21}} \propto \nu^3.$$

Hence, the ratio of spontaneous emission and stimulated emission is proportional to ν^3. This suggests that as the energy differential between the two energy levels widens, the probability of spontaneous emission dominates the probability of induced emission more and more.

5.6 Population Inversion

Consider an energy state E containing N atoms per unit volume. This number is called population and is given by Boltzmann's equation

Lasers

$$N = N_0 e^{-E/KT}, \qquad (5.11)$$

where N_0 is the number of atoms in ground state ($E = 0$), K is Boltzmann's constant, and T is an absolute temperature.

From equation (5.11), it is obvious that population is highest in the ground state and rapidly declines as we move to higher energy states.

Consequently, if N_1 is the population of atoms in the lower energy state E_1 and N_2 is the population of atoms in the greater energy state E_2, we have

$$N_1 = N_0 e^{-E_1/KT}$$

$$N_2 = N_0 e^{-E_2/KT}$$

$$\frac{N_2}{N_1} = \frac{e^{-\frac{E_2}{KT}}}{e^{-\frac{E_1}{KT}}} = e^{-\frac{E_2-E_1}{KT}}$$

$$N_2 = N_1 e^{-\frac{E_2-E_1}{KT}}. \qquad (5.12)$$

As $E_2 > E_1$, $N_2 < N_1$, therefore, if an electromagnetic wave is incident on the substance, there is net absorption of radiation. Usually, the population decreases with an increase of energy of the state. If N_1, N_2, and N_3 are the population in the energy states E_1, E_2, and E_3, respectively, such that $E_1 < E_2 < E_3$, then

$$N_1 > N_2 > N_3.$$

If the process of stimulated emission dominates over the process of spontaneous emission, then it may be possible that $N_2 > N_1$. If this happens, the state is called population inversion. In the state of population inversion, the upper energy levels are more populated than the lower energy levels. To achieve population inversion, the external energy is supplied to excite the atoms of the substance. This process is known as pumping.

5.7 Threshold Condition for Laser Action

For laser action, two threshold conditions should be satisfied:

i. The probability of spontaneous emission that produced incoherent radiations should be much smaller than the probability of stimulated radiations, that is, $A_{21} \ll B_{21} u(\nu)$.

ii. The emission rate should be larger than the absorption rate. This can be achieved by population inversion, that is, the number of atoms in the higher energy state exceeds with that in the lower energy state as shown in Figure 5.3. The population inversion can be achieved by pumping methods.

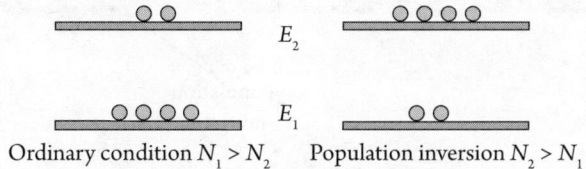

Ordinary condition $N_1 > N_2$ Population inversion $N_2 > N_1$

Fig. 5.3 Population inversion.

5.8 Pumping

By stimulating the medium with the appropriate type of energy, the population inversion may be realized. Pumping is the name of this procedure.

A laser can be pumped using a variety of techniques to create the population inversion required for the occurrence of stimulated emission. The most popular pumping techniques include the following.

1. **Optical Pumping:** The pumping is referred to as optical pumping if optical energy is used to the medium to induce population inversion. In optical pumping, the luminous energy typically originates from the light source in brief light bursts, as seen in Figure 5.4. T. H. Maiman employed this technique for the first time in the ruby laser, and solid state lasers currently use it as well. The same sort of helical xenon flash lamp that is used in photography is simply filled with the laser material.

2. **Electric Discharge:** The pumping by electric discharge is preferred in gaseous ion laser (e.g., Argon–ion laser). In discharge tube when a potential difference is applied between cathode and anode. The electrons emitted from the cathode are accelerated toward anode; some of these electrons collide with the atoms of active medium to ionize the medium and raise it to higher energy level as shown in Figure 5.5. This generates the required population inversion. This process is also called direct electron excitation.

3. **Inelastic Atom–Atom Collision:** One class of atoms are brought to their excited states during an elastic discharge. With other types of atoms, these atoms clash elastically. Inversion in population is provided by these atoms. He-Ne laser serves as an example.

4. **Direct Conversion:** Light-emitting diodes (LEDs) directly transform electrical energy into radiant radiation. In semiconductor lasers, population inversion via direct collision is demonstrated.

5. **Chemical Conversion:** There is no need for additional energy sources in a chemical laser conversion since energy is produced by a chemical process. In order to accomplish population inversion, this process is employed to pump a CO_2 laser.

To achieve stimulated emission of radiation, the steps are shown in Figure 5.6.

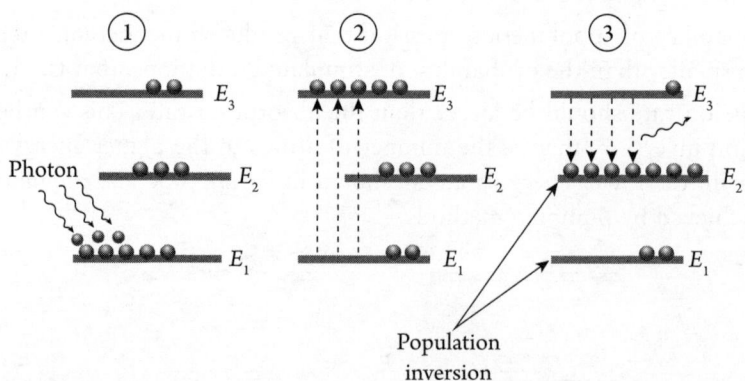

Fig. 5.4 Optical pumping process: Step 1–Photon absorption by atoms; Step 2–Transition of atoms; Step 3–Population inversion achieved.

Lasers

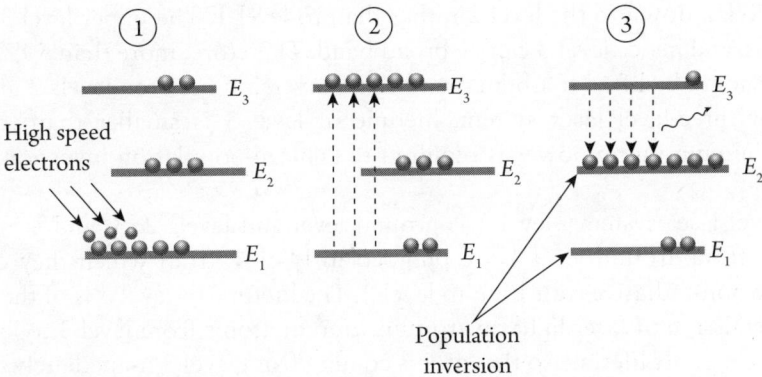

Fig. 5.5 Electric discharge process.

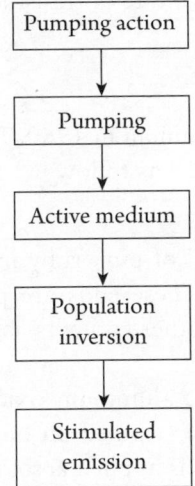

Fig. 5.6 Flow chart to achieve stimulated emission.

5.9 CONCEPT OF THREE- AND FOUR-LEVEL LASER SYSTEMS

When pumping action takes place between two energy levels, we raise a large number of atoms from the ground state to the excited state. Resorting to continued pumping, more and more atoms are transferred to the excited state till a situation is reached where the population density of two levels becomes equal while pumping. This state of equal population takes place when the rate of absorption becomes equal to that of stimulated emission. This state is called a saturated state and further increase in population in both energy levels is not possible. Thus, population inversion is not possible in two energy-level system.

Now, consider a three-level laser system and assume that all the levels are non-degenerate. Pump is applied to transfer atoms from 1 to 3, while laser is produced due to transition from level 2 to 1. In this system, pump lifts atoms from level 1 to 3 and from level 3 to 2, there is non-radiative transition. The level 2 is a metastable state. Thus, using pump, the atoms are transferred from 1 to 2 through level 3. If the relaxation from 3 to 2 is very fast, then most of the atoms relax down to the level 2 rather than to level 1. The upper level 3 is not one of the laser levels, and hence level 3 can be broad band. Therefore, more than 50% of the atoms from level 1 should be lifted to obtain population inversion between levels 2 and 1.

In case of three-level laser system, lifetime of level 3 is smaller than the lifetime of level 2, so minimum pump power is required to achieve population inversion, as shown in Figure 5.7.

In four-level laser systems, level 1 is ground level and levels 2, 3, and 4 are the excited levels. Atoms from ground level 1 are pumped to level 4, from which they quickly decay through some non-radiative transition to level 3. The lifetime of level 3 is of the order of 10^{-3} sec. There is emission of laser light due to transition of atoms from level 3 to level 2. Level 2 must have a very short lifetime so that atoms coming from level 3 immediately drop down to level 1, ready for being pumped to level 4. The population inversion can be obtained between level 3 and level 2 even for a very small pump power because the rate of relaxation of atoms from level 2 to level 1 is faster than the rate of transition of atoms from level 3 to level 2.

5.10 RUBY LASER

Ruby laser was developed by T. H. Maiman in 1960. This is a three-level solid-state laser. Its construction and working are described as follows.

Construction

Ruby laser consists of a single crystal of pink ruby in the form of a cylindrical rod whose opposite ends are flat and parallel. These ends are polished in such a way that one end acts as fully silvered mirror and the other end acts as partially silvered mirror as shown in Figure 5.8.

Ruby crystal is basically synthetic aluminum oxide (Al_2O_3) with 0.05% of chromium dioxide (Cr_2O_3). A xenon flash tube is winded on outer surface of ruby rod that provides optical pumping, when a high voltage is applied across it.

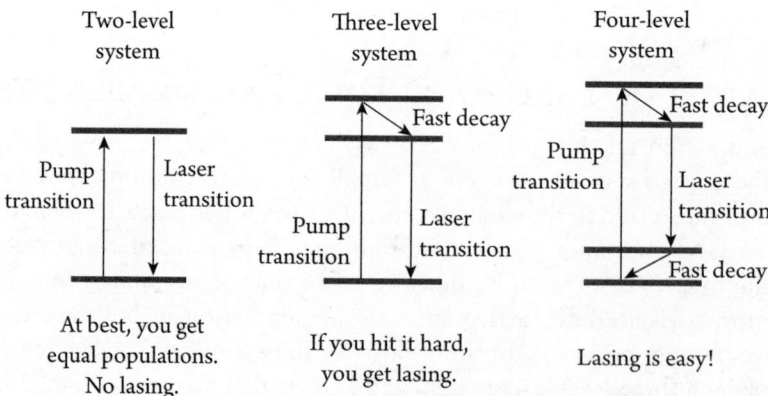

Fig. 5.7 Concept of two-, three-, and four-level laser systems.

Lasers

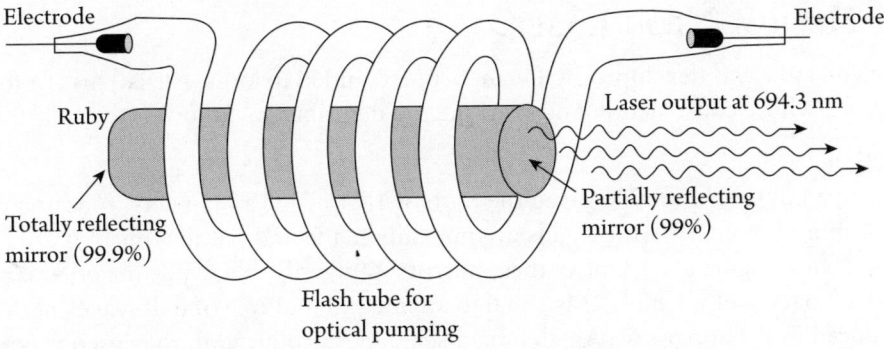

Fig. 5.8 Ruby laser setup.

Working

In a ruby rod, some of the Al^{+3} ions are replaced by Cr^{+3} ions. These impurity ions give pink color of ruby laser. The chromium doped ruby has three energy level ground state (G), metastable stable (M), and higher/excited state (H), which are used for laser action, as shown in Figure 5.9.

Initially, the chromium ions are in ground energy level G. The xenon flash lamp excites the chromium ions to a high-energy level H by emitting the photons of wavelength 5600 Å. These photons are absorbed by chromium ions for transitions $G \rightarrow H$. Some of these Cr^{+3} ions fall to the metastable energy level M by losing energy due to collisions with each other. This gives rise to radiation less transition $H \rightarrow M$. The metastable energy level M provides long lifetime to Cr^{+3} ions as compared to high-energy level H. Therefore, the number of Cr^{+3} ions in level M goes on increasing, while due to pumping, the number of Cr^{+3} ions in ground level G goes on decreasing. Thus, population inversion is established between the metastable level M and the ground level G.

When some Cr^{+3} ions spontaneously return to ground level G from metastable level M, they emit photons of red light having wavelength 6943 Å. These photons are reflected back and forth between the mirrored ends of ruby rod, stimulating other Cr^{+3} ions to jump from metastable level M to the ground level G by emitting photons of wavelength 6943Å. These new photons further stimulate the remaining Cr^{+3} ions. Thus, a photon multiplication takes place by a chain of stimulated emission. Finally, a large pulse of monochromatic, coherent, and highly unidirectional red laser light emerges from partially silvered end of the ruby rod.

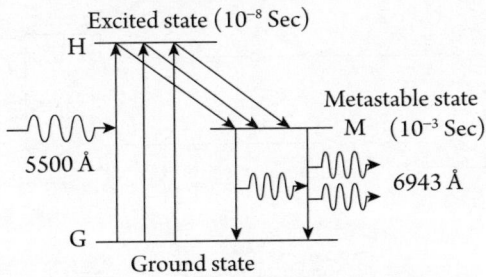

Fig. 5.9 Energy levels of ruby laser.

5.11 HELIUM-NEON LASER

Helium–neon laser was developed by Javan, Bennett, and Harriot in 1961. This is a four-level gaseous laser, whose construction and working are described as follows.

Construction

It consists of a mixture of He-Ne gas in the ratio of 10:1 inside a long glass tube at a pressure of about 1 mm of mercury, whose ends are optically plane and parallel mirrors are fitted on them, as shown in Figure 5.10. One of these mirrors is fully silvered, while the other is partially silvered. The spacing of the mirrors is equal to an integral multiple of half wavelength of laser light produced in the apparatus. An electric discharge is produced in the gaseous mixture of helium and neon by electrodes connected to a high frequency electric source.

Working

When the electric discharge is passed through the gaseous mixture of helium and neon, the electrons accelerating in the tube collide with helium and neon atoms due to which they excite to their metastable energy levels M_1 and M_2 having energies 20.61 eV and 20.66 eV, respectively. Some of the excited helium atoms transfer their energy to neon atoms in collisions, with 0.05 eV of additional energy being provided by the kinetic energy of the atoms. The purpose of the helium atoms is to achieve a population inversion in the neon atoms. The laser transition occurs when Ne atoms fall from metastable level M_2 of energy 20.66 eV to an excited energy level E of energy 18.70 eV by emitting photons of wavelengths 632.8 nm. From the excited energy level E, the neon atoms transfer to a lower metastable level M_3 by spontaneous emission and then they transit to ground energy level.

This process is continuous with time; thus, He-Ne laser beam is continuous, as shown in Figure 5.11.

5.12 SUPERIORITY OF HE-NE LASER OVER RUBY LASER

1. He-Ne gas laser produces a continuous beam while ruby laser produces beam in ultra-short pulses.
2. The He-Ne laser beam is highly monochromatic and highly directional because of the fact that in gaseous lasers the crystalline imperfection, thermal distortion, and scattering are almost absent.

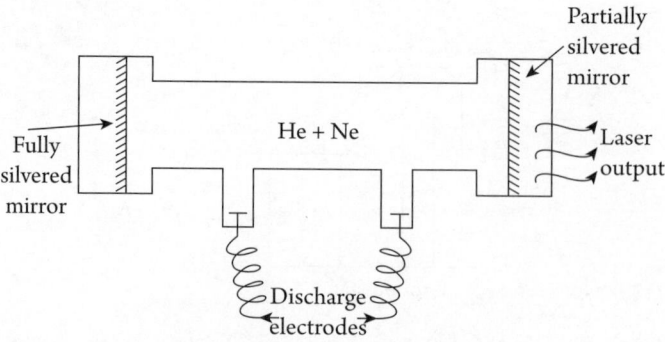

Fig. 5.10 He-Ne lasers setup.

Lasers

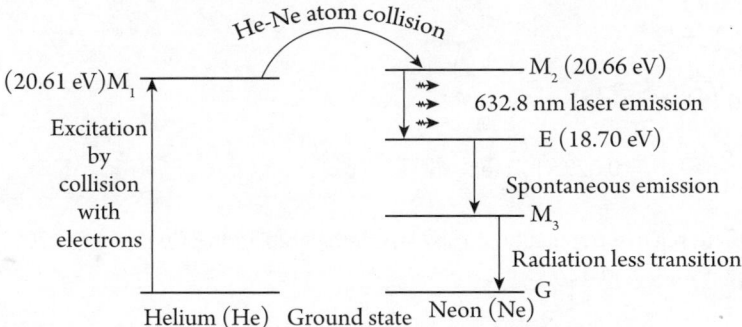

Fig. 5.11 Energy levels of He-Ne laser.

3. He-Ne lasers are capable of operating continuously without any need of cooling. Thus, He-Ne laser is superior to ruby laser.

5.13 Applications of Lasers

The lasers are generally used in almost every field including fundamental research. The common applications of lasers are as follows:

1. Metallic rods can be melted and joined by means of a laser beam, that is, laser welding.
2. The laser beam is used to vaporize unwanted material during the manufacturing of electronic circuits on semiconductor chips, that is, laser etching.
3. Lasers are used to detect and destroy enemy missiles during war.
4. The narrow red laser beam is used in supermarkets to read the bar codes.
5. Lasers are also being employed for separating the various isotopes of an element.
6. Lasers are used in the production of three-dimensional images of an object in holography.
7. High power lasers are used in CD players, laser printers, laser copiers, facsimile machine and so on.
8. CO_2 gas lasers of about 100-Watt output are helpful in surgery because they seal small blood vessels.
9. Laser beam is used in plasmon fundamental research.
10. Lasers are used in nanotechnology and fabricating nanomaterials.

Solved Problems

Ex. 1: Calculate the frequency of radiation of CO_2 laser having the energy difference between two levels is 0.121 eV.

Solution:

The energy difference between the energy levels is given by

$\Delta E = h\nu \Rightarrow \nu = \dfrac{\Delta E}{h}$.

Given $\Delta E = 0.121 \text{ eV} = 0.121 \times 1.6 \times 10^{-19}$ J.

$\nu = \dfrac{0.121 \times 1.6 \times 10^{-19}}{6.63 \times 10^{-34}} = 0.029 \times 10^{15} \text{ sec}^{-1}$

Ex. 2: Calculate relative population if ruby laser emits radiation of wavelength 7000 Å and has two states at 27°C and 227°C.

Solution:

The relative population of the two states at temperature T kelvin is given by

$\dfrac{N_2}{N_1} = e^{-\frac{E_2 - E_1}{KT}}$.

Since $E_2 - E_1 = h\nu = \dfrac{hc}{\lambda}$.

Given $\lambda = 7000 \times 10^{-10}$ m, $h = 6.63 \times 10^{-34}$ J.sec, $c = 3 \times 10^8$ m/sec,

$E_2 - E_1 = \dfrac{6.63 \times 10^{-34} \times 3 \times 10^8}{7000 \times 10^{-10} \times 1.6 \times 10^{-19}} = 1.77$ eV.

At temperature $T = 27°C = 27 + 273 = 300$ K, Boltzmann constant $K = 8.6 \times 10^{-5}$ eV,

$\dfrac{N_2}{N_1} = e^{-\frac{1.77}{8.6 \times 10^{-5} \times 300}} = e^{-69}$.

At temperature $T = 227°C = 227 + 273 = 500$ K, Boltzmann constant $K = 8.6 \times 10^{-5}$ eV,

$\dfrac{N_2}{N_1} = e^{-\frac{1.77}{8.6 \times 10^{-5} \times 500}} = e^{-41.1}$.

Ex. 3: In a ruby laser, total number of Cr^{+3} ions are 2.8×10^{19}. If the laser emits radiation of wavelength 7000 Å, calculate the energy of the laser pulse.

Solution:

The energy of the laser pulse is given by

$E = nh\nu = \dfrac{nhc}{\lambda}$,

where n is the number of active atoms/ions. Given $= 2.8 \times 10^{19}$ ions, $\lambda = 7000 \times 10^{-10}$ m,

$E = \dfrac{2.8 \times 10^{19} \times 6.63 \times 10^{-34} \times 3 \times 10^8}{7000 \times 10^{-10}} = 7.95$ J.

Ex. 4: A certain ruby laser emits 1 J pulses of light whose wavelength is 696 nm. Calculate the minimum number of Cr^{+3} ions in the ruby rod?

Solution:

The energy of the laser pulse is given by

$E = nh\nu = \dfrac{nhc}{\lambda}$,

where n is the number of active atoms/ions. Given $= 1$ J, $\lambda = 696 \times 10^{-9}$ m,

$$I = \frac{n \times 6.63 \times 10^{-34} \times 3 \times 10^{8}}{696 \times 10^{-9}} \Rightarrow n = 3.59 \times 10^{18} \text{ ions.}$$

Ex. 5: Calculate the energy and momentum of a photon of a laser beam of wavelength 6328 Å.

Solution:

The energy of a photon is given by

$$E = h\nu = \frac{hc}{\lambda}.$$

Given $\lambda = 6328 \times 10^{-10}$ m, $h = 6.63 \times 10^{-34}$ J.sec, $c = 3 \times 10^{8}$ m/sec,

$$E = \frac{6.63 \times 10^{-34} \times 3 \times 10^{8}}{6328 \times 10^{-10}} = 3.14 \times 10^{-19} \text{ J}$$

The momentum of a photon is given by

$$p = \frac{h}{\lambda} = \frac{6.63 \times 10^{-34}}{6328 \times 10^{-10}} = 1.05 \times 10^{-27} \text{ kg.m/sec.}$$

Previous Year Questions (University Examination)

1. What do you mean by laser? Write its full name.
2. Distinguish between spontaneous and stimulated radiation.
3. Define transition probabilities.
4. Discuss Einstein's coefficient of spontaneous and stimulated radiation.
5. Discuss the necessary condition to achieve laser radiation.
6. Explain the principal of laser action.
7. What is a metastable state?
8. What do you mean by pumping method? Which method is used in ruby and He-Ne laser?
9. "A gas laser has several advantages over the solid-state lasers." Explain this statement.
10. What is the difference between ordinary and laser radiation?
11. Write down the use of lasers in medical and industrial fields.
12. Give the four properties of lasers.
13. What are Einstein's coefficients? Derive Einstein relation between them.
14. What is spontaneous and stimulated emission of radiation? What do you mean by population inversion?
15. Explain the principle of optical pumping and stimulated emission of radiation.
16. Discuss the properties of laser radiation and mention some of its important applications.
17. Explain the action of a helium-neon laser. How it is superior to a ruby laser?
18. Describe the construction, action, and working of ruby laser.
19. Calculate the population ratio of the two states in the He-Ne laser that produces light of wavelength 6000 Å at 27°C.

20. A laser beam of a wavelength 692.8 nm and aperture 10×10^{-3} m from the He-Ne laser can be focused on an area equal to the square of its wavelength. If the laser source radiates energy at 20 mW, calculate
   (a) The angular spread of the beam
   (b) The intensity of focused beam

Multiple Choice Questions

1. The LASER acronym is
   (a) The name of a scientist
   (b) Light atom by state emission of radar
   (c) Light amplification by stimulated emission of radiation
   (d) Light atomic for spontaneous emission of radar

2. The term for the emission of photons without any external influence is
   (a) Radiation absorption   (b) Excited emission
   (c) Spontaneous emission   (d) Light tuning

3. The basic idea behind lasers is
   (a) Light emission by stimulation   (b) Radiation absorption
   (c) Light tuning   (d) Absorption

4. The Laser beam properties are
   (a) Highly monochromatic   (b) Highly coherent
   (c) Low divergence   (d) All of the above

5. Ruby laser's population inversion is caused by
   (a) Collision of atoms   (b) Optical pumping
   (c) Direct conversion   (d) Chemical excitation

6. He-Ne laser's population inversion is brought about by
   (a) Pumping by optical energy   (b) Phonon excitation
   (c) Inelastic atomic collision (d) Discharge method

7. The Ruby laser is
   (a) Pulsed laser   (b) Continuous laser
   (c) Gas laser   (d) Chemical laser

8. The output beam of He-Ne laser is
   (a) Continuous   (b) Pulsed
   (c) Discrete   (d) All of the above

9. A combination ratio of He and Ne gas is used in He-Ne lasers is
   (a) 2:5   (b) 10:1
   (c) 1:5   (d) 5:5

10. Which of the following energy states is the least stable?
    (a) Open state   (b) Ground state
    (c) Real state   (d) None

11. The requirement of Laser beam emission is
    (a) Active medium
    (b) Population inversion
    (c) Metastable energy level
    (d) All of these

12. Ruby laser cannot compete with He-Ne Laser. The cause is
    (a) Output beam is continuous in He-Ne Laser
    (b) He-Ne Laser is highly monochromatic
    (c) No cooling is required in He-Ne Laser
    (d) All of these

13. The He-Ne laser is
    (a) Solid laser
    (b) Gaseous laser
    (c) Semiconductor laser
    (d) Dye laser

Bibliography

1. Silfvast, W. (2008). *Laser Fundamentals*, 2nd ed. Cambridge University Press.
2. Young, M. (2000). *Optics and Lasers*, 5th ed. Springer.
3. Thyagarajan, K. and Ghatak. A. (2010). *Lasers: Fundamentals and Applications*, 2nd ed. Springer.
4. Paschotta, R. (2008). *Encyclopedia of Laser Physics and Technology*, 2 vol. Wiley-VCH.
5. Diels, J. C. and Arissian, L. (2011). *Lasers: The Power and Precision of Light*. Wiley-VCH.
6. Quimby, R. (2008). *Photonics and Lasers: An Introduction*. Wiley-Interscience.

Keys

1. (c) 2. (c) 3. (a) 4. (d) 5. (b) 6. (c) 7. (a) 8. (b) 9. (b) 10. (c)
11. (d) 12. (d) 13. (b)

CHAPTER 6

Dielectric Materials

6.1 Introduction

The nonconducting materials such as paper, wood, glass, ceramics, polymers and so on do not have free charge carriers, that is, electrons or holes. Therefore, they prevent the flow of electrical current and heat through them.

When the main function of nonconducting materials is to provide electrical isolation then they are called **insulators**.

When the main function of nonconducting materials is for charge storage then it is called **dielectric**.

The dielectrics are polarized under the influence of an external electric field.

6.2 Dielectric Constant

Let us consider two parallel plates separated by a distance "d" connected with a dc supply of voltage V, as shown in Figure 6.1(a). Now the circuit is disconnected, and the dielectric is inserted between the plates, as shown in Figure 6.1(b).

Then, the voltage across the capacitor is reduced from V to V'. The change in voltage across the plates can be related by a factor as

$$V = \varepsilon' V',$$
$$V' = \frac{V}{\varepsilon_r},$$
$$\varepsilon_r = \frac{V}{V'}. \tag{6.1}$$

Since $V' < V$, the relative permittivity or dielectric constant $\varepsilon_r > 1$.

The capacitance without dielectric is given as

$$C = \frac{Q}{V}.$$

The capacitance with dielectric is given as

$$C' = \frac{Q}{V'}.$$

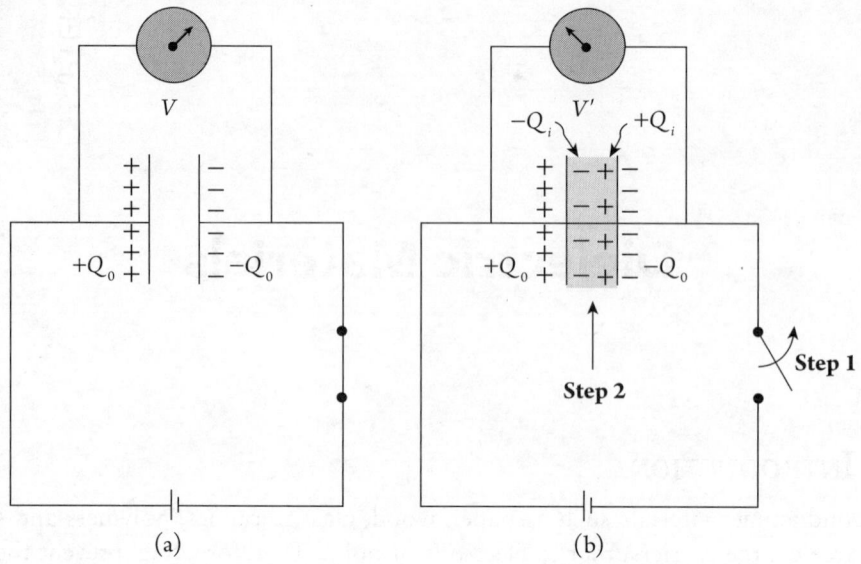

Fig. 6.1 Parallel plate capacitor (a) without dielectric, (b) with dielectric.

Now, put the value of C and C' in equation (6.1), the relative permittivity or dielectric constant is

$$\varepsilon_r = \frac{C'}{C}, \tag{6.2}$$

If "A" is the area of the plates, then

$$C = \frac{\varepsilon_0 A}{d} \text{ or, } C' = \frac{\varepsilon A}{d},$$

where ε is the permittivity of the material.

From equation (6.2),

$$\varepsilon_r = \frac{\varepsilon}{\varepsilon_0}.$$

Thus, the dielectric constant is the ratio of permittivity of the material to the permittivity of free space.

6.3 Polar and Nonpolar Molecules

Nonpolar molecules are classified as such when the centers of gravity of positive and negative charges in a molecule coincide. It does not have any permanent dipoles, as shown in Figure 6.2(a). Examples of nonpolar molecules are H_2, N_2, O_2, and so on.

When the center of gravity of positive charge and negative charge do not coincide, the molecule is known as polar molecules. It has permanent dipoles with electric dipole moment, as shown in Figure 6.2(b). Examples of polar molecules are H_2O, HCl, and so on.

Dielectric Materials

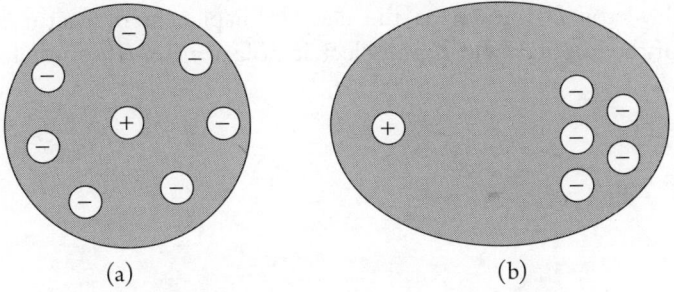

Fig. 6.2 (a) Nonpolar molecule, (b) Polar molecule.

6.4 DIELECTRIC POLARIZATION

Dielectric polarization is the displacement of charge particles under the action of an electric field.

Consider a polar dielectric placed between two parallel plates of a capacitor. The permanent dipoles of the polar dielectrics are randomly oriented in the absence of the electric field known as unpolarized dielectric as shown in Figure 6.3. When an external electric field "E" is applied, a torque is exerted on the dipoles causing them to align with the field. Depending on the temperature and the strength of the applied electric field, the molecules will align themselves with the electric field. In general, alignment rises when the external electric field increases and the dielectric temperature reduces.

The polarization of the material ($\vec{P}$) is directly proportional to the electric field ($\vec{E}$) and can be written as

$$\vec{P} \propto \vec{E}$$

or, $\vec{P} = \varepsilon_0 \chi \vec{E}$, (6.3)

where χ is the proportionality constant known as dielectric susceptibility of the material. It describes how easily a dielectric can be influenced or polarized by an external electric field. It is a dimensionless quantity.

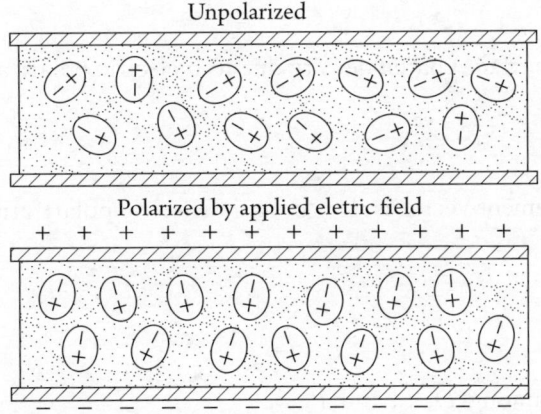

Fig. 6.3 Unpolarized and polarized dielectric due to external electric field.

An additional vector D known as the electric displacement vector is defined as the combined effect of the applied field E and electric polarization (P) on dielectric material, as follows:

$$\vec{D} = \varepsilon_0 \vec{E} + \vec{P}. \tag{6.4}$$

For vacuum $\vec{P} = 0$,

$$\vec{D} = \varepsilon_0 \vec{E}.$$

For a dielectric medium,

$$\vec{D} = \varepsilon_0 \vec{E} + \vec{P}$$

$$\vec{D} = \varepsilon_0 \vec{E} + \varepsilon_0 \chi \vec{E} \quad \text{[From equation (6.3)]}$$

$$\vec{D} = \varepsilon_0 \vec{E}(1+\chi) \quad \text{[Since } (1+\chi) = \varepsilon_r\text{]}$$

$$\vec{D} = \varepsilon_r \varepsilon_0 \vec{E}$$

$$\vec{D} = \varepsilon \vec{E} \quad \text{[Since } \varepsilon_r = \frac{\varepsilon}{\varepsilon_0}\text{]}.$$

This is the expression for electric displacement vector $\vec{D}$ for dielectric medium.

6.5 Relation among $\vec{D}$, $\vec{E}$, and $\vec{P}$

The effective electric field across capacitor is given by

$$\vec{E} = \vec{E_0} - \vec{E'}.$$

Since $\vec{E} = \dfrac{\sigma}{\varepsilon_0}$,

$$\therefore E = \frac{\sigma}{\varepsilon_0} + \frac{\sigma_p}{\varepsilon_0}.$$

Here, σ_p is bound surface charge density due to dielectric polarization and σ is the free charge density.

$$\varepsilon_0 \vec{E} = \sigma - \sigma_p.$$

The electric displacement vector is $\vec{D} = \sigma$ and dielectric polarization is $\vec{P} = \sigma_p$; then, the polarization is given by

$$\varepsilon_0 \vec{E} = \vec{D} - \vec{P}$$

$$\vec{D} = \varepsilon_0 \vec{E} + \vec{P}.$$

This is the relation among $\vec{D}$, $\vec{E}$, and $\vec{P}$.

Dielectric Materials

6.6 Relation between Dielectric Constant and Electrical Susceptibility

The effective electric field across capacitor is given by

$$\vec{E} = \vec{E_0} - \vec{E'}. \quad (6.5)$$

Since $\vec{E} = \dfrac{\sigma}{\varepsilon_0}$,

$$\therefore E = \dfrac{\sigma}{\varepsilon_0} + \dfrac{\sigma_p}{\varepsilon_0}.$$

Here, σ_p is bound surface charge density due to dielectric polarization and σ is the free charge density.

$$\varepsilon_0 \vec{E} = \sigma - \sigma_p.$$

The electric displacement vector is $\vec{D} = \sigma$ and dielectric polarization is $\vec{P} = \sigma_p$; then, the polarization is given by

$$\varepsilon_0 \vec{E} = \vec{D} - \vec{P}.$$

Since $\vec{D} = \varepsilon \vec{E} = \varepsilon_r \varepsilon_0 \vec{E}$,

$$\varepsilon_0 \vec{E} = \varepsilon_r \varepsilon_0 \vec{E} - \vec{P}$$

$$\vec{P} = \varepsilon_r \varepsilon_0 \vec{E} - \varepsilon_0 \vec{E}$$

$$\vec{P} = (\varepsilon_r - 1)\varepsilon_0 \vec{E}. \quad (6.6)$$

Since $\vec{P} = \varepsilon_0 \chi \vec{E}$ comparing with equation (6.6), we get

$$(\varepsilon_r - 1) = \chi$$

$$\varepsilon_r = 1 + \chi.$$

This is the relation between dielectric constant and electrical susceptibility.

6.7 Polarizability

Let us consider a dielectric slab placed in a uniform external electric field and then the charges are displaced to produce electric dipoles known as induced electric dipoles. The electric dipole moment is directly proportional to the applied electric field.

$$\vec{p} \propto \vec{E}.$$

$$\vec{p} = \alpha \vec{E}.$$

The constant of proportionality α is called polarizability.

The dielectric polarization is defined as dipole moment per unit volume. So, the polarization of the material P is given by

$$\vec{P} = N\vec{p}.$$

$$\vec{P} = N\alpha\vec{E}.$$

Here N is the number of dipoles (atoms) per unit volume.

6.8 Types of Polarization (Polarizability)

There are four important types of polarization.

1. Electronic polarization
2. Ionic polarization
3. Orientation polarization
4. Space-charge polarization

6.9 Electronic Polarization

Electronic polarization occurs in nonpolar dielectrics. In this type of polarization, the atom is initially unpolarized in the absence of an electric field. Electronic polarization occurs very quickly (10^{-15} – 10^{-14} sec.) after the external electric field $\vec{E}$ is applied, causing an atom's electron cloud to move away from its heavy fixed nuclei by a distance smaller than the dimension of the atom or molecule depicted in Figure 6.4.

Electronic polarization is independent of temperature and described by

$$\vec{P}_e = N\alpha_e \vec{E},$$

where α_e is the electronic polarizability and N is the number of atoms per unit volume.

Since we know that

$$\varepsilon_r = 1 + \chi,$$

$$\therefore \vec{P} = \varepsilon_0 \chi \vec{E} \Rightarrow \chi = \frac{\vec{P}}{\varepsilon_0 \vec{E}}$$

$$\varepsilon_r = 1 + \frac{\vec{P}_e}{\varepsilon_0 \vec{E}}$$

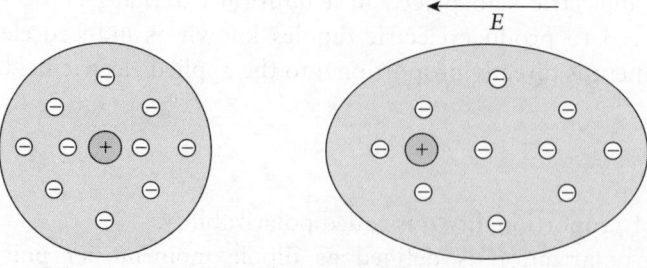

Fig. 6.4 Electronic polarization.

Dielectric Materials

$$\varepsilon_r = 1 + \frac{N\alpha_e \vec{E}}{\varepsilon_0 \vec{E}}$$

$$\varepsilon_r = 1 + \frac{N\alpha_e}{\varepsilon_0}.$$

Hence, dielectric constant is dependent on the electric polarizability and independent of external applied electric field.

In case of monoatomic gas,

$$\alpha_e = 4\pi\varepsilon_0 R^3$$

$$\varepsilon_r = 1 + \frac{N\alpha_e}{\varepsilon_0} = 1 + \frac{N 4\pi\varepsilon_0 R^3}{\varepsilon_0}$$

$$\varepsilon_r = 1 + 4\pi N R^3,$$

where R is the radius of an atom.

6.10 Ionic Polarization

Ionic polarization occurs in ionic crystals. Let us consider NaCl molecule that is bound through ionic bond. The inter-ionic distance of these ions varies with the application of electric field. When the applied electric is in the direction of ionic bonds, the inter-ionic distance decreases, while an electric field opposite to it causes an increase in inter-ionic distance, as shown in Figure 6.5.

The ionic polarization of the material is given by

$$\vec{P}_i = N\alpha_i \vec{E},$$

where α_i is the ionic polarizability and N is the number of atoms per unit volume. Ionic polarization is also independent of temperature.

For most of the materials, the ionic polarizability is less than the electronic polarizability as

$$\alpha_i = \frac{1}{10}\alpha_e,$$

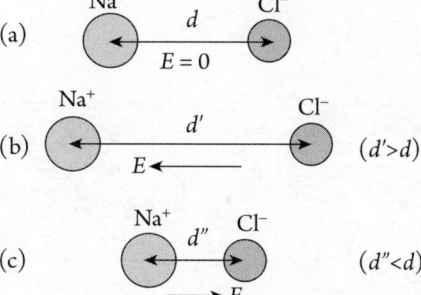

Fig. 6.5 Ionic polarization mechanism.

6.11 Orientation Polarization

Polar dielectrics with permanently dipole-momentous molecules experience orientation polarization. As seen in Figure 6.6, the permanent dipoles are orientated at random to make the dielectric slab electrically neutral in the absence of any external electric field. When an external field is applied, then a torque is exerted on these permanent dipoles to align in the direction of applied electric field E, as shown in Figure 6.6. Such type of polarization is called orientation polarization. It is dependent on temperature. The built-up time for orientation polarization is of the order of 10^{-10} sec.

The orientation polarization is given by

$$\vec{P}_o = N\alpha_o \vec{E}.$$

$$\therefore \alpha_o = \frac{p^2}{3kT}.$$

$$\therefore \vec{P}_o = \frac{Np^2 \vec{E}}{3kT}.$$

Where α_o is the orientation polarizability, p is the electric dipole moment, k is the Boltzmann's constant, and T is an absolute temperature.

6.12 Space-charge Polarization

When an external electric field is applied on a dielectric material, the accumulation of the charges at the electrodes occurs, as shown in Figure 6.7. Therefore, the tendency for redistribution of charges in the dielectric medium in the presence of an applied electric field is known as space charge polarization.

The different polarization mechanism in a dielectric material by an application of an external electric field is summarized in Figure 6.8.

6.13 Total Polarization

When the material experiences all the three types of polarization, then the total sum of polarizability is the sum of electronic, ionic, and orientation polarizabilities, that is,

$$\alpha = \alpha_e + \alpha_i + \alpha_o.$$

The total polarization P is given by

$$\vec{P} = \vec{P}_e + \vec{P}_i + \vec{P}_o.$$

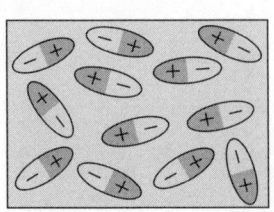 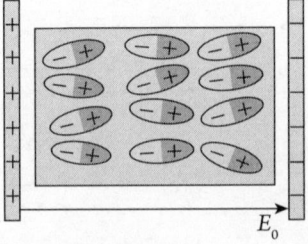

Fig. 6.6 Orientation polarization.

Dielectric Materials

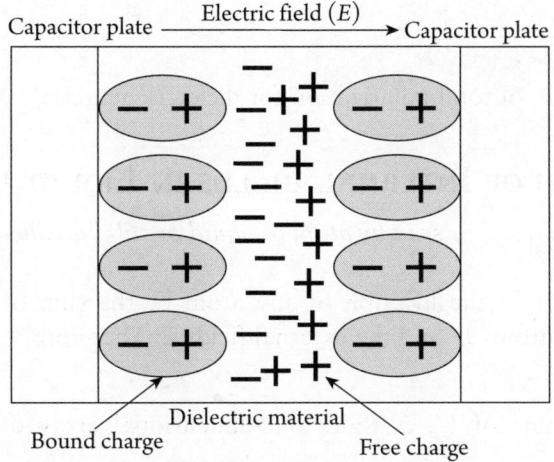

Fig. 6.7 Space-charge polarization.

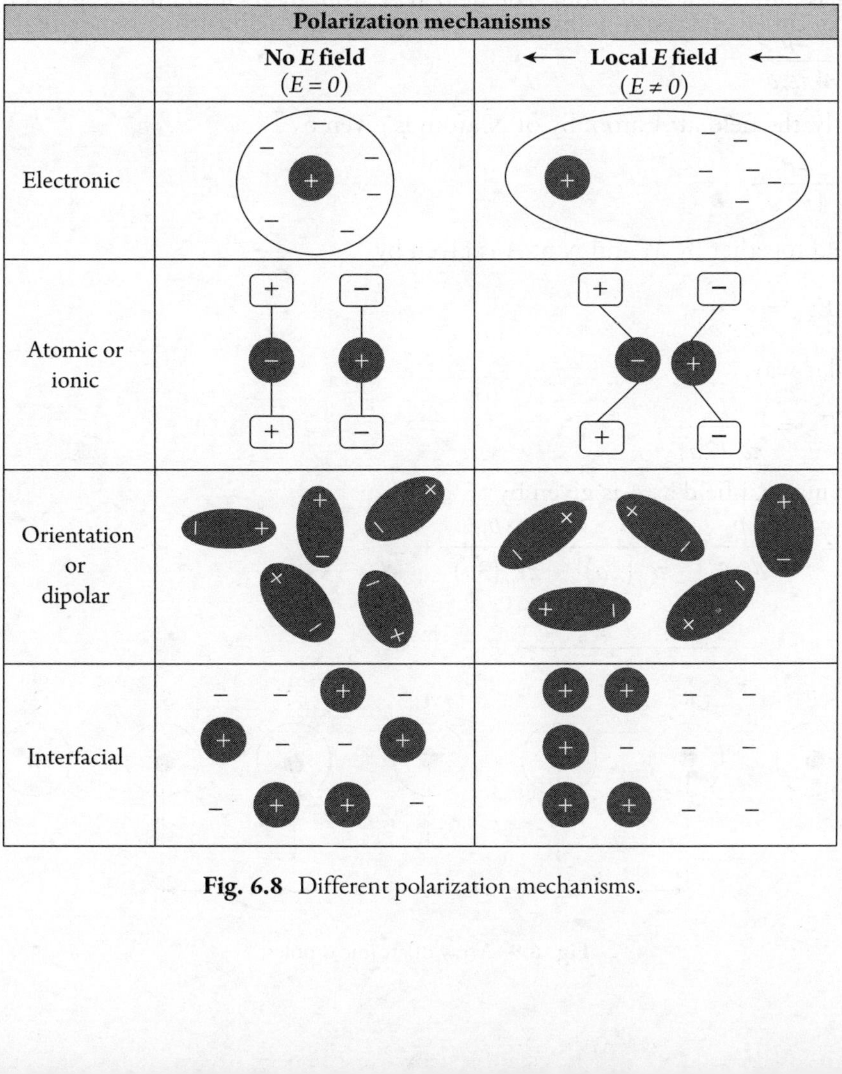

Polarization mechanisms		
	No E field ($E = 0$)	**← Local E field ←** ($E \neq 0$)
Electronic		
Atomic or ionic		
Orientation or dipolar		
Interfacial		

Fig. 6.8 Different polarization mechanisms.

$$\vec{P} = NE\left[\alpha_e + \alpha_i + \frac{p^2}{3kT}\right].$$

This is an expression of total polarization for dielectric material.

6.14 EQUATION OF INTERNAL FIELDS IN LIQUID AND SOLIDS

The electric field at atom site or seen by atom in liquid or solid is called internal field or local field.

The internal field E_i at the location of any atom be the sum of fields created by the neighboring polarized atoms E' and the external fields E. Therefore,

$$E_i = E + E'. \tag{6.7}$$

To evaluate the value of E', consider one-dimensional array of atoms, as shown in Figure 6.9.

An array of equispaced atomic dipoles with dipole moment "p_i" (induced dipole moment) separated by a distance "a". Let an electric field be applied from left to right; then dipole moment p_i is induced in each atom. The field at A atom by p_i of M atom is given by

$$E_{AM} = \frac{2p_i}{4\pi\varepsilon_0 a^3}. \tag{6.8}$$

Similarly, the field at A atom by of N atom is given by

$$E_{AN} = \frac{2p_i}{4\pi\varepsilon_0 a^3}. \tag{6.9}$$

The field together by M and N at A is given by

$$E_{AM} + E_{AN} = \frac{p_i}{\pi\varepsilon_0 a^3}.$$

In similar way,

$$E_{AR} + E_{AS} = \frac{p_i}{\pi\varepsilon_0 (2a)^3}.$$

So, the internal field at A is given by

$$E_i(A) = E + \frac{p_i}{\pi\varepsilon_0 a^3} + \frac{p_i}{\pi\varepsilon_0 (2a)^3} + \frac{p_i}{\pi\varepsilon_0 (3a)^3} + \ldots$$

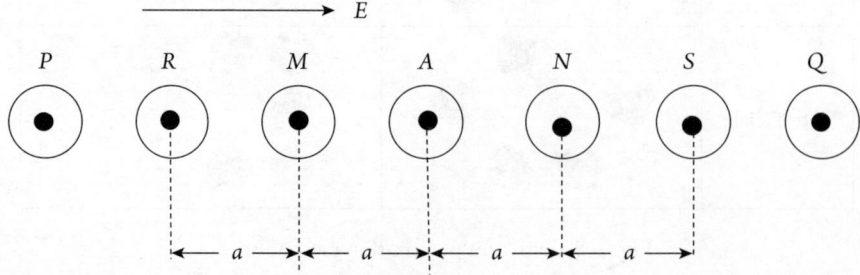

Fig. 6.9 Array of atomic dipoles.

Dielectric Materials

$$E_i(A) = E + \frac{p_i}{\pi\varepsilon_0 a^3} \sum_{n=1}^{n=\infty} \frac{1}{n^3}. \tag{6.10}$$

Since the value of term is $\sum_{n=1}^{n=\infty} \frac{1}{n^3} = 1.2$ by harmonic progression,

thus $E_i(A) = E + \frac{1.2 p_i}{\pi\varepsilon_0 a^3}.$ (6.11)

From equation (6.11), the total field exceeds the applied external field. In three-dimensional dielectric case, $\frac{1.2}{\pi}$ is replaced by γ constant, which depends on the structure of the crystal and is known as the internal field constant, while $\left(\frac{1}{a^3}\right)$ is replaced by N (the number of atoms per unit volume). Thus, equation (6.11) becomes

$$E_i(A) = E + \frac{\gamma N p_i}{\varepsilon_0}$$

$$E_i(A) = E + \frac{\gamma P}{\varepsilon_0} \quad [\because P = N p_i]. \tag{6.12}$$

In case of crystal of cubic symmetry, the internal field constant $\gamma = \frac{1}{3}$,

$$E_i(A) = E + \frac{P}{3\varepsilon_0}. \tag{6.13}$$

The field equation given by equation (6.13) is called Lorentz electric field.

6.15 Clausius–Mossotti Equation

Clausius–Mossotti equation relates the polarizability of the material to the dielectric constant of the dielectric medium. Let us consider solid nonpolar dielectric placed in a uniform electric field; then, the polarization of the material is given by

$$P_e = N\alpha_e E_i.$$

$$\alpha_e = \frac{P_e}{NE_i}.$$

On substituting the value of internal field E_i for cubic symmetry with $\left(\gamma = \frac{1}{3}\right)$, we get

$$\alpha_e = \frac{P}{N\left[E + \dfrac{P}{3\varepsilon_0}\right]}.$$

$$\alpha_e = \frac{P}{N\left[\dfrac{P}{\varepsilon_0(\varepsilon_r - 1)} + \dfrac{P}{3\varepsilon_0}\right]} \quad [\because P = \varepsilon_0(\varepsilon_r - 1)].$$

$$\alpha_e = \cfrac{1}{N\left[\cfrac{1}{\varepsilon_0(\varepsilon_r - 1)} + \cfrac{1}{3\varepsilon_0}\right]}.$$

$$\frac{N\alpha_e}{3\varepsilon_0} = \cfrac{1}{\left[\cfrac{1}{(\varepsilon_r - 1)} + \cfrac{1}{3}\right]}.$$

$$\frac{N\alpha_e}{3\varepsilon_0} = \frac{\varepsilon_r - 1}{\varepsilon_r + 2}.$$

This is an expression of Clausius–Mossotti equation, which is valid for nonpolar dielectric of cubic symmetry.

6.16 Relation between Dielectric Constant and Refractive Index of Dielectric Material

According to Maxwell's equations of electromagnetic wave theory, the velocity of electromagnetic waves in the medium is given by

$$v = \frac{1}{\sqrt{\mu\varepsilon}}.$$

The refractive index is given by

$$n = \frac{C}{v} \quad \left[\because C = \frac{1}{\sqrt{\mu_0 \varepsilon_0}}\right].$$

Therefore, $n = \sqrt{\dfrac{\mu\varepsilon}{\mu_0 \varepsilon_0}}$.

For nonmagnetic materials $\mu = \mu_0$,

$$n = \sqrt{\frac{\mu_0 \varepsilon}{\mu_0 \varepsilon_0}} = \sqrt{\frac{\varepsilon}{\varepsilon_0}} \quad \left[\because \varepsilon_r = \frac{\varepsilon}{\varepsilon_0}\right].$$

$$n = \sqrt{\varepsilon_r}.$$

This is the relation between dielectric constant and refractive index of dielectric material. On substituting this value in Clausius–Mossotti equation, we have

$$\frac{N\alpha_e}{3\varepsilon_0} = \frac{n^2 - 1}{n^2 + 2}.$$

This equation is known as Lorentz–Lorentz equation.

6.17 Frequency Dependence of Dielectric Constant

When a nonpolar dielectric is subjected to an alternating electric field, the polarization of the material follows field reversal. The total polarization of the material depends on the

Dielectric Materials

frequency of an alternating field applied on the dielectric. Relaxation time and relaxation frequency are terms used to describe how long it typically takes a dipole to reorient itself in the direction of the applied electric field (f). When $\tau < f$, there will be no polarization in the material. When $\tau > f$, there is polarization in the material.

In audio frequency region, all types of polarization occur in dielectric, as shown in Figure 6.10. In radio wave or microwave frequency regions, the permanent dipoles are unable to follow the field reversal. Therefore, the orientation polarization is vanished from the dielectric material ($\vec{P}_o = 0$). In infrared region, ionic polarization will be zero inside the material ($\vec{P}_i = 0$). At optical frequency region, only electronic polarization will occur in the dielectric material. At ultraviolet region, all polarization becomes zero and dielectric constant becomes unity, that is, $\varepsilon_r = 1$.

6.18 Dielectric Loss

Dielectric loss may be defined as the loss of energy in the form of heat by a dielectric medium due to internal friction developed in switching of dipoles to their normal state under the action of an alternating field.

Consider a dielectric substance that is placed between the capacitor's plates and is subject to the alternating electric field. The result is the same as a resistance being present alongside the capacitor, as seen in Figure 6.11.

The angle between I and Ic is denoted by δ and is called dielectric loss angle shown in Figure 6.12. The tangent of this angle δ, that is, $\tan \delta$ is called the loss tangent.

The dielectric loss is the power loss due the resistance given by

Dielectric loss = $V \times I_R$

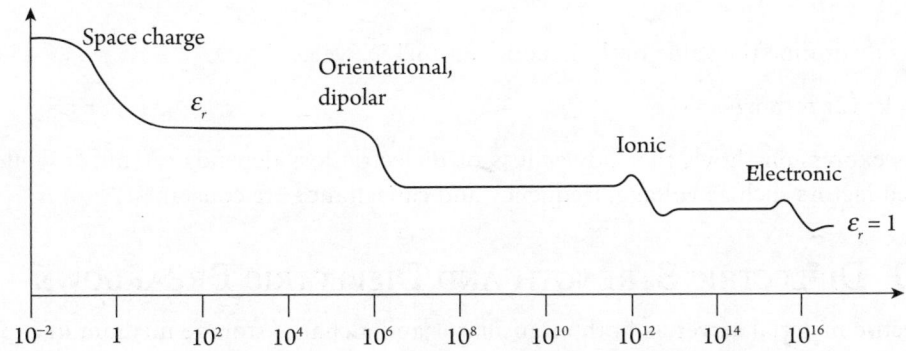

Fig. 6.10 Dielectric constant variation in a.c. field.

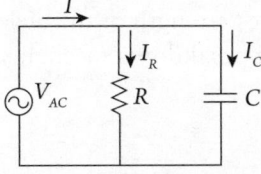

Fig. 6.11 Circuit diagram to measure dielectric loss.

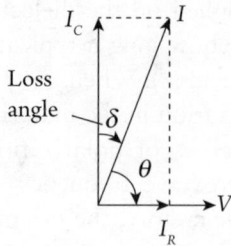

Fig. 6.12 Phasor diagram.

$P = V \times I \cos(90° - \delta)$

$P = V \times I \sin \delta.$ (6.14)

From Phasor diagram, we have

$I_c = I \times \cos \delta$

$I = \dfrac{I_c}{\cos \delta}.$

On substituting the value of I in equation (6.14), we get

$P = V \times \dfrac{I_c \sin \delta}{\cos \delta} = VI_c \tan \delta.$ (6.15)

The capacitive current is given by

$I_c = \dfrac{V}{X_c} \qquad \left[\because X_c = \dfrac{1}{2\pi fc} \right].$

$I_c = V(2\pi fc).$

On substituting the value of I_c in equation (6.15), we get

$P = V^2 (2\pi fc) \tan \delta.$ (6.16)

This expression shows that power loss or dielectric loss depends on $\tan \delta$, while other provided factors such as voltage, frequency and capacitance are constants.

6.19 Dielectric Strength and Dielectric Breakdown

A dielectric material serves as both an insulator and a charge storage medium in capacitors since it is a perfect nonconducting material. The potential difference across an electric field grows in proportion to the applied electric field. When a spark happens and the dielectric stops acting as an insulator, a limit has been reached. This voltage's limiting value, called "breakdown voltage," measures the "strength of the dielectric."

There are mainly three kinds of breakdown mechanisms as follows:

- Avalanche Breakdown
- Thermal Breakdown
- Defect Breakdown

Dielectric Materials

Avalanche Breakdown: Dielectric has a large band gap (more than 5 eV) and therefore when a high electric field is applied, the valence band electrons get enough energy to cross the band gap and become stimulated to the conduction band. The mobile electrons of conduction band also acquire high acceleration and collide with other electrons of the conduction band. As a result, an avalanche of conduction electrons results from the release of more and more electrons into the conduction band. The material is considered to have reached the point of breaking down field when it eventually becomes strongly conducting.

Thermal Breakdown: Dielectric losses are energy losses that happen when high frequency a.c. is applied to a medium that is dielectric. Heat energy is released from this energy. The material heats up and occasionally melts if the dissipation is ineffective because of the substance's weak conductivity. This is known as thermal breakdown.

Defect Breakdown: This is present in dielectrics that have processes, fractures, and so on. When there are significant electric fields, gas discharge will happen in small cracks and pores and cause the dielectric to breakdown.

6.20 Various Kinds of Dielectric Materials

Similar to the magnetic materials, the dielectric materials too are classified as given below on the basis of their behavior in the external applied electric field. These are:

- Ferroelectric materials
- Para electric materials
- Dielectric materials do not exist

Ferroelectric Materials: In some special type of polar dielectrics, the permanent electric dipoles in the material are oriented in a specific direction, even in absence of electric field. These dipolar dielectrics are known as ferroelectrics. Examples are Rochelle salt, Perovskiet group (titivates and niobates such as $BaTiO_3$), dihydrogen phosphates, and arsenates (KH_2PO_4).

Following are some characteristic properties of Ferroelectrics:

- All Ferroelectric materials exhibit Piezo-electric effect.
- Ferroelectric dielectrics exhibit "hysteresis", that is, polarization P is nonlinear with E, as shown in Figure 6.13.

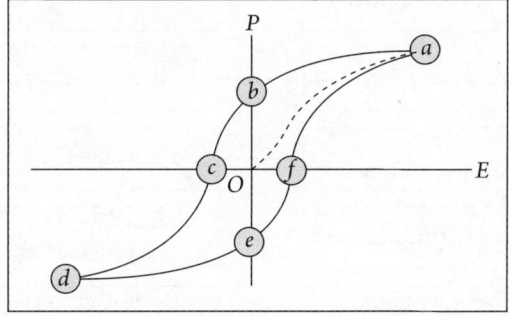

Fig. 6.13 P versus E curve for ferroelectric material.

- They acquire spontaneous polarization below a certain temperature.
- Their polarization is reversible.
- These exhibit ferroelectric transition temperature.

6.21 Hysterisis in Ferroelectric Materials

When a ferroelectric material is subjected to a time-varying electric field, the polarization (P) lags behind the electric field. This is known as hysteresis. The nature of variation of P versus E is as follows:

OM - Remanent polarization
OT - Coercive field
Ps - Saturation polarization

As P versus E curve is nonlinear for a ferroelectric material, its dielectric constant is not a constant of E. Normally, the dielectric constant of a ferroelectric is measured at low fields. The value of dielectric constant of ferroelectric materials is very high (10^4–10^5) compared to ordinary dielectric constants, which are only around 10.

6.22 Applications of Ferroelectric Materials in Devices

Ferroelectric materials have several device applications due to their special characteristics. The following Figure 6.14 summarizes the various special properties of ferroelectrics and the corresponding device application.

6.23 Loss Factor and Its Significance

The product of ϵ_r and $\tan \delta$ is known as loss factor and $\tan \delta$ is also sometimes known as loss tangent. The loss factor $\epsilon_r \tan \delta$ is very significant as it characterizes the usefulness of

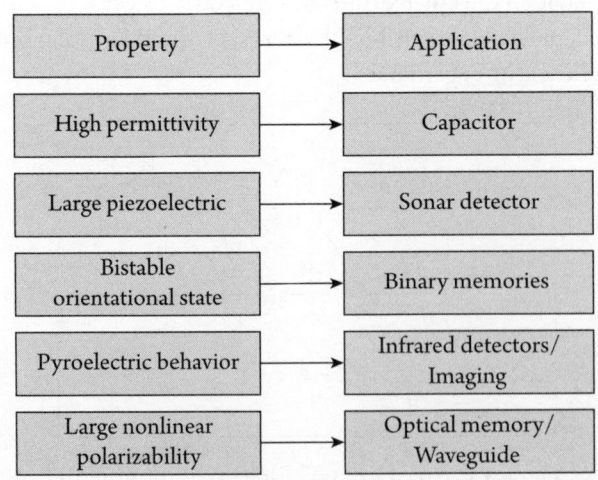

Fig. 6.14 Applications of various special properties of ferroelectrics.

a dielectric material as insulator. In every case, a low loss tangent is desirable. Classification of dielectrics is also done on the basis of dielectric loss. These are

i. Low–loss dielectrics
ii. High–loss dielectrics

6.24 Electrostriction Effect

On applying the electric field to a dielectric material, it gets polarized either by shifting of the electron cloud (electronic polarization), by shifting of positive and negative ions (ionic polarization), or by orientation of atomic dipoles (orientation polarization). The dimensions of the dielectric material undergo a slight change due to shifting of ions or orientation of dipoles. Therefore, a strain is developed in the dielectric material. This is observed in all dielectrics and is known as *electrostriction effect*.

6.25 Direct Piezoelectric Effect and Inverse Piezoelectric Effect

Certain dielectric materials are also there in which an electric potential appears across a crystal when the dimensions of the crystal are changed by the mechanical force. This is known as the Piezoelectric effect. The reverse is also possible: if a varying potential is applied to the proper axis of the crystal, the dimension of the crystal gets changed, or the crystal gets deformed. These phenomena are named as inverse Piezoelectric effect. The materials showing the above effect are Rochelle salt, tourmaline, lead ziconate, barium titanates, lead titanates, and so on. All ferroelectric materials exhibit Piezoelectric effect.

Applications of Piezo-electric Materials: The following are the applications of Piezo-electric materials:

1. Crystal oscillators
2. Gas ignitors
3. Displacement transducers
4. Piezo-electric transformers
5. Ultrasonics
6. Delay lines
7. Microbalances.

6.26 Pyro-electric Material

This class is subset of Piezoelectric materials. These materials are spontaneously polarized, but they do not respond to an electric field like ferroeclectics because they require a very high electric field for orienting the dipoles. The field required is so high that the material undergoes dielectric breakdown before it can get polarized.

Solved Problems

Ex. 1: Determine the displacement vector D and electric permittivity in a dielectric material having a dielectric constant of 2.5 and polarization $P = 2.1 \times 10^{-8}$ C/m².

Solution:
The polarization P of the material is related to its dielectric constant K or ε_r. Now,

$P = \varepsilon_0 E(\varepsilon_r - 1)$ and $\varepsilon_0 E = D$.

So, $D = \dfrac{P}{\varepsilon_r - 1}$.

Here, $\varepsilon_r = 2.5$ and $P = 2.1 \times 10^{-8}$ C/m², then

$D = \dfrac{2.1 \times 10^{-8}}{2.5 - 1} = 1.4 \times 10^{-8}$ C/m²

and $\varepsilon = \varepsilon_r \varepsilon_0 = 1.4 \times 8.85 \times 10^{-12} = 12.39 \times 10^{-12}$ C² N⁻¹ m⁻².

Ex. 2: The electrical susceptibility of material is 22.13. Find the dielectric constant and electric permittivity.

Solution:
The dielectric constant of the material is

$\varepsilon_r = 1 + \chi_e = 1 + 22.13 = 23.13$.

$\therefore \varepsilon = \varepsilon_r \varepsilon_0 = 23.13 \times 8.85 \times 10^{-12} = 2.04 \times 10^{-10}$ C² N⁻¹ m⁻².

Ex. 3: At 0°C and 1 atmosphere pressure, the dielectric constant of helium is 1.000074. Determine the induced dipole moment on each helium atom and electric polarization when the gas is subjected to electric field strength of 200 V/m.

Solution:
Here, we have $\varepsilon_r = 1.000074$, $N_a = 6.02 \times 10^{23}$, $V = 22.4$ liters $= 22.4 \times 10^{-3}$ m³, and $E = 200$ V/m.

$P = \varepsilon_0 E(\varepsilon_r - 1) = Np$.

$\therefore p = \dfrac{\varepsilon_0 E(\varepsilon_r - 1)}{N} = \dfrac{\varepsilon_0 E(\varepsilon_r - 1)V}{N_a} \qquad \left[\because N = \dfrac{N_a}{V} \right].$

Thus, induced dipole moment on each helium atom is

$p = \dfrac{8.85 \times 10^{-12} \times 200 \times (1.000074 - 1) \times 22.4 \times 10^{-3}}{6.02 \times 10^{23}} = 48.84 \times 10^{-40}$ Cm.

The electric polarization is

$P = Np = \dfrac{N_a p}{V} = \dfrac{6.02 \times 10^{23} \times 48.84 \times 10^{-40}}{22.4 \times 10^{-3}} = 13.12 \times 10^{-14}$ Cm⁻².

Ex. 4: If the capacitor's plates are filled with an insulator with a dielectric constant of 7, determine the electrical susceptibility.

Dielectric Materials

Solution:

Here, we have $\chi_e = (\varepsilon_r - 1) = (7 - 1) = 6$.

Ex. 5: The permittivity of diamond is 1.46×10^{-10} c^2 N^{-1} m^{-2}. Calculate the electrical susceptibility of it.

Solution:

The relative dielectric constant is given by

$$\varepsilon_r = \frac{\varepsilon}{\varepsilon_0} = \frac{1.46 \times 10^{-10}}{8.85 \times 10^{-12}} = 16.48.$$

The electrical susceptibility is given by

$$\chi_e = (\varepsilon_r - 1) = 16.48 - 1 = 15.48.$$

Previous Year Questions (University Examination)

1. What is a dielectric? How it different from an insulator?
2. What is meant by polarization of substance? Mention the different mechanisms of polarization in a dielectric.
3. What is dielectric constant? Derive an expression for it.
4. What is dielectric loss? Derive an expression for it.
5. Establish the relation between $\vec{D}$, $\vec{E}$, and $\vec{P}$.
6. Establish the relation between dielectric constant and electrical susceptibility.
7. Explain frequency dependence of dielectric constant.
8. Explain electronic and ionic polarizability. Show that the polarizability of an atom of radius r is $4\pi\varepsilon_0 r^2$.
9. What is meant by ionic polarization? How does it vary with temperature?
10. Explain polarization in a dielectric material. Discuss its various kinds.
11. Distinguish between polar and nonpolar dielectrics. Show that the polarizability of a nonpolar dielectric is equal to $4\pi\varepsilon_0 r^2$.
12. Explain the behavior of a dielectric in d.c. folds. Derive a relation between polarization P and external electric field E.
13. State and prove Gauss' law in a dielectric material. Derive a relation between E, D, and P.
14. Explain the following terms:
15. (i) Electric field   (ii) Electric displacement vector D (iii) Polarization vector P
16. Also discuss the physical significance of these vectors.
17. Discuss Langevin–Debye theory of orientational polarization. Also prove that susceptibility of polar dielectric is inversely proportional to temperature.
18. Define relative permittivity and electric susceptibility, and explain how these are related to each other.
19. Derive an expression for the electric field strength within a molecule in a dielectric. Then, obtain Clasuius–Mossotic equation. How can the diameter of an atom be estimated with its help?

20. What is meant by Lorentz local field? Derive an expression for the internal field in solids and liquids.
21. Discuss the behavior of dielectric in alternating electric field. Explain the effect of frequency upon the dielectric constant of a material.
22. Draw a sketch of frequency dependence of different contributions to the polarizability of a dielectric.
23. Explain the terms: dielectric loss and loss angle.
24. Explain the phenomenon of Piezo-electricity and discuss some applications of Piezoelectric crystals.
25. What is a dipolar relaxation? Show that the dipolar relaxation leads to complex dielectric constant of the material and also deduce an expression for it.
26. What are the ferroelectric materials? Discuss some applications of ferroelectric materials and their characteristics.
27. What is meant by ferroelectric hysteresis? Show that dielectric loss is given by area of hysteresis loop.
28. What is meant by breakdown phenomenon in dielectric material? Discuss various kinds of breakdown mechanisms.
29. Compare the characteristics of Piezo-electric, Pyro-electric, and ferroelectric crystals and discuss some of their applications.
30. Write short notes on:
    (i) Complex dielectric constant   (ii) Dielectric loss

Multiple Choice Questions

1. The dielectric material, which is
   (a) Metal
   (b) An insulating material
   (c) Semiconductor
   (d) All of the above
2. Dielectric and insulators vary in that
   (a) Working
   (b) Magnitude
   (c) Same
   (d) All of the above
3. Electric displacement vector and electric field have the following relationship:
   (a) $D = \varepsilon / E$
   (b) $D = E / \varepsilon$
   (c) $D = \varepsilon E$
   (d) All of the above
4. Polarization in a dielectric is
   (a) Tangential function of applied field
   (b) Sinusoidal function of applied field
   (c) Linear function of applied field
   (d) All of the above
5. The polarization in dielectric at frequencies 10^{12} Hz is
   (a) No polarization
   (b) Quad polarization
   (c) Orientational
   (d) All polarization
6. The material in which orientational polarization exist is
   (a) Metal
   (b) Polar material
   (c) Semiconductor
   (d) Superconductor

7. Electronic polarization takes place in
   - (a) Metals
   - (b) Nonpolar solids
   - (c) Semiconductors
   - (c) All of the above
8. The dielectric loss is due to
   - (a) Internal resistance in switching dipoles
   - (b) Attenuated
   - (c) Material capacitance
   - (d) None of the above
9. In a perfect dielectric, the loss of dielectric is
   - (a) 0
   - (b) ∞
   - (c) 5
   - (d) All of the above
10. The relation between electric susceptibility and relative permittivity is
    - (a) $\varepsilon_r = 1 + \chi$
    - (b) $\varepsilon_r = 1/\chi$
    - (c) Both (a) and (b)
    - (d) None of the above
11. When a monoatomic gas atom is placed in uniform electric field E, the resulting induced dipole moment is proportional to
    - (a) E
    - (b) E^2
    - (c) $1/E$
    - (d) None of these
12. The orientation polarizability per molecule in a polyatomic gas is proportional to
    - (a) T
    - (b) T^2
    - (c) $1/T$
    - (d) None of these

Bibliography

1. Juan, M., ed. (2013). *Dielectric Materials for Electrical Engineering*. John Wiley & Sons.
2. Mailadil, T. S. (2002). *Dielectric Materials for Wireless Communication*. Elsevier.
3. Kwan Chi, K. (2002). *Dielectric Phenomena in Solids*. Elsevier.
4. Friedrich, K. and Schönhals, A., eds. (2002). *Broadband Dielectric Spectroscopy*. Springer Science & Business Media.
5. Aldo, P. (2007). *Dielectric Resonator Antenna Handbook*. Artech.
6. Dietrich, M. (2013). *Theory of Dielectric Optical Waveguides*. Elsevier.

Keys

1. (b) 2. (a) 3. (c) 4. (c) 5. (d) 6. (b) 7. (b) 8. (a) 9. (a) 10. (a)
11. (a) 12. (c)

CHAPTER 7

Semiconducting Materials

7.1 BAND THEORY OF SOLIDS

The band theory of solids is different from the others because the atoms are arranged very close to each other such that the energy levels of the outermost orbital electrons are affected. But the energy level of the innermost electrons is not affected by the neighboring atoms.

In general, if there is n number of atoms, then there will be n discrete energy levels in each energy band. In such a system of n number of atoms, the molecular orbitals are called energy bands shown in Figure 7.1.

7.2 CLASSIFICATION OF SOLIDS ON THE BASIS OF BAND THEORY

The solids can be classified on the basis of band theory. The parameter that differentiates the solids among insulator, conductor, and semiconductor is known as energy band gap and represented by (E_g), as shown in Figure 7.2. When the energy band gap (E_g) between conduction band and valence band is greater than 5 eV (electron-volt) then the solid is classified as insulator. When the energy band gap (E_g) between conduction band and valence band is 0 eV (electron-volt), that is, overlapping of bands occurs then the solid is classified as conductor. When the energy band gap (E_g) between conduction band and valence band is approximately equals to 1 eV (electron-volt) then the solid is classified as semiconductors.

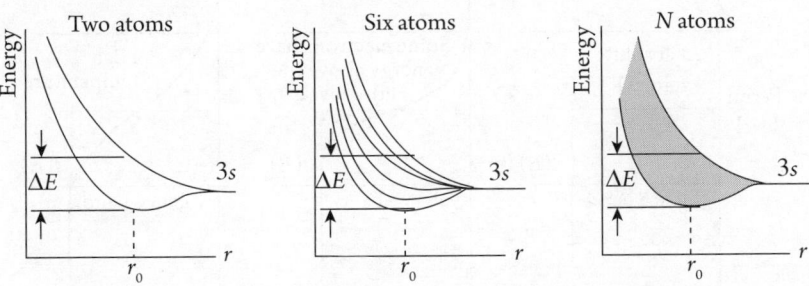

Fig. 7.1 Formation of bands due to different numbers of atoms.

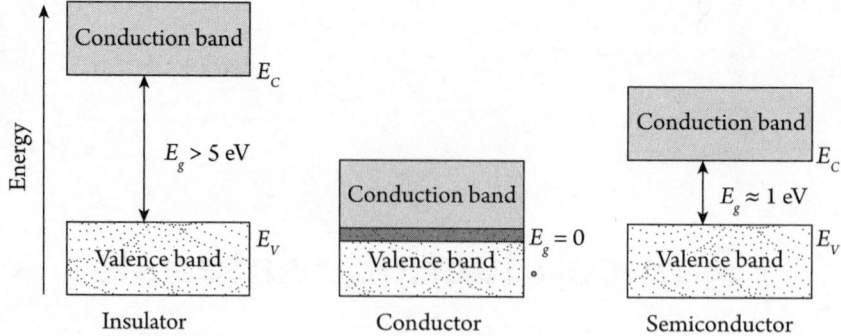

Fig. 7.2 Pictorial diagram of solids classification on the basis of band theory.

7.3 FERMI ENERGY

The energy difference between the greatest and lowest occupied single-particle states in a quantum system of noninteracting fermions at absolute zero temperature is referred to as the Fermi energy in quantum mechanics.

The Fermi energy is defined as the Fermi level at absolute zero temperature (273.15°C). It is the greatest kinetic energy an electron can have at absolute zero degrees Celsius. Each solid has a constant Fermi energy.

7.4 FERMI ENERGY LEVEL

The Fermi level is the maximum energy level an electron may occupy when it is at absolute zero temperature.

Since the electrons are all in the lowest energy state at absolute zero, the Fermi level is located between the valence band and conduction band. The Fermi level may be thought of as the sea of fermions (or electrons) above which no electrons exist because there is insufficient energy at absolute zero Kelvin. Pauli's exclusion principle, which stipulates that two fermions cannot occupy the same quantum state, is what causes this energy level to exist. Therefore, if a system contains many fermions, each fermion is linked with a unique set of magnetic quantum numbers. The intensity of Fermi increases with temperature, as shown in Figure 7.3.

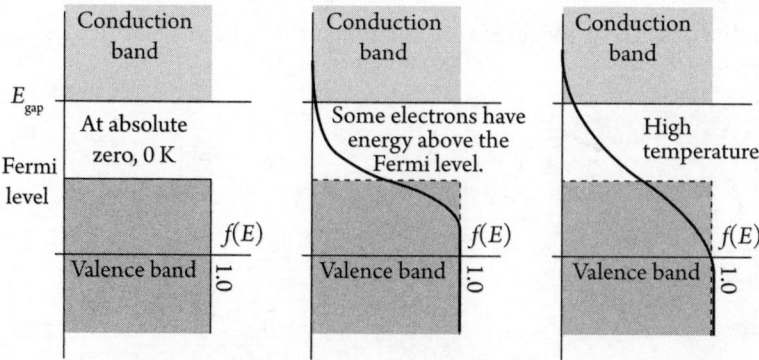

Fig. 7.3 Fermi energy level at different temperatures.

Semiconducting Materials

7.5 Density of States

The density of states is defined as the number of energy states present in a unit energy range. The density of state for energy E is denoted by $N(E)$ and is given by

$$N(E) = \frac{dn}{dE}. \tag{7.1}$$

Since Fermi energy is given by

$$E_f = \frac{h^2}{2m}\left(\frac{3n}{8\pi}\right)^{\frac{2}{3}}$$

or, $\left[E_f\right]^{\frac{3}{2}} = \left(\frac{h^2}{2m}\right)^{\frac{3}{2}} \frac{3n}{8\pi},$

$$\therefore n = \left(\frac{8\pi}{3}\right)\left[E_f\right]^{\frac{3}{2}}\left(\frac{2m}{h^2}\right)^{\frac{3}{2}}.$$

Put the value of n in equation (7.1), we get

$$N(E) = \frac{d}{dE}\left\{\frac{8\pi}{3}[E]^{\frac{3}{2}}\left(\frac{2m}{h^2}\right)^{\frac{3}{2}}\right\}.$$

$$N(E) = \frac{4\pi}{h^3}[2m]^{\frac{3}{2}}[E]^{\frac{1}{2}}. \tag{7.2}$$

This is an expression of density of state for energy E.

7.6 Density of State for Electrons in Conduction Band

The density of states at the bottom of the conduction bands is given by

$$N(E) = \frac{4\pi}{h^3}\left[2m_e\right]^{\frac{3}{2}}\left[E - E_c\right]^{\frac{1}{2}}. \tag{7.3}$$

7.7 Density of State for Holes in Valence Band

The density of states at the top of the valence band is given by

$$N(E) = \frac{4\pi}{h^3}\left[2m_h\right]^{\frac{3}{2}}\left[E_v - E\right]^{\frac{1}{2}}. \tag{7.4}$$

7.8 Fermi–Dirac Distribution Function

Density of states indicates that how many states exist at a given energy E. The Fermi–Dirac distribution function or Fermi function $f(E)$ specifies the probability of an electron to occupy an available state at an energy E under equilibrium conditions. It is a probability distribution function given as

$$F(E) = \cfrac{1}{1+e^{\frac{E-E_f}{kT}}},$$

where E_f is the Fermi energy level, k is the Boltzmann constant, and T is an absolute temperature.

7.9 Free Carrier Density (Electron–Hole Concentration)

1. **Concentration of Electrons in Conduction Band:**

Consider the following:

a. Electrons in the conduction band behave as free particle with an effective mass m_e.
b. The number of conduction electrons per cubic meter whose energy lies between E and $E + dE$.

$$dn_e = N(E)F(E), \tag{7.5}$$

where $N(E) = \dfrac{4\pi}{h^3}[2m_e]^{\frac{3}{2}}[E-E_c]^{\frac{1}{2}}$

and $F(E) = \cfrac{1}{1+e^{\frac{E-E_f}{kT}}}.$

Now, the electrons in the conduction band may have energies lying from E_c to ∞, as shown in Figure 7.4.

Then, the concentration of electrons in the conduction band from equation (7.5) is given by

$$n_e = \int_{E_c}^{\infty} N(E)F(E).$$

On substituting the values of $N(E)$ and $F(E)$, we get

$$n_e = \int_{E_c}^{\infty} \frac{4\pi}{h^3}[2m_e]^{\frac{3}{2}}[E-E_c]^{\frac{1}{2}} \cdot \cfrac{1}{1+e^{\frac{E-E_f}{kT}}} dE. \tag{7.6}$$

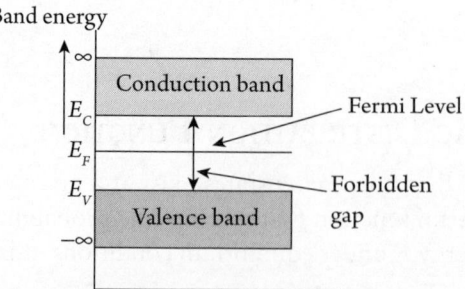

Fig. 7.4 A schematic diagram of band energies and Fermi level position.

Semiconducting Materials

Since $E \geq E_c, E - E_f \gg kT$.

Then, $F(E) = \dfrac{1}{1+e^{\frac{E-E_f}{kT}}} \cong \dfrac{1}{e^{\frac{E-E_f}{kT}}} = e^{\frac{E_f - E}{kT}}$.

$$n_e = \dfrac{4\pi}{h^3}\left[2m_e\right]^{\frac{3}{2}} \int_{E_c}^{\infty} \left[E - E_c\right]^{\frac{1}{2}} \cdot e^{\frac{E_f - E}{kT}} \, dE.$$

$$n_e = \dfrac{4\pi}{h^3}\left[2m_e\right]^{\frac{3}{2}} \int_{E_c}^{\infty} \left[E - E_c\right]^{\frac{1}{2}} \cdot e^{\frac{E_f - E + E_c - E_c}{kT}} \, dE.$$

$$n_e = \dfrac{4\pi}{h^3}\left[2m_e\right]^{\frac{3}{2}} e^{\frac{E_f - E_c}{kT}} \int_{E_c}^{\infty} \left[E - E_c\right]^{\frac{1}{2}} \cdot e^{\frac{E_c - E}{kT}} \, dE. \tag{7.7}$$

Let $\dfrac{E - E_c}{kT} = x; \dfrac{1}{kT} dE = dx \Rightarrow dE = kT dx.$

$$n_e = \dfrac{4\pi}{h^3}\left[2m_e\right]^{\frac{3}{2}} e^{\frac{E_f - E_c}{kT}} \int_{0}^{\infty} [x]^{\frac{1}{2}} \cdot [kT]^{\frac{1}{2}} e^{-x} kT \, dx.$$

$$n_e = \dfrac{4\pi}{h^3}\left[2m_e\right]^{\frac{3}{2}} [kT]^{\frac{3}{2}} e^{\frac{E_f - E_c}{kT}} \int_{0}^{\infty} [x]^{\frac{1}{2}} \cdot e^{-x} \, dx.$$

$$n_e = \dfrac{4\pi}{h^3}\left[2m_e\right]^{\frac{3}{2}} [kT]^{\frac{3}{2}} e^{\frac{E_f - E_c}{kT}} \left(\dfrac{\sqrt{\pi}}{2}\right).$$

$$n_e = 2\left[\dfrac{2\pi m_e kT}{h^2}\right]^{\frac{3}{2}} \cdot e^{\frac{E_f - E_c}{kT}}. \tag{7.8}$$

This is an expression for concentration of electrons in the conduction band.

2. Concentration of Holes in Valence Band

A hole represents the vacancy created by the movement of electron from the valence band to conduction band due to thermal energy.

The number of valence holes per cubic meter whose energy lies between E and $E + dE$. Concentration of holes is given by

$$dn_h = N(E)\left[1 - F(E)\right], \tag{7.9}$$

where $N(E) = \dfrac{4\pi}{h^3}\left[2m_h\right]^{\frac{3}{2}} \left[E_v - E\right]^{\frac{1}{2}}.$

And $\left[1 - F(E)\right] = 1 - \dfrac{1}{1+e^{\frac{E-E_f}{kT}}} = \dfrac{1+e^{\frac{E-E_f}{kT}} - 1}{1+e^{\frac{E-E_f}{kT}}} = \dfrac{e^{\frac{E-E_f}{kT}}}{1+e^{\frac{E-E_f}{kT}}}.$

In the above case, $E_f - E \gg kT$.

So, in the above expression, $e^{\frac{E-E_f}{kT}}$ can be neglected in comparison to 1.

$$[1-F(E)] = e^{\frac{E-E_f}{kT}}. \quad (7.10)$$

For valence band, the density of states can be written as

$$n_h = \int_{-\infty}^{E_v} N(E)[1-F(E)].$$

$$n_h = \int_{-\infty}^{E_v} \frac{4\pi}{h^3}[2m_h]^{\frac{3}{2}}[E_v - E]^{\frac{1}{2}} \cdot e^{\frac{E-E_f}{kT}} dE. \quad (7.11)$$

Let $\dfrac{E_v - E}{kT} = x;\ \dfrac{1}{kT}dE = -dx \Rightarrow dE = -kTdx.$

$$n_h = \frac{4\pi}{h^3}[2m_h]^{\frac{3}{2}} e^{\frac{E_v-E_f}{kT}} \int_{\infty}^{0} [x]^{\frac{1}{2}} \cdot [kT]^{\frac{1}{2}} e^{-x}(-kT)dx.$$

$$n_h = \frac{4\pi}{h^3}[2m_h]^{\frac{3}{2}} [kT]^{\frac{3}{2}} e^{\frac{E_v-E_f}{kT}} \int_{0}^{\infty} [x]^{\frac{1}{2}} \cdot e^{-x} dx.$$

$$n_h = \frac{4\pi}{h^3}[2m_h]^{\frac{3}{2}} [kT]^{\frac{3}{2}} e^{\frac{E_v-E_f}{kT}} \left(\frac{\sqrt{\pi}}{2}\right).$$

$$n_h = 2\left[\frac{2\pi m_h kT}{h^2}\right]^{\frac{3}{2}} \cdot e^{\frac{E_v-E_f}{kT}}. \quad (7.12)$$

This is an expression for concentration of holes in the valence band.

7.10 Fermi Level for Intrinsic Semiconductors

For intrinsic semiconductors, we know that

Number of electrons = Number of holes

$$n_e = n_h$$

$$2\left[\frac{2\pi m_e kT}{h^2}\right]^{\frac{3}{2}} \cdot e^{\frac{E_f-E_c}{kT}} = 2\left[\frac{2\pi m_h kT}{h^2}\right]^{\frac{3}{2}} \cdot e^{\frac{E_v-E_f}{kT}}.$$

$$[m_e]^{\frac{3}{2}} \cdot e^{\frac{E_f-E_c}{kT}} = [m_h]^{\frac{3}{2}} \cdot e^{\frac{E_v-E_f}{kT}}.$$

$$e^{\frac{E_f-E_c-E_v+E_f}{kT}} = \left[\frac{m_h}{m_e}\right]^{\frac{3}{2}}.$$

Semiconducting Materials

$$e^{\frac{2E_f - E_c - E_v}{kT}} = \left[\frac{m_h}{m_e}\right]^{\frac{3}{2}}.$$

Now taking log on both sides, we get

$$\frac{2E_f - E_c - E_v}{kT} = \frac{3}{2}\log\frac{m_h}{m_e}.$$

$$2E_f - E_c - E_v = \frac{3}{2}kT\log\frac{m_h}{m_e}$$

$$E_f = \frac{E_c + E_v}{2} + \frac{3}{4}kT\log\frac{m_h}{m_e}.$$

For intrinsic semiconductors, the effective mass is $m_e \cong m_h$.

$$\therefore E_f = \frac{E_c + E_v}{2}.$$

Thus, Fermi energy level lies midway between conduction band and valence band, as shown in Figure 7.5.

7.11 Fermi Level for Extrinsic Semiconductors

1. For *n*-type Semiconductors:

In case of *n*-type semiconductor, the number of ionized donor atoms is equal to the number of electrons.

$$n_e = N_d = 2\left[\frac{2\pi m_e kT}{h^2}\right]^{\frac{3}{2}} \cdot e^{\frac{E_f - E_c}{kT}}. \tag{7.13}$$

Let us consider $N_c = 2\left[\dfrac{2\pi m_e kT}{h^2}\right]^{\frac{3}{2}}$; then put in equation (7.13), we get

$$N_d = N_c \cdot e^{\frac{E_f - E_c}{kT}}.$$

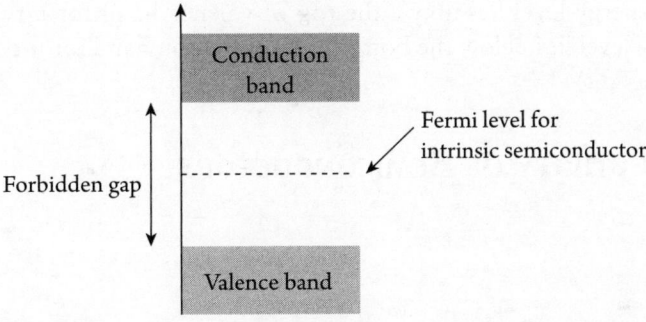

Fig. 7.5 A schematic diagram of Fermi level position for an intrinsic semiconductor.

$$\frac{N_c}{N_d} = e^{\frac{-(E_f - E_c)}{kT}}.$$

Taking log on both sides, we get

$$\log\left(\frac{N_c}{N_d}\right) = \frac{(E_c - E_f)}{kT}.$$

$$E_f = E_c - KT \log\left(\frac{N_c}{N_d}\right). \tag{7.14}$$

Thus, Fermi energy level lies below the bottom of the conduction band.

2. For *p*-type Semiconductors

In case of *p*-type semiconductor, the number of ionized acceptor atoms is equal to the number of holes.

$$n_h = N_a = 2\left[\frac{2\pi m_h kT}{h^2}\right]^{\frac{3}{2}} \cdot e^{\frac{E_v - E_f}{kT}}. \tag{7.15}$$

Let us consider $N_v = 2\left[\dfrac{2\pi m_h kT}{h^2}\right]^{\frac{3}{2}}$. Put in equation (7.15), we get

$$N_a = N_v \cdot e^{\frac{E_v - E_f}{kT}}.$$

$$\frac{N_v}{N_a} = e^{\frac{-(E_v - E_f)}{kT}}.$$

Taking log on both sides, we get

$$\log\left(\frac{N_v}{N_a}\right) = \frac{(E_f - E_v)}{kT}.$$

$$E_f = E_v + KT \log\left(\frac{N_v}{N_a}\right). \tag{7.16}$$

Thus, Fermi energy level lies above the top of the valence band.

Hence, Fermi energy level lies above the top of valence band for *p*-type semiconductor, while Fermi energy level lies below the bottom of conduction band for *n*-type semiconductor, as shown in Figure 7.6.

7.12 Conductivity of Semiconductors

We know that $J = \sigma E$

$$\sigma = \frac{J}{E},$$

$$\therefore J = ne\mu E.$$

Semiconducting Materials

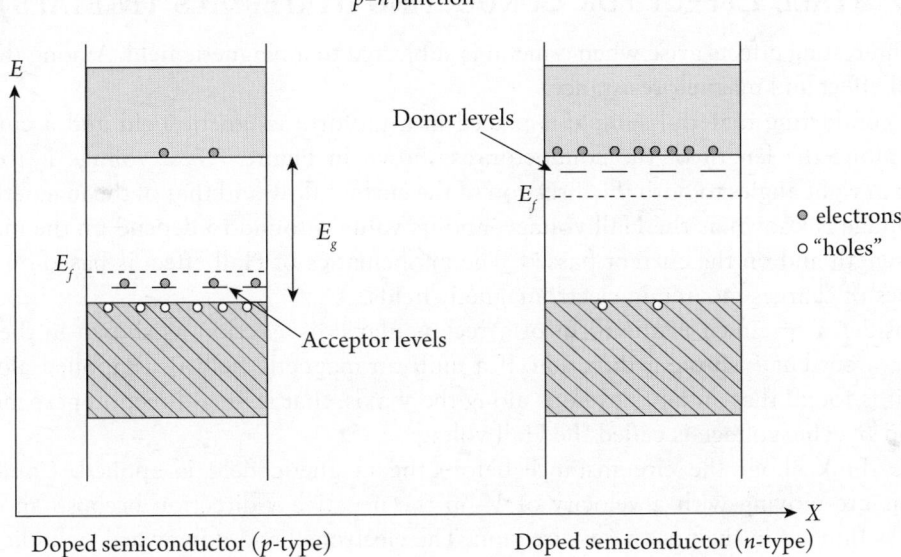

Fig. 7.6 Fermi energy level position in *p*-type and *n*-type semiconductors.

Here, μ is the mobility and E is applied electric field.

$$\therefore \sigma = ne\mu$$

This is the general formula of conductivity.

In case of semiconductors, the carrier density is of two types n_e and n_h. Similarly, mobility is also different μ_e and μ_h.

So, conductivity for semiconductors is expressed as

$$\sigma = \sigma_n + \sigma_p.$$

$$\sigma = n_e e \mu_e + n_h e \mu_h.$$

$$\sigma = e(n_e \mu_e + n_h \mu_h).$$

For intrinsic semiconductors, conductivity is given as

$$n = p = n_i.$$

$$\therefore \sigma = en_i(\mu_e + \mu_h).$$

The conductivity for extrinsic semiconductors can be given as follows:

For *n*-type, conductivity is given as

$$n \gg p.$$

$$\sigma_{n-type} = n_e e \mu_e.$$

For *p*-type, conductivity is given as

$$n \ll p.$$

$$\sigma_{p-type} = n_h e \mu_h.$$

7.13 Hall Effect for Conducting Materials (Metals)

Many interesting effects arise when a metal is subjected to a magnetic field. Among them are the Hall effect and magnetoresistance.

If a conducting material sample is placed in a uniform magnetic field and a current is passed along the length of the conductor, as shown in Figure 7.7, a voltage is found to develop at right angles to both the direction of the current flow and that of the magnetic field. This voltage is known as the Hall voltage, and its value is found to depend on the magnetic field strength and on the current passed. The mathematics of Hall effect is based on simple dynamics of charges moving in electromagnetic fields.

Consider a specimen in the form of a rectangular cross section as shown in the figure carrying a current I_x in the x-direction. If a uniform magnetic field B_z is applied along the z-axis, it is found that an emf develops along the y-axis, that is, in a direction perpendicular to I_x and B_z. This voltage is called the Hall voltage.

Let's think about the circumstance before the magnetic field is applied. Conduction electrons are moving with a velocity of V_x in the negative x-direction because an electric current is flowing in the positive x-direction. The electrons bend downward, as indicated in the illustration, when a magnetic field is supplied due to the F_L Lorentz force. As a result, the bottom surface develops a net negative charge as electrons amass there. Because there aren't enough electrons on the top surface, a net positive charge develops there at the same time. The downward electric field produced by this contrast of positive and negative surface charges is known as the Hall field.

Lorentz force F_L, which produces the charge accumulation in the negative y-direction, has the value

$$F_L = ev_x B_z.$$

The surface charge field now generates a force that resists the Lorentz force. Up until the Lorentz force is totally cancelled by the Hall force, the accumulation process continues. Thus, in the study state $F_H = F_L$,

$$eE_H = ev_x B_z.$$

$$E_H = v_x B_z. \tag{7.17}$$

The current density j_x is given by the equation as

$$j_x = -nev_x. \tag{7.18}$$

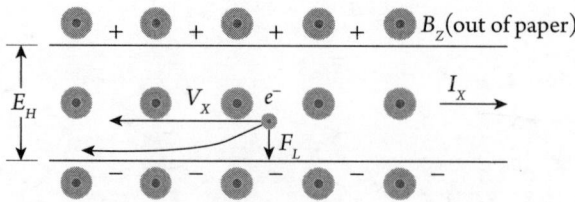

Fig. 7.7 Origin of Hall effect and Hall field.

Semiconducting Materials

Dividing equations (7.17) by (7.18), we get

$$\frac{E_H}{j_x} = -\frac{B_z}{ne}.$$

$$E_H = -\left(\frac{1}{ne}\right) j_x B_z.$$

As a result, the Hall field is proportional to both the current and the magnetic field. The constant of proportionality, that is, $\frac{E_H}{j_x B_z}$ is known as the Hall constant and is usually denoted by R_H, that is,

$$\frac{E_H}{j_x B_z} = -\left(\frac{1}{ne}\right) = R_H. \tag{7.19}$$

Now, the Hall constant for Hall coefficient R_H is defined as the ratio of the electric field strength reduce per unit current density to the transverse magnetic field. It will be noted that R_H depends on the sign of e, and if E_H is in a certain direction for a flow of negative charges, then it will be in the opposite sense for the same current when it is produced by a flow of positive charges in a reverse direction. In the monovalent metals, R_H is negative, which is consistent with the fact that the current is produced by a flow of negatively charged particles. The magnitude of R_H is then such that there is of the order of one moving charge per atom.

In more complicated metals, particularly those in which there is a band overlap, R_H can be positive (example in zinc and cadmium) and here it is assumed that most of the conduction occurs by the motion of positive holes. From equation (7.19),

$$R_H = \frac{E_H}{j_x B_z} = \left(\frac{V_y}{y}\right)\left(\frac{yz}{I_x}\right)\frac{1}{B_z}. \tag{7.20}$$

Unit of $R_H = \dfrac{volt - m}{\dfrac{amp - weber}{m^2}} = Vm^3 A^{-1} wb^{-1}$.

The general expression for current density is

$$j_x = -nev_x.$$

Therefore, electrical conductivity is

$$\sigma = \frac{j_x}{E_x} = \frac{nev_x}{E_x}.$$

The drift velocity produced for unit electric field is called the mobility of charge carriers. that is, $\sigma = \dfrac{j_x}{E_x} = ne\mu_e \Rightarrow \mu_e = \left(\dfrac{1}{ne}\right)\sigma.$

$$\mu_e = \left(\frac{E_H}{j_x B_z}\right)\frac{j_x}{E_x} \quad [\because \mu_e = R_H \sigma].$$

Table. 7.1 Hall coefficient and mobility for some metals at 300 K.

S. No.	Metal	R_H (Vm^{3}A^{-1}wb^{-1})	μ (m^2 V^{-1}s^{-1})
1	Silver	−0.84	0.0056
2	Copper	−0.55	0.0032
3	Gold	−0.71	0.0030
4	Sodium	−2.50	0.0052
5	Aluminum	−0.31	0.0012
6	Lithium	−1.70	0.0018
7	Zinc	+0.30	0.0060
8	Cadmium	+0.60	0.0080

$$\mu_e = \left(\frac{V_y}{B_z y}\right)\left(\frac{x}{V_x}\right) = \left(\frac{V_y}{V_x}\right)\left(\frac{x}{y}\right)\left(\frac{1}{B_z}\right). \tag{7.21}$$

From this, the mobility of electrons may be determined. Measurement of the Hall voltage helps one to determine the following:

1. The sign of the current carrying charges can be determined.
2. The number of charge carriers present in unit volume can be calculated from the magnitude of R_H.
3. The mobility of the charge carriers may be obtained directly from the measurement of the hall voltage. The hall coefficient and mobility of some selected metals are given in Table 7.1.

7.14 Hall Effect in Semiconductors

Finding out if a substance is *n*-type or *p*-type is frequently essential. This information cannot be determined by measuring a specimen's conductivity since it cannot differentiate between positive hole and electron conduction. The two different types of carriers may be distinguished using the Hall effect, which also makes it possible to calculate the charge carrier density.

An electric field is created within a conductor (metal or semiconductor) when it is put in a transverse magnetic field in a direction normal to both the current and the magnetic field, as illustrated in Figure 7.8. Hall effect is the name of both the phenomenon and the voltage it produces.

The majority of the current flow is made up of electrons that are moving from right to left, assuming the material is an *n*-type semiconductor. This is in line with the illustration's depiction of the left to right direction of the conventional current. If v is the speed of electrons traveling at right angles to the magnetic field, then each electron feels a downward force with a magnitude of be Bev. The electron stream is then deflected downward, which causes a buildup of negative charge on the slab's bottom face. It should be determined from the top to the bottom of the specimen, with the bottom face being negative, if a difference is possible.

Semiconducting Materials

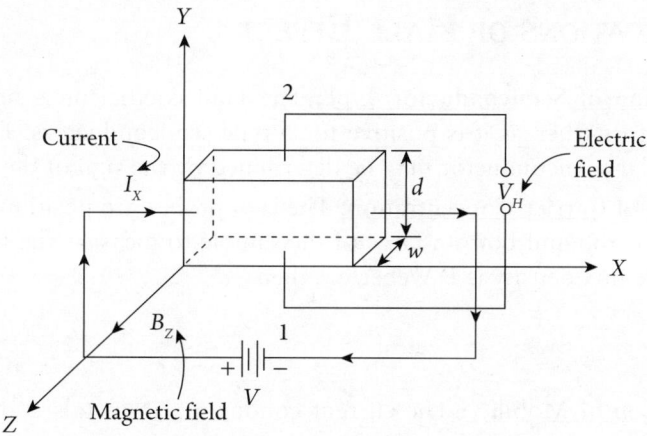

Fig. 7.8 Hall effect and Hall field in semiconductors.

There is a field EH in the negative y direction as a result of this potential difference. There is a field EH in the negative y direction as a result of this potential difference.

$$eE_H = eBv \Rightarrow E_H = Bv.$$

If J_x is the current density in the x-direction, then

$$J_x = nev,$$

where n is the concentration of current carriers.

Thus, $E_H = \dfrac{BJ_x}{ne}$.

The Hall effect is described by means of the Hall coefficient R_H defined in terms of the current density J_x by the relation,

$$E_H = R_H J_x B \Rightarrow R_H = \dfrac{E_H}{J_x B}.$$

That is, $R_H = -\dfrac{1}{ne}$

Negative sign is used because the electric field developed is in the negative y-direction.

$$R_H = -\dfrac{E_H}{J_x B} = -\dfrac{1}{ne}.$$

All the three quantity, E_H, B and J_x, can be measured; therefore, the Hall coefficient and carrier density can also be determined.

In case of p-type specimen when the current is entirely by holes. In this case,

$$R_H = \dfrac{E_H}{J_x B} = \dfrac{1}{pe},$$

where p is the positive hole density.

7.15 Applications of Hall Effect

1. **Determination of Semiconductor Type:** The Hall coefficient is negative for *n*-type semiconductors whereas it is positive for *p*-type semiconductors. Therefore, the type of a particular semiconductor may be determined by the sign of the Hall coefficient.
2. **Calculation of Carrier Concentration:** The two probes are positioned as usual at the centers of the top and bottom faces of the sample to measure the Hall voltage V_H. If the magnetic flux density is B Weber/m², then

$$n = \frac{1}{R_H e}.$$

3. **Determination of Mobility:** The current conduction in metals is due to one type of carrier, i.e., electrons; therefore, conductivity is given as:

$\sigma = ne\mu_n.$

$$\mu_n = \frac{\sigma}{ne} = \sigma R_H = \sigma \left(\frac{V_H b}{I_x B} \right).$$

The mobility μ_n can be determined by measuring σ.

4. **Measurement of Magnetic Flux Density:** Since Hall voltage V_H is proportional to the magnetic flux density B for a given current I_x to the sample, the Hall effect can be used as the basis for the design of a magnetic flux density meter.
5. **Measurement of Power in an Electromagnetic Wave:** The magnetic field H and the electric field E are at right angles in an electromagnetic wave traveling through empty space. As a result, if a semiconductor sample is positioned parallel to E, a semiconductor current I will result. A transverse magnetic field H applied on a semiconductor produces a Hall voltage across the sample. The Hall voltage is proportional to the magnitude of the Poynting vector (the product of E and H) of the electromagnetic wave. Thus, the power flow in electromagnetic waves may be determined using the Hall effect.
6. **Hall Effect Multiplier:** If the magnetic flux density B is produced by passing a current I through an air core coil, B will be proportional to I'. The Hall voltage is thus proportional to the product of I and I'. This forms the basis of the multiplier.

7.16 Compound Semiconductors

A compound semiconductor is made up of chemical elements from two or more distinct periodic table groups, such as III-V. In contrast to silicon, compound semiconductors offer special material qualities that enable the development of photonic, high-speed, and high-power device technologies. These properties include a direct energy bandgap, high breakdown electric fields, and high electron mobility. Processing is more than 100 times faster in compound semiconductors than in silicon, thanks to the substantially faster electron motion. Below are a few instances and uses of compound semiconductors.

Semiconducting Materials

Disordered or glassy varieties of crystalline semiconductor materials are known as amorphous semiconductors. They are network formations with a predominance of covalent bonding, similar to nonconducting glasses. The diamond structure of crystalline silicon is an organized arrangement of fused six-membered silicon rings, all of which are in the "chair" conformation. The silicon atoms' local bonding environment is tetrahedral. Although the structure of amorphous silicon (a-Si) lacks long-range organization, the silicon atoms are mostly tetrahedrally coordinated. There are five- and seven-membered rings in addition to six-membered rings, as well as certain "dangling bond" locations where the closest neighbors for Si atoms are just three, as shown in Figure 7.9.

Amorphous silicon (a-Si) and amorphous selenium (a-Se), two of the most extensively investigated amorphous semiconductors. Glassy forms of Si and Se may both be produced, often by sputtering or low-temperature evaporation. Similar to a-Si, most of the atoms in a-Se have their "normal" valence locally, but the structure has several flaws and abnormalities. Electrons in dangling bonds in amorphous semiconductors have orbital energies in the center of the gap, making them effectively nonbonding states. These dangling bond sites exist over a wide range of energies and are far from one another, resulting in little orbital overlap, as shown in Figure 7.10. Anderson localization is the term for the localization of electrons in these mid-gap states. Since electrons near the Fermi level (at the center of the gap) remain immobile in the lattice, amorphous silicon is insulating. Only the electrons in strongly bonding or antibonding states are delocalized because of the mobility gap created by these localized states. As a result, unaltered a-Si is not a particularly useful semiconductor. However, the

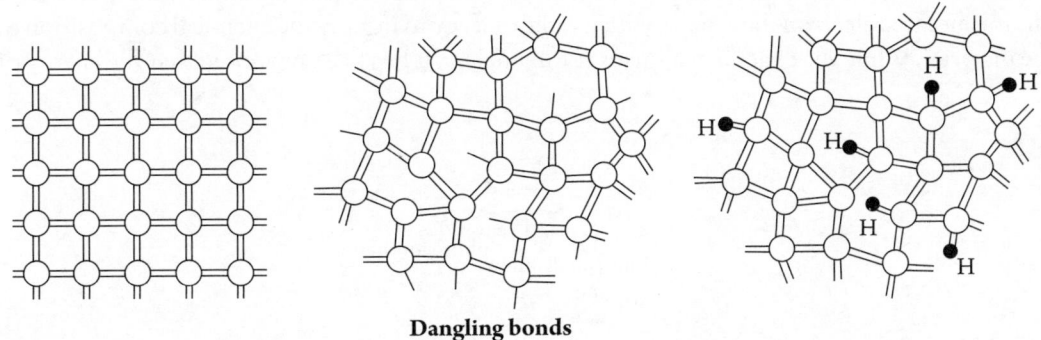

Dangling bonds

Fig. 7.9 Schematic illustration of the structures of crystalline silicon (left), amorphous silicon (middle), and amorphous hydrogenated silicon (right).

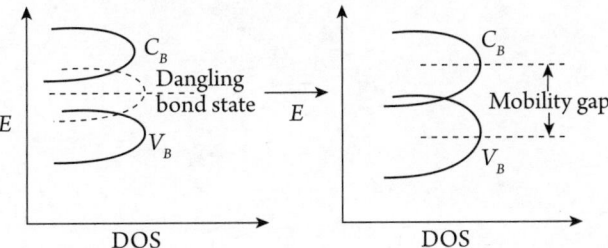

Fig. 7.10 Energy versus DOS for an amorphous semiconductor. Disorder and dangling bonds result in localized mid-gap states.

under-coordinated Si atoms are bound to hydrogen atoms by hydrogenating the material as it is being created (usually in a plasma of H atoms). Due to this, filled bonding and empty antibonding orbitals are produced, both of which have energies that are outside the mobility gap. In the mobility gap, hydrogenation therefore reduces the density of states. Hydrogenated amorphous silicon (a-Si:H) is insulating in the dark but is a good photoconductor because light absorption creates electrons and holes in mobile states that are outside the mobility gap.

In xerography, amorphous Se's photoconductivity is utilized. A corona discharge from a wire causes static electricity to build up on a conductive drum that has been coated with the insulating compound a-Se. The lighted a-Se sections of the drum become conductive when subjected to a sequence of light and dark (the picture to be copied), and the static charge is released from those regions of the drum. To create the copy, carbon-containing toner particles are transferred and bound to the paper in the parts that were not exposed to light by static electricity. The very low conductivity of a-Se in the dark and its strong conductivity under illumination determine the process' speed and high resolution of pattern transfer, as displayed in Figure 7.11.

Thin film solar cells made of affordable materials employ amorphous hydrogenated silicon. The mobility gap is around 1.7 eV, which is higher than the value of the crystalline Si bandgap (1.1 eV). Since a-Si:H is a direct-gap material, thin films made of it perform well as light absorbers. Large-area sheets of a-Si:H solar cells may be vapor-deposited. Cells made of p+Si-a-Si:H-n+Si have a 10% power conversion efficiency. Amorphous Si solar cells, however, eventually lose effectiveness when exposed to light.

The Staebler–Wronski effect, which accounts for this efficiency loss, requires photogenerated electron–hole pairs with enough energy to modify the chemical composition of the material. While the exact mechanism is still unclear, it has been proposed that the energy of

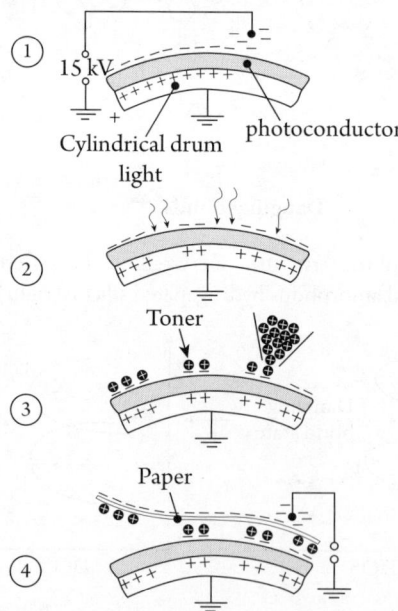

Fig. 7.11 Charging of amorphous Se and pattern transfer in the xerographic cycle.

electron–hole recombination breaks a weak Si–Si bond, and that one of the resulting dangling bonds abstracts a H atom, leaving a passivated Si-H center and a permanent dangling bond. The effect is minimized by hydrogenating a-Si and can be partially reversed by annealing.

In heterojunction intrinsic thin-layer (HIT) solar cells shown in Figure 7.13, crystalline silicon is combined with thin layers of amorphous silicon. The amorphous-crystalline interface has a potential energy barrier that deflects electrons and holes away from it because the mobility gap of a-Si is larger than the bandgap of c-Si. Only holes can tunnel through the barrier at the p+ contact, but only electrons can do so at the n+ contact. In comparison to traditional c-Si p-n junction cells, the photovoltage and photocurrent in solar cells are increased due to the passivation of surface imperfections that are locations of electron–hole recombination. Production of HIT cells with up to 23% power conversion efficiency has been disclosed by Panasonic and Sanyo in electronic devices, as shown in Figure 7.12.

Fig. 7.12 A calculator that runs on solar and battery power.

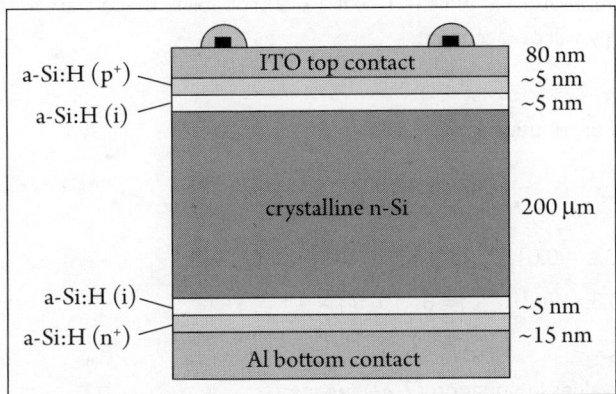

Fig. 7.13 Layered structure of a HIT solar cell. The layers are not drawn to scale. A thick crystalline n-silicon layer is the light absorber, and photogenerated holes, which are the minority carriers, are reflected away from the aluminum back contact by the thin intrinsic a-Si layer there.

7.17 Applications of Semiconductors

Semiconductors are used in a variety of applications, including diodes, transistors, and FETs, JFETs, and MOSFETs. They play a crucial role in computer hardware, such as microprocessors, ICs, and memory cells, and are essential for creating RAM for computers. Additionally, semiconductors are found in transmitters and receivers, as well as in transducers and sensors. They serve as amplifiers and are integral to medical instruments like EEGs and ECGs. Semiconductors also enable the development of digital electronic devices, lasers, and optoelectronic devices, such as LEDs, photodiodes, and solar cells. Moreover, they are used in smart refrigerators, digital cameras, smart mobile phones, and washing machines.

Solved Problems

Ex. 1: In a sample of intrinsic germanium at room temperature, the mobility of electrons and holes is 0.40 m² / V – sec and 0.23 m² / V – sec, respectively. If the electron and hole densities are each equal to 1.5×10^{23} m^{-3}, find out the electrical conductivity and resistivity of germanium.

Solution:

Here, given $\mu_e = 0.40$ m² / V – sec, $\mu_h = 0.23$ m² / V – sec. Now, $n_e = n_h = 1.5 \times 10^{23}$ m$^{-3} = n_i$.
The conductivity of intrinsic semi-conductor is given by

$$\sigma = en_i(\mu_e + \mu_h) = 1.6 \times 10^{-19} \times 1.5 \times 10^{23} \times (0.40 + 0.23)$$

$\sigma = 1.512 \times 10^4 \; \Omega m^{-1}$.

Resistivity of intrinsic semiconductor is given by

$$\rho = \frac{1}{\sigma} = \frac{1}{1.512 \times 10^4} = 0.66 \times 10^{-4} \; \Omega^{-1} \text{ m}.$$

Ex. 2: Find the temperature at which there is 1.0% probability that a state with energy 0.5 eV above Fermi energy will be occupied.

Solution:

According to the Fermi distribution law,

$$F(E) = \frac{1}{1 + e^{\frac{E-E_f}{kT}}}. \tag{7.22}$$

We have $F(E) = 1.0\% = 0.01$.

$E - E_F = 0.5$ eV $= 0.5 \times 1.6 \times 10^{-19}$ J $= 8 \times 10^{-20}$ J.

$k = 1.38 \times 10^{-23}$ J/K.

Substituting theses values in equation (7.22), we get

$$0.01 = \frac{1}{1 + e^{\frac{8 \times 10^{-20}}{1.38 \times 10^{-23} \times T}}} \Rightarrow 1 + e^{\frac{5797.1}{T}} = 100 \Rightarrow \frac{5797.1}{T} = 4.595.$$

or $T = 1261.6$ K.

Semiconducting Materials

Ex. 3: Find the probability with which an energy level 0.02 eV below Fermi level will be occupied at room temperature of 300 K and at 1000 K.

Solution:

According to the Fermi distribution function,

$$F(E) = \frac{1}{1+e^{\frac{E-E_f}{kT}}}. \tag{7.23}$$

We have $E - E_F = -0.02\,\text{eV}$.

$k = 1.38 \times 10^{-23}\,\text{J/K} = 8.625 \times 10^{-5}\,\text{eV/K}$.

For $T = 300$ K,

$$F(E) = \frac{1}{1+e^{\frac{-0.02}{8.625 \times 10^{-5} \times 300}}} = 0.684.$$

For $T = 1000$ K,

$$F(E) = \frac{1}{1+e^{\frac{-0.02}{8.625 \times 10^{-5} \times 1000}}} = 0.56.$$

Ex. 4: For n-type semiconductor with energy gap $E_g = 0.7\,\text{eV}$, calculate the concentration of n-type charge carriers at 300 K.

Solution:

The n-type charge carrier concentration is

$$n_e = 2\left[\frac{2\pi m_e kT}{h^2}\right]^{\frac{3}{2}} \cdot e^{-\frac{E_g}{2kT}}.$$

We have $m_e = 9.1 \times 10^{-31}$ k, $k = 1.38 \times 10^{-23}$ J/K, $T = 300$ K.

$h = 6.63 \times 10^{-34}$ Js, $E_g = 0.7\,\text{eV} = 0.7 \times 1.6 \times 10^{-19}$ J.

So, $n_e = 2\left[\dfrac{2\pi \times 9.1 \times 10^{-31} \times 1.38 \times 10^{-23} \times 300}{(6.63 \times 10^{-34})^2}\right]^{\frac{3}{2}} \cdot e^{-\frac{0.7 \times 1.6 \times 10^{-19}}{2 \times 1.38 \times 10^{-23} \times 300}}$

$n_e = 3.34 \times 10^{19}$ /m³.

Ex. 5: Calculate the conductivity of pure silicon at room temperature when the concentration of carriers is 1.6×10^{10} /cm³. Given that $\mu_e = 1300\,\text{cm}^2/\text{V}-\text{s}$ and $\mu_h = 700\,\text{cm}^2/\text{V}-\text{s}$.

Solution:

We know that the conductivity for intrinsic semiconductors,

$$\sigma_i = n_i e(\mu_e + \mu_h)$$

We have

$n_i = 1.6 \times 10^{10}$ /cm³, $e = 1.6 \times 10^{-19}$ C, $\mu_e = 1300\,\text{cm}^2/\text{V}-\text{s}$, $\mu_h = 700\,\text{cm}^2/\text{V}-\text{s}$.

So, $\sigma = 1.6 \times 10^{10} \times 1.6 \times 10^{-19}(1300 + 700) = 5.17 \times 10^{-6}$ mho/cm.

Ex. 6: Calculate the number of donor atoms which must be added to an intrinsic semiconductor to obtain the resistivity as 10^{-6} ohm cm. Assume $\mu_e = 1300$ cm^2/V–s.

Solution:

We know that $\rho = \dfrac{1}{\sigma} = \dfrac{1}{n_e e \mu_e}$.

Now, $\rho = 10^{-6}$ ohm cm, $e = 1.6 \times 10^{-19}$ C, $\mu_e = 1300$ cm^2/V–s.

$$n_e = \dfrac{1}{10^{-6} \times 1.6 \times 10^{-19} \times 1300} = 6.25 \times 10^{22} / \text{cm}^2.$$

Ex. 7: In order to obtain the resistivity 10^{-6} ohm cm, calculate the number of donor atoms which must be added to an intrinsic semiconductor (assume $\mu_e = 100 \dfrac{\text{cm}^2}{\text{V}-\text{sec}}$).

Solution:

We know that

$$\rho = \dfrac{1}{\sigma} = \dfrac{1}{n_e e \mu_e}.$$

Given, $\rho = 10^{-6}$ ohm cm, $e = 9.1 \times 10^{-19}$ C, $\mu_e = 100 \dfrac{\text{cm}^2}{\text{V}-\text{sec}}$,

$$n_e = \dfrac{1}{10^{-6} \times 9.1 \times 10^{-19} \times 100} = 6.25 \times 10^{22} / \text{cm}^3.$$

Previous Year Questions (University Examination)

1. What do you understand by intrinsic and extrinsic semiconductors? Derive an expression for the densities of free electrons and holes in an intrinsic semiconductor.
2. Show that the Fermi level in an intrinsic semiconductor lies halfway between the top of the valence band and bottom of the conduction band.
3. Show that the Fermi level lies below the bottom of conduction band and above the top of valence band in extrinsic semiconductors.
4. Derive an expression for conductivity in semiconductors.
5. Derive an expression for free carrier density in conduction band and valence band of intrinsic semiconductors.
6. Consider a compensated GaAs semiconductor at $T = 300$ K doped at $N_d = 5 \times 10^{15}$ cm^{-3}. Calculate the thermal equilibrium electron and hole concentration.
7. Silicon is doped at $N_d = 10^{15}$ cm^{-3} and $N_a = 0$. (a) Plot the concentration of electrons versus temperature over the range $300 \leq T \leq 600$ K. (b) Calculate the temperature at which the electron concentration is equal to 1.1×10^{15} cm^{-3}.
8. What are different types of semiconductor materials? Discuss their applications. Describe the merits and characteristic properties of semiconducting materials?

Semiconducting Materials

9. Enumerate different types of semiconductors. Intrinsic semiconductors are not suitable for applications in electronic devices – why? Show that the Fermi level for intrinsic germanium lies in the middle of its forbidden gap.
10. Differentiate between intrinsic and extrinsic semiconductors. Discuss in brief the mechanism of conduction in n-type and p-type semiconductors.
11. What is semiconductor? Discuss their classification. Differentiate between n-type and p-type semiconductors and how are they produced?
12. Explain how electrons and holes both conduct in a pure silicon crystal. How is conductivity influenced by mobility of electrons and holes?
13. Differentiate between n-type and p-type semiconductors. Name various impurities used to prepare them from intrinsic elements. Draw energy diagrams for both the above types, show salient levels on them, and explain the mechanism of conduction in them.
14. Describe "effective mass" and "density of states", and discuss their importance in designing of the solid state devices.
15. Describe the effects of temperature on (a) carrier concentration, and (b) mobility of carriers. Also discuss the effects of doping on mobility.
16. Explain direct and indirect semiconductors. Discuss their characteristic features. Enlist their examples also.
17. What do you mean by degenerate semiconductors? What are its different types? Also discuss the effects of heavy doping?
18. Compare the following:
    (a) Direct and indirect bandgap semiconductor
    (b) Degenerate and nondegenerate semiconductor
19. What is Hall effect? Explain its origin and derive the relation between Hall coefficient and carrier density?
20. What are the basic semiconductors parameters that are measured to characterize the semiconductor material?

Multiple Choice Questions

1. The forbidden energy gap in an insulator is
   (a) 6 eV (b) 3 eV
   (c) 0 eV (d) 1 eV
2. The forbidden energy gap in an insulator is
   (a) 6 eV (b) 3 eV
   (c) 0 eV (d) 1 eV
3. The resistivity of conductors
   (a) Increases with temperature (b) Decreases with temperature
   (c) Is independent of temperature (d) None of these

4. The resistivity of insulators
   - (a) Is low
   - (b) Is high
   - (c) Remains constant
   - (d) None of these
5. Position of Fermi level in intrinsic semiconductors is
   - (a) Above the valence band
   - (b) Below the conduction band
   - (c) Midway of the valence and conduction band
   - (d) None of these
6. Position of Fermi level in p-type semiconductors is
   - (a) Above the valence band
   - (b) Below the conduction band
   - (c) Midway of the valence and conduction band
   - (d) None of these
7. Position of Fermi level in n-type semiconductors is
   - (a) Above the valence band
   - (b) Below the conduction band
   - (c) Midway of the valence and conduction band
   - (d) All of these
8. For n-type semiconductors,
   - (a) $n_e \gg n_h$
   - (b) $n_e = n_h$
   - (c) $n_e \ll n_h$
   - (d) None of these
9. For p-type semiconductors,
   - (a) $n_e \gg n_h$
   - (b) $n_e = n_h$
   - (c) $n_e \ll n_h$
   - (d) None of the above
10. For intrinsic semiconductors,
    - (a) $n_e \gg n_h$
    - (b) $n_e = n_h$
    - (c) Both (a) and (b)
    - (d) None of the above
11. The solids having negative temperature coefficients are
    - (a) Conductors
    - (b) Semiconductors
    - (c) Insulators
    - (d) Dielectrics

BIBLIOGRAPHY

1. Berger, L. I. (2020). *Semiconductor Materials*. CRC Press.
2. Dieter, K. S. (2015). *Semiconductor Material and Device Characterization*. John Wiley & Sons.
3. Man, S. T. (2008). *Introduction to Semiconductor Materials and Devices*. John Wiley & Sons.
4. Ben, G. Y. (2003). *Semiconductor Materials: An Introduction to Basic Principles*. Springer Science & Business Media.
5. Maria, C. T. (2002). *II-VI Semiconductor Materials and Their Applications*, vol. 12. CRC Press.
6. Donald, A. N. (2003). *Semiconductor Physics and Devices: Basic Principles*. McGraw-Hill.

7. Sergio, P. (2015). *Physical Chemistry of Semiconductor Materials and Processes*. John Wiley & Sons.
8. Leah, B. and McHale, L. J., eds. (2011). *Handbook of Luminescent Semiconductor Materials*. CRC Press.
9. Parans, M. P., Wong-Ng, W. and Bhattacharya, N. R., eds. (2016). *Semiconductor Materials for Solar Photovoltaic Cells*, vol. 218. Springer International Publishing.

Keys

1. (b) 2. (d) 3. (a) 4. (b) 5. (c) 6. (a) 7. (b) 8. (a) 9. (c) 10. (b) 11. (b)

CHAPTER 8

Nanomaterials

8.1 Nano

The term "nano" is derived from a Greek word that means "dwarf" (small) and is represented by the symbol "*n*." As a unit prefix, it signifies "one billionth," denoting a factor of 10^{-9} or 0.000000001. It is primarily used with the metric system, as illustrated in Figures 8.1 and 8.2. For example, one nanometer is equal to 1×10^{-9} m, and one nanosecond is equal to

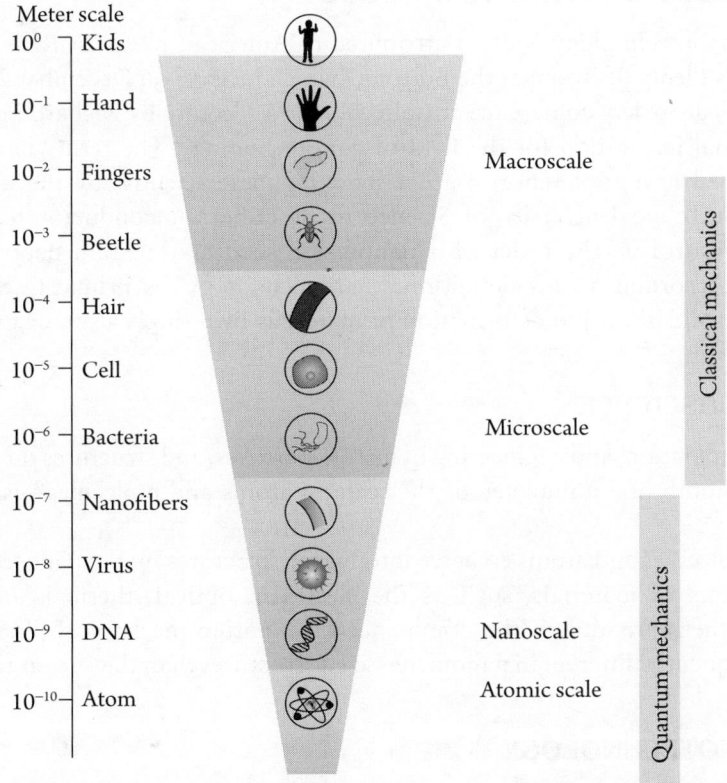

Fig. 8.1 Scale of measurement.

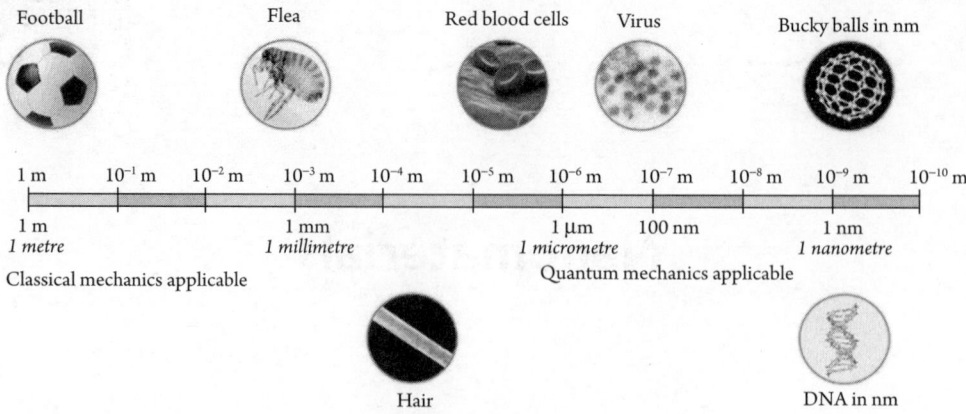

Fig. 8.2 Scale of measurement.

1×10^{-9} sec. It is frequently encountered in science and electronics, particularly for prefixing units of time and length.

8.2 History of Nanotechnology

The origin of nanotechnology is often attributed to American physicist Richard Feynman's speech, "There's Plenty of Room at the Bottom," which he gave on December 29, 1959, at an American Physical Society conference at Caltech. A 1959 lecture by Richard Feynman served as the intellectual inspiration for the field of nanotechnology. The term "nanotechnology" was initially used in a conference in 1974 by a Japanese scientist by the name of Norio Taniguchi from Tokyo University of Science to describe semiconductor techniques with characteristic control on the order of a nanometer, such as thin film deposition and ion beam milling. According to his definition, "nanotechnology" is primarily the processing, separation, consolidation, and deformation of materials by a single atom or molecule.

8.3 Nanoscience

The study, manipulation, and engineering of matter, particles, and structures on the nanometer scale—one millionth of a millimeter, or the scale of atoms and molecules—is referred to as nanoscience.

The way molecules and atoms coalesce into bigger structures on the nanoscale determines important aspects of materials, such as the electrical, optical, thermal, and mechanical properties. Furthermore, due to the dominance of quantum mechanical phenomena, these features are frequently different in nanometer-sized structures than they are at the macroscale.

8.4 Nanotechnology

The technology of design, synthesis, characterization, and applications of materials on nanoscale is called nanotechnology.

The application of nanoscience, or nanotechnology, results in the usage of novel nanomaterials and nanoscale components in practical products. In the end, nanotechnology will enable us to develop novel materials and products with enhanced properties, as well as new nanoelectronic components, "smart" medications and sensors, and even electronic-biological interfaces.

These newborn scientific disciplines are situated at the interface between physics, chemistry, materials science, microelectronics, biochemistry, and biotechnology. Control of these disciplines therefore requires an academic and multidisciplinary scientific education.

8.5 NANOMATERIALS

The materials are tiny (in at least one dimension) and range in size from 1 to 100 nm, termed as nanomaterials.

Nanomaterial is defined as the "material with any external dimension in the nanoscale or having internal structure or surface structure in the nanoscale," with nanoscale defined as the "length range approximately from 1 nm to 100 nm." This includes both nano-objects, which are discrete pieces of material, and nanostructured materials, which have internal or surface structure on the nanoscale; a nanomaterial may be a member of both these categories.

8.6 TYPES OF NANOMATERIALS

Materials having at least one dimension in the nanoscale are termed quantum wells. Examples of quantum wells include layers, thin films, and surface coatings. Materials having two dimensions in the nanoscale are often called quantum wires. Examples of quantum wires include nanowires and nanotubes. Materials having all three dimensions in the nanoscale are termed quantum dots. Examples of quantum dots include nanoparticles and nanodots.

8.7 PROPERTIES OF NANOMATERIALS

The general properties of nanomaterials are as follows:

1. Nanomaterials are hard.
2. Nanomaterials are exceptionally strong.
3. Nanomaterials are ductile at high temperature.
4. Nanomaterials are chemically very active.
5. Nanomaterials are wear resistant.

8.8 REASON BEHIND PROPERTY CHANGE AT NANOSCALE

There are two main reasons behind the property change at the nanoscale:

1. **Increase in Surface Area-to-Volume Ratio**
   The properties of materials are drastically changed at nanoscale due to increased surface area to volume ratio.

Let us consider a large sphere of radius R, volume V, and surface area S. Then, the surface area to volume ratio is given by

$$\frac{S}{V} = \frac{4\pi R^2}{\frac{4}{3}\pi R^3} \Rightarrow \frac{S}{V} = \frac{3}{R}.$$

This means that when radius decreases from microscale to nanoscale, the surface area to volume ratio increases drastically.

Hence, nanomaterials are chemically very reactive.

2. **Quantum Confinement Effect**

    The quantum effects are begin to dominate at lower nanoscale that causes the change in optical, electrical, magnetic, and mechanical properties.

    Some examples at nanoscale are as follows:

    (a) Opaque materials can become transparent, e.g., cooper.
    (b) Inert material can become catalyst, e.g., platinum.
    (c) Stable material can turn combustible, e.g., aluminum.
    (d) Solids can turn into liquids at room temperature, e.g., gold.
    (e) Insulator becomes conductor, e.g., silicon.

8.9 Material Fabrication at Nanoscale (Nanomaterials)

Generally, the two approaches used to fabricate materials at nanoscale are as follows:

Top-down Approach

Breaking down the bulk material into nanoscale structures or particles is a top-down strategy.

The methods employed for creating the micron-sized particles in Figure 8.3 are extended by top-down synthesis methods. Top-down strategies are by nature simpler and rely on either the removal or division of bulk material or the miniaturization of bulk production methods to create the required structure with the right attributes. The top-down strategy's main flaw is the irregularity of the surface structure.

For instance, lithographically produced nanowires have a rough surface and may have several impurities and structural flaws. High-energy wet ball milling, electron beam lithography, atomic force manipulation, gas-phase condensation, aerosol spray, and so on are a few examples of such methods.

Bottom-up Approach

The alternate strategy is called "bottom-up," which has the potential to produce less waste and is thus more cost-effective.

In a bottom-up approach, nanomaterial is built up from lower-dimensional particle sizes, i.e., by adding atom-by-atom, then molecule by molecule, and then cluster-by-cluster.

Many of these methods are either still being developed or are only now starting to be applied in the manufacturing of nanopowders for commercial purposes. Some of the well-known bottom-up methods used to create the luminescent nanoparticles depicted in Figure 8.3 include the revere-micelle route, the organometallics chemical route, sol-gel synthesis, colloidal precipitation, hydrothermal synthesis, template assisted sol-gel, electro-deposition, and so on. There is minimal difference between the characteristics of solid matter

Nanomaterials

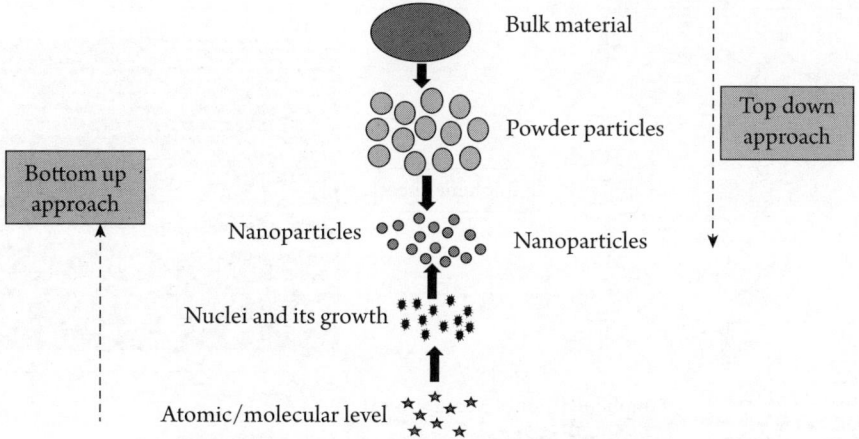

Fig. 8.3 Classification of nanomaterial's synthesis.

particles with visible scale particle sizes and those that can be seen with a standard optical microscope. However, the characteristics of the materials differ dramatically from those at larger scales when particles with diameters of around 1–100 nanometers are produced (where the particles can only be "seen" with strong specialized microscopes). At this scale, alleged quantum phenomena control the behavior and characteristics of particles. In this scale range, material properties are scale-dependent.

As a result, attributes including melting point, fluorescence, electrical conductivity, magnetic permeability, and chemical reactivity vary as a function of particle size when particle size is changed to be nanoscale. The special characteristics that exist at the nanoscale are demonstrated by nanoscale gold. The color of gold at the nanoscale is different from the yellow we are used to seeing; it might seem red or purple. The movement of the electrons in gold is constrained at the nanoscale. Gold nanoparticles behave differently from gold particles of a greater size in terms of how they respond to light because this mobility is constrained. The idea of "tenability" of qualities is an intriguing and potent outcome of quantum processes at the nanoscale.

In other words, by adjusting the size of the particle, a scientist can actually fine-tune a material property of interest (for example, by altering fluorescence color; a particle's fluorescence color can be used to identify the particle, and different materials can be "labelled" with fluorescent markers for various purposes). Tunneling is a powerful quantum phenomenon that occurs at the nanoscale and gives rise to devices like the scanning tunneling microscope and flash memory for computers.

8.10 Categories of Nanomaterials

Nanomaterials are generally fall into two categories:

1. **Fullerenes**

   A fullerene is an allotrope of carbon whose molecule consists of carbon atoms connected by single and double bonds, conceptually a graphene sheet (isolated one-atom thick layer of graphite) rolled into tubes or spheres, as shown in Figures 8.4 and 8.5.

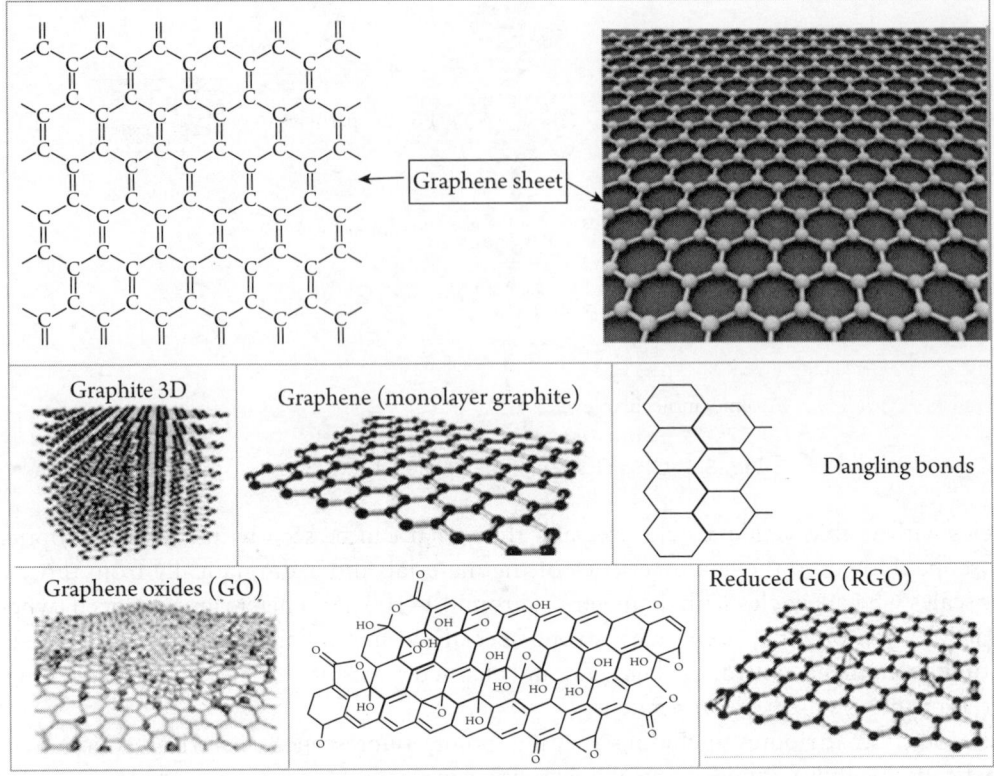

Fig. 8.4 Graphene sheets of different dimensions.

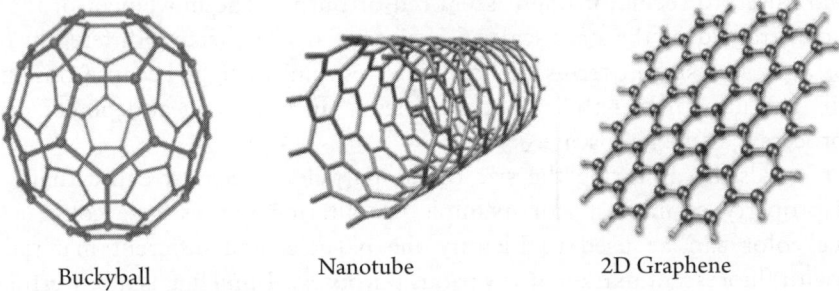

Fig. 8.5 Graphene sheet and its different rolling shapes.

Fullerenes with a closed mesh topology are informally denoted by their empirical formula C_n, where n is the number of carbon atoms.

The graphene sheet that is one-atom thick layer of graphite is shown in Figure 8.4. The figure illustrates the three-dimensional graphene sheet, two-dimensional graphene sheet, graphene oxide, and reduced graphene oxide, which are currently used in nanomaterials research.

2. **Nanoparticles**

A nanoparticle or ultrafine particle is usually defined as a particle of matter that is between 1 and 100 nanometers (nm) in diameter.

As a result of their smaller size, nanoparticles are usually distinguished from microparticles (1–1000 m), "fine particles" (sized between 100 and 2500 nm), and "coarse particles" (ranging from 2500 to 10000 nm). These properties, such as colloidal properties and optical or electric properties, are driven by very different physical or chemical properties.

8.11 BUCKYBALLS

Buckyballs, also known as Buckminsterfullerene, is a type of fullerene with the formula C_{60}, as shown in Figure 8.6. It features a fused-ring cage-like structure consisting of twenty hexagons and twelve pentagons that resembles a football. Three bonds connect each carbon atom. It is a black solid that turns into a violet solution when it is dissolved in hydrocarbon solvents. Despite much research, there aren't many practical uses for the chemical.

8.12 CREATION OF BUCKYBALLS

8.12.1 Arc Discharge Method

Buckyballs can be created by vaporizing carbon placed between two carbon electrodes. As illustrated in Figure 8.7, buckyballs and carbon soot are produced when an arc forms between two carbon electrodes positioned very close to one another in a reaction chamber with low pressures of helium, neon, or argon. Benzene and other solvents are used to separate them.

8.12.2 Properties of Buckyballs

1. Buckyballs can be used in many applications because of its unusual hollow structures.
2. Buckyballs are extremely stable and can bear very high temperature and pressure.
3. Buckyballs produce stable spherical structure after reaction with other atoms.
4. Buckyballs can trap other atoms due to its hollow cage-like structure.

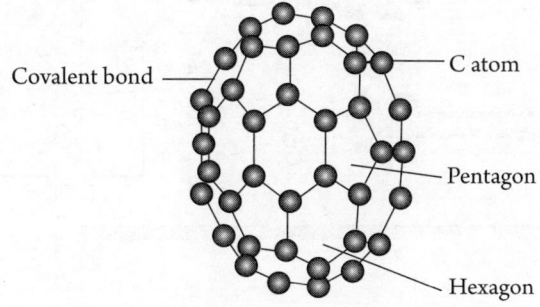

Fig. 8.6 Buckyballs.

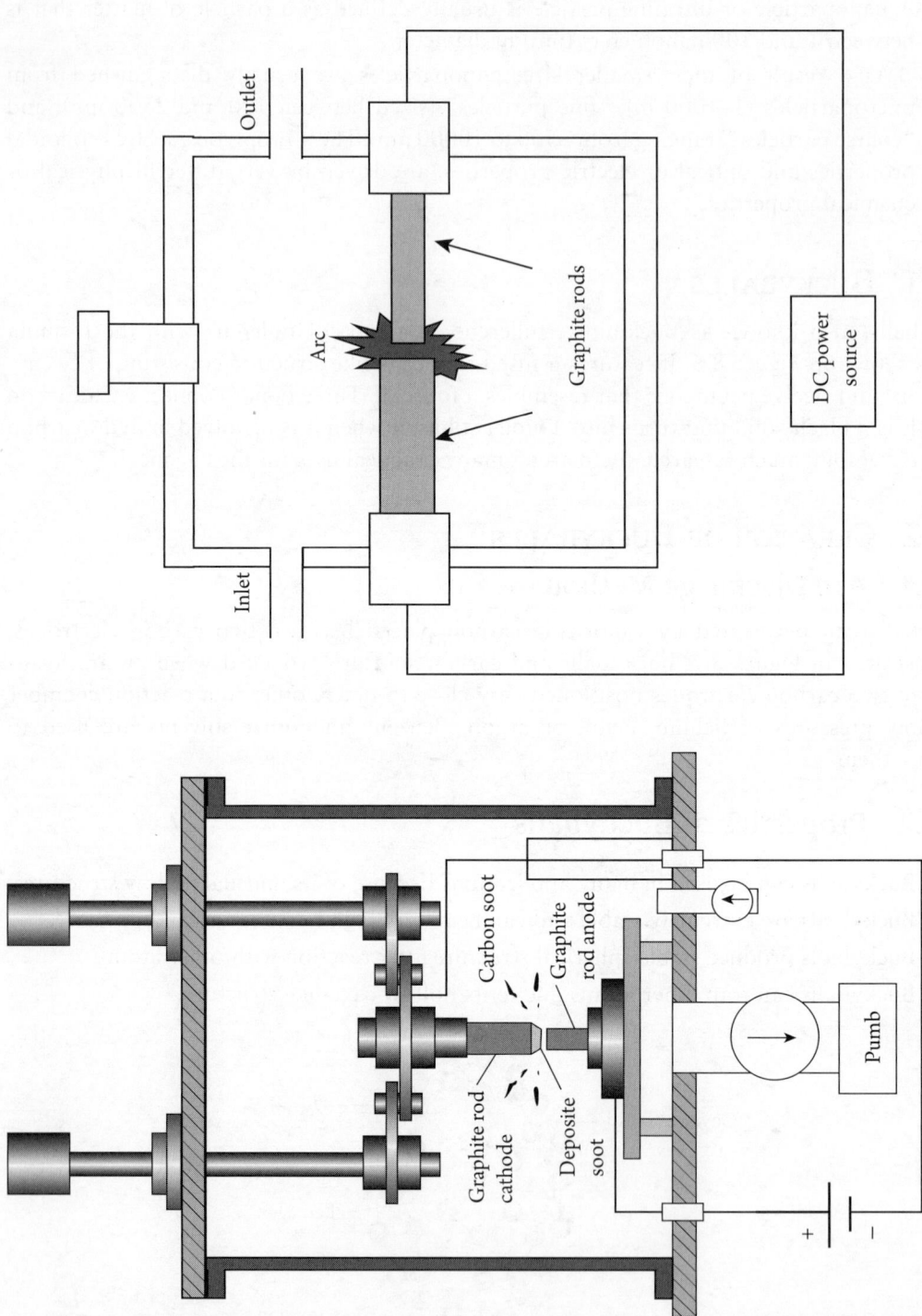

Fig. 8.7 Arc discharge experimental setup.

8.13 Applications of Buckyballs

Some potential applications of buckyballs include:

1. Superconductors (buckyballs doped with potassium or cesium).
2. Lubricants.
3. Catalysts due to their high reactivity.
4. Drug delivery systems, pharmaceuticals, and targeted cancer therapies.
5. Hydrogen storage as almost every carbon atom in C_{60} can absorb a hydrogen atom without disrupting the buckyballs structure, making it more effective than metal hydrides. This could lead to applications in fuel cells.
6. Optical devices.
7. Chemical sensors.
8. Photovoltaic.
9. Polymer electronics such as organic field effect transistors (OFETs).
10. Antioxidants.
11. Polymer additives.
12. Cosmetics, where they "mop up" free radicals.
13. Diamonds and fullerenes have been used as precursors to produce diamond films.

8.14 Carbon Nanotubes

Carbon nanotubes, long, thin cylinders of carbon, were discovered in 1991 by Sumio Iijima. These are large macromolecules that are unique for their size, shape, and remarkable physical properties. They can be thought of as a sheet of graphite (a hexagonal lattice of carbon) rolled into a cylinder, as shown in Figure 8.8. They are less than 100 nanometers in diameter and can be as thin as 1 or 2 nanometers, while they can be up to several millimeters in length.

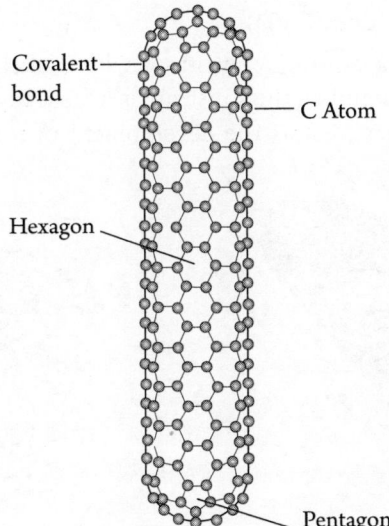

Fig. 8.8 Carbon nanotube.

8.15 Types of Carbon Nanotubes

Nanotubes are generally of two types:

1. **Single-walled Carbon Nanotubes (SWNT/SWCNT)**

   Single-walled carbon nanotubes are defined as one-dimensional cylindrically shaped allotropes of carbon that have a high surface area and aspect ratio (length to diameter ratio). They are made of one-atom-thick nano carbon sheet that forms a tube shape, as shown in Figure 8.9.

2. **Multi-walled Carbon Nanotubes (MWNT/MWCNT)**

   Figure 8.10 illustrates how numerous single-walled carbon nanotubes are nestled inside of one another to produce multi-walled carbon nanotubes (MWCNTs), a unique kind of carbon nanotube. MWCNTs are, however, still categorized as a one-dimensional type of carbon.

8.16 Production (Synthesis) of Carbon Nanotubes

Commercially, nanotubes are being produced with a number of methods. Two of these methods are discussed here.

1. **High-pressure Carbon Mono-oxide Deposition (HiPCO)**

   As seen in Figure 8.11, this technique uses a heated chamber through which carbon monoxide gas and tiny clusters of iron atoms flow. Iron clusters serve as catalysts for carbon monoxide molecules that come into contact with them. This promotes the dissociation of a carbon monoxide molecule into an oxygen and a carbon atom. A lattice of nanotubes is created when this carbon atom forms a connection with other carbon atoms. In order to create carbon dioxide, the oxygen atom joins forces with another carbon monooxide molecule. This carbon dioxide gas then rises to the surface and floats away.

2. **Chemical-Vapor Deposition (CVD)**

   This technique involves coating a substrate with nickel, cobalt, or iron-based metal catalyst particles. In a chamber, the substrate is heated to around 700°C. Two gases are blasted into the chamber to start the development of the nanotube. One is a process

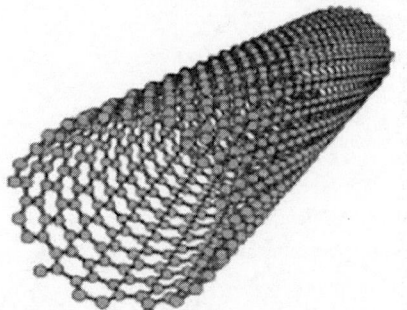

Fig. 8.9 Single-walled carbon nanotube.

Fig. 8.10 Multi-walled carbon nanotube.

Nanomaterials

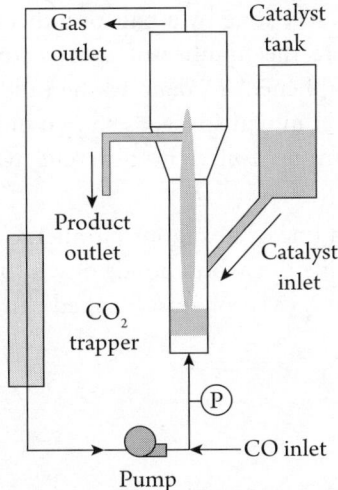

Fig. 8.11 High-pressure carbon mono-oxide deposition setup.

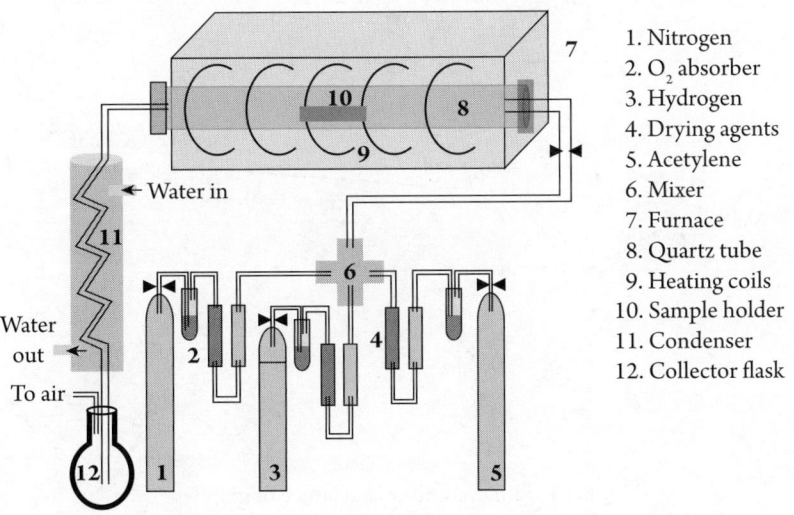

1. Nitrogen
2. O_2 absorber
3. Hydrogen
4. Drying agents
5. Acetylene
6. Mixer
7. Furnace
8. Quartz tube
9. Heating coils
10. Sample holder
11. Condenser
12. Collector flask

Fig. 8.12 Chemical vapor deposition setup.

gas like hydrogen, nitrogen, or ammonia. Methane and other carbon-containing gases are additional. One carbon atom and four hydrogen atoms make up methane. The methane molecules' bonds between the carbon and hydrogen atoms are severed by the chamber's high temperature. These carbon atoms join with additional carbon atoms after adhering to catalyst particles. As a result, nanotubes are created. Figure 8.12 depicts the CVD configuration.

8.17 Structure of Carbon Nanotubes

When viewed with a transmission electron microscope, carbon nanotubes appear as planes. A SWNT appears as two planes, whereas in a MWNT, more than two planes are observed.

The naming scheme of SWNT is done by a pair of indices (*n, m*) called the chiral vector.

The integers *n* and *m* denote the number of unit vectors along two directions in the honeycomb crystal lattice of graphene. "*T*" denotes the tube axis, and **a**$_1$ and **a**$_2$ are the two unit vectors of the graphene sheet in real space, as shown in Figure 8.13.

As the graphene sheets can be rolled in different ways, this leads to three types (Figure 8.14) of nanotubes:

a. Armchair (*n, n*): There is a line of hexagons parallel to the axis of nanotube.
b. Zig-zag (*n*, 0): There is a line of carbon bonds down to the center. Here, *m* = 0.
c. Chiral (*n, m*): They exhibit a twist or spiral (called chirality) around the nanotube.

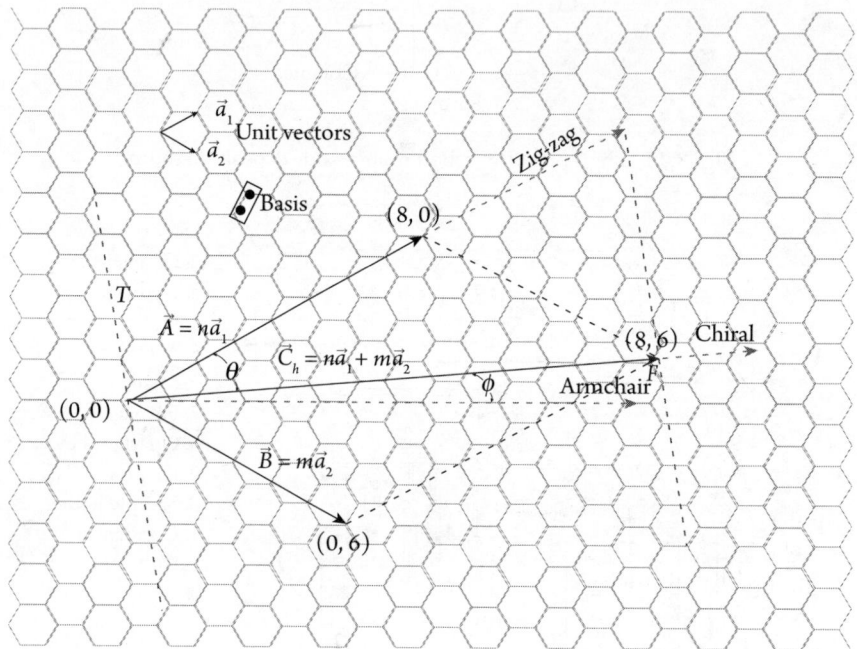

Fig. 8.13 Honeycomb crystal lattice of graphene.

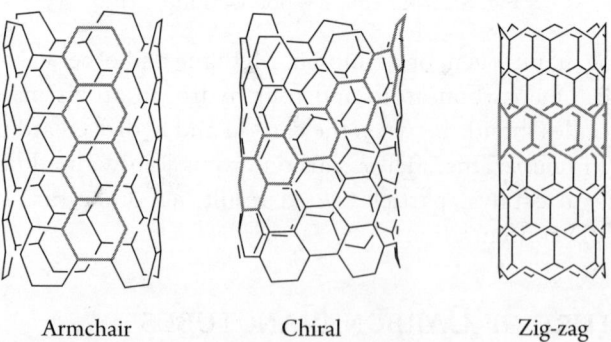

Fig. 8.14 Different structures of carbon nanotubes.

Nanomaterials

8.18 Properties of Nanotubes

1. Nanotubes can be either metallic or semiconductor, depending on their size, energy gaps, diameter, and helicity of the tubes.
2. A SWNT is considered metallic if the value of (n–m) is divisible by three. Otherwise, the nanotube is semiconducting.
3. Ultra-small SWNTs (diameter 4 Å) exhibit superconductivity below 20 K.
4. Nanotubes are very strong, with a high Young's modulus, and are extremely flexible.
5. High thermal conductivity.
6. High sensitivity to gas absorption.

8.19 Applications of Nanotubes

Applications of nanotechnology are shown in Figure 8.15. The figure illustrates that nanotechnology is used in almost all fields, including efficient energy storage, defense and security, metallurgy and materials, electronic devices, optical engineering and communication, biomedical and drug delivery applications, agriculture and food processes, cosmetics and paints, biotechnology and biomaterials, textiles, computer science, civil engineering, and electrical sciences. Consequently, nanotechnology is applied across all engineering disciplines.

Previous Year Questions (University Examination)

1. What is nanotechnology? What are the changes in the properties that take place in a material when its size is reduced to nanoscale?
2. What are nanoscience and nanotechnology? Also, explain the reasons for property change at nanoscale.
3. What are carbon nanotubes? Explain their synthesis, properties, and uses.
4. What are buckyballs? Explain their synthesis, properties, and uses.
5. What do you mean by quantum wells, quantum wires, and quantum dots?

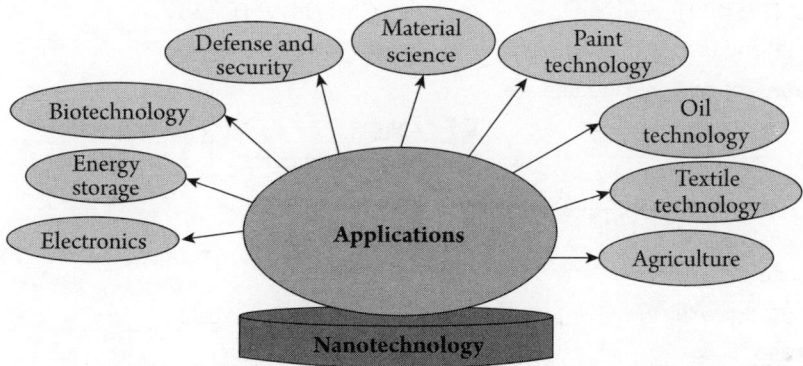

Fig. 8.15 Applications of nanotechnology in different fields.

Multiple Choice Questions

1. Nanomaterials are in the range of
   - (a) 100–200 nm
   - (b) 1–100 nm
   - (c) 1000 nm
   - (d) 1 m
2. The properties of nanomaterials can be explained by
   - (a) Classical mechanics
   - (b) Quantum mechanics
   - (c) Statistical mechanics
   - (d) None of these
3. Fullerene is
   - (a) Molecule
   - (b) Chemical compound
   - (c) Metal particles
   - (d) None of these
4. The property of bulk material changes at nanoscale due to
   - (a) High surface-to-volume ratio
   - (b) Quantum confinement effect
   - (c) Both (a) and (b)
   - (d) None of these
5. Carbon nanotubes are
   - (a) Graphene sheet rolled into sphere
   - (b) Graphene sheet rolled into tube
   - (c) Graphene sheet rolled into cone
   - (d) None of these
6. Which of the following is carbon nanotube
   - (a) Armchair
   - (b) Arc discharge
   - (c) Chemical discharge
   - (d) None of these
7. The synthesis method of buckyballs is
   - (a) High-pressure carbon mono-oxide deposition
   - (b) Chemical vapor deposition
   - (c) Arc discharge method
   - (d) All of these
8. Buckyballs are a cluster of
   - (a) 30 carbon atoms
   - (b) 60 carbon atoms
   - (c) 90 carbon atoms
   - (d) None of these
9. Types of carbon nanotubes are
   - (a) SWCNT/SWNT
   - (b) MWCNT/MWNT
   - (c) Both (a) and (b)
   - (d) None of the above
10. The structures of carbon nanotube are
    - (a) Armchair
    - (b) Zig-zag
    - (c) Chiral
    - (d) All of the above
11. The known third form of the carbon is
    - (a) Fullerene
    - (b) Diamond
    - (c) Graphite
    - (d) None of these
12. Which of the following forms of pure carbon is similar to a soccer ball?
    - (a) Fullerene
    - (b) Diamond
    - (c) Graphite
    - (d) None of these

13. The drastic property change at nanoscale is due to
    (a) Seeback effect
    (b) Electromagnetic effect
    (c) Quantum mechanical effect
    (d) None of these

BIBLIOGRAPHY

1. Vance, M. E., Kuiken, T., Vejerano, E. P., McGinnis, S. P., Hochella Jr, M. F., Rejeski, D., and Hull, M. S. (2015). "Nanotechnology in the Real World: Redeveloping the Nanomaterial Consumer Products Inventory." *Beilstein Journal of Nanotechnology*, 6.1: 1769–1780.
2. Saleh, T. A. and Gupta, V. K. (2016). *Nanomaterial and Polymer Membranes: Synthesis, Characterization, and Applications*. Elsevier.
3. Schäfer, A. I. and Fane, A. G., eds. (2021). *Nanofiltration: Principles, Applications, and New Materials*. John Wiley & Sons.

Keys

1. (b)   2. (b)   3. (a)   4. (c)   5. (b)   6. (a)   7. (c)   8. (b)   9. (b)   10. (d)
11. (a)   12. (a)   13. (c)

CHAPTER 9

Superconducting Materials

9.1 INTRODUCTION

All metals and alloys exhibit a reduction in electrical resistance as they cool. As the temperature drops, atoms' thermal vibrations become less intense, and conduction electrons scatter less frequently. The resistivity should decrease toward zero as the temperature approaches zero Kelvin for a perfect pure metal, where the only thing standing in the way of an electron's travel is the thermal vibrations of the lattice. This zero resistance, which a hypothetical perfect specimen would acquire if it could be cooled to absolute zero, is the phenomenon of superconductivity. Any real specimen of metal cannot be perfectly pure and will contain some impurities. As a result, in addition to being scattered by the thermal vibrations of the lattice atoms, the electrons are also dispersed by impurities, and this impurity scattering is largely temperature independent. As a result, at the lowest temperature, there will be some residual resistance. The residual resistivity of a metal increases with the degree of impurity.

The phenomenon of superconductivity was first discovered by Dutch physicist H. Kamerling Onnes of Leiden University in 1911 during the investigation of the variation of electrical resistance of mercury in the newly available range of low temperatures, in the neighborhood of temperature of liquid helium (or 4.2 K). He observed that the resistance of mercury suddenly falls from 0.08 ohm at about 4 K to less than 3×10^{-6} ohm over a very small temperature of 0.01 K.

9.2 SUPERCONDUCTIVITY

When a substance is cooled below a certain temperature (called the transition temperature or critical temperature), the electrical resistance of the substance suddenly drops to zero. This phenomenon is known as superconductivity. Substances that exhibit the property of superconductivity are known as superconductors.

Or

The disappearance of a material's electrical resistance below a specific temperature is referred to as superconductivity, and the substance is referred to as a superconductor.

Or

When the specimen is cooled to a low enough temperature, up to a temperature in the liquid helium region, many metals and alloys have electrical resistivities that abruptly decrease to zero. Superconductivity is the name given to this phenomenon.

9.3 Superconductivity and Transition Temperature

It is a well-known fact that the resistivity of pure metals decreases with decrease in the temperature. The electrical resistivity of many metals and alloys drops suddenly to zero when they are cooled to a low temperature nearly equal to liquid helium temperature. This phenomenon was observed for the first time by Dutch physicist H. Kamerling Onnes of Leiden University in 1911. The material exhibiting this remarkable characteristic is known as superconductor. The temperature at which the normal material changes into superconducting state is called the critical temperature T_c. Other substances that exhibit this phenomenon are aluminum, silver, cadmium, LED, gallium, iridium, and so on. There are so many other alloys and ceramics that behave as superconductors. It may be possible that individual elements may not be superconductor, but their alloy can be superconductor. Figure 9.1 shows the comparison of resistance of normal metal and superconductor. From Figure 9.1, it is clear that the critical temperature separates the superconducting state from the normal states.

Above critical temperature T_c, the specimen is in the normal state with finite resistivity but below T_c, it changes into superconducting state with infinite conductivity. This critical temperature and its range are different for different materials and are affected very much by impurity in the material. A small amount of impurity can change from steep fall of resistivity to gradual fall. Further, this transition of material is reversible, and the material transforms to normal state when temperature rises above the critical temperature.

9.4 Temperature Dependence of Resistivity in Superconducting Materials

Metals are good conductors of electricity as they have plenty of free electrons. However, they offer resistance to the flow of charges that is current. The fluctuation of a metal's resistance with temperature is seen in Figure 9.1. The metals provide what is known as residual resistance even at 0 K.

The resistivity of a superconductor is a function of temperature.

The resistance of the superconductor in nonsuperconducting state decreases with the decrease in temperature as in the case of normal metal. However, at a particular temperature T_c, the resistivity abruptly drops to zero. This temperature T_c is called the critical temperature. Therefore, the temperature at which normal material turns into a superconductor is called the critical temperature (T_c). The critical temperature is different for different conductors.

9.5 Effect of Critical Magnetic Field

In 1913, Kamerling Onnes observed that superconductivity is destroyed if a sufficient strong magnetic field is applied. In other words, the superconducting material restores its normal resistance when a strong magnetic field is applied. The minimum magnetic field that is

Superconducting Materials

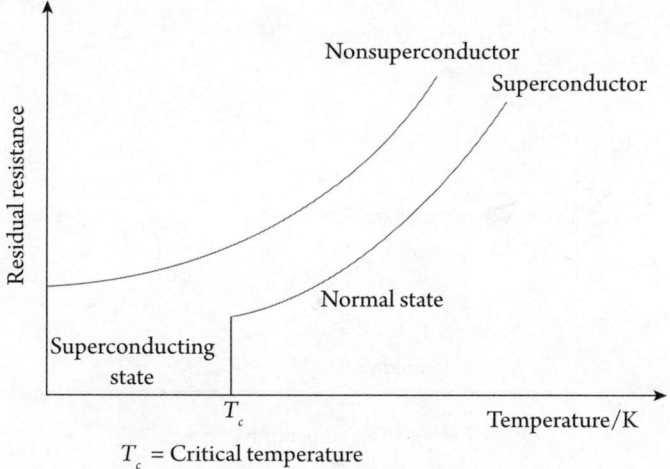

Fig. 9.1 The dependence of resistivity of a superconductor and nonsuperconductor.

necessary to regain the normal resistivity is called critical magnetic field H_c. If the applied magnetic field exceeds the critical value H_c, the superconducting state is destroyed. The variation of critical magnetic field with temperature is shown in Figure 9.2.

As seen from Figure 9.2(a), the curve is nearly parabolic and can be represented by the following relation:

$$H_c(T) = H_c(0)\left[1 - \left(\frac{T}{T_c}\right)^2\right], \tag{9.1}$$

In equation (9.1), $H_c(T)$ is the maximum critical field strength at temperature T, $H_c(0)$ is the maximum critical field strength occurring at absolute zero (characteristic of a material) and T_c is the critical temperature. In general, the higher value of T_c, the higher is the value of H_c.

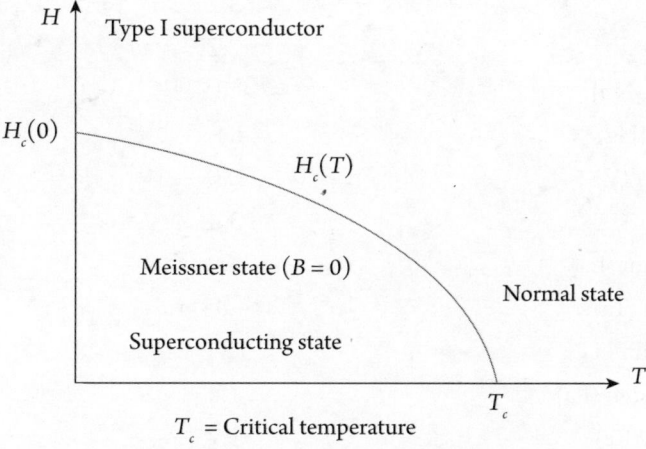

Fig. 9.2 (a) Critical magnetic fields in type I superconductor.

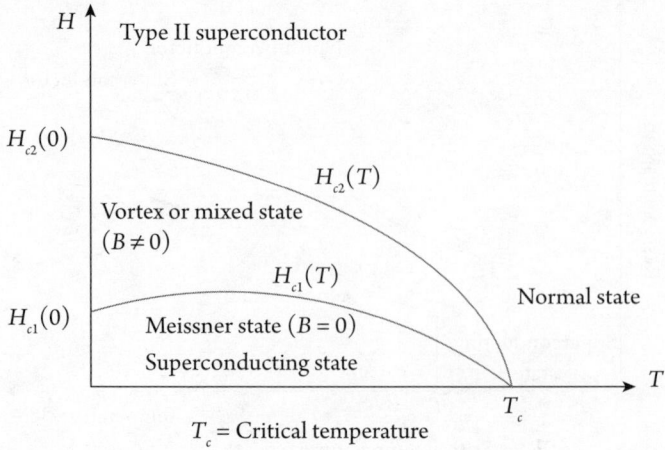

Fig. 9.2 (b) Critical magnetic fields in type II superconductor.

Some superconducting materials with transition temperature and critical magnetic field value are given in Table 9.1.

9.6 CRITICAL CURRENT AND CURRENT DENSITY

A superconducting material loses its superconducting property when a high electric current is applied to it. Critical current and critical current densities are the terms used to describe the current necessary for this.

Table 9.1 Some superconducting materials with transition temperature and critical magnetic field value

S. No.	Materials	T_c (K)	H_c (Gauss)
1.	Aluminum (Al)	1.1	99
2.	Tin (Sn)	3.7	305
3.	Lead (Pb)	7.2	803
4.	Niobium (Nb)	9.25	1980
5.	Mercury (Hg)	4.15	411
6.	Vanadium (V)	5.38	1020
7.	Indium (In)	3.4	293
8.	Lanthanum (La)	4.9	798
9.	Tantalum (Ta)	4.48	830
10.	Technetium (Tc)	7.77	1410
11.	Protactinium (Pa)	1.4	–
12.	Rhenium (Re)	1.7	198
13.	Thallium (Tl)	2.39	171

Superconducting Materials

If a superconducting wire of radius r carries current I then critical current I_c is given by $I_c = 2\pi r H_c$.

At $I = I_c$, superconductivity will be destroyed. In presence of applied transverse magnetic field H, the value of critical current decreases and expressed as

$$I_c = 2\pi r (H_c - 2H).$$

This is called Silsbee's rule. The critical current I_c will decrease linearly with increase of applied field until it reaches zero at $= H_c/2$. If the applied field is zero, then $I_c = 2\pi r H_c$.

The current density is given by

$$J_c = \frac{I_c}{A} = \frac{I_c}{\pi r^2} A/m^2.$$

9.7 MEISSNER EFFECT (FLUX EXCLUSION)

In 1913, Meissner observed that if a superconductor is cooled below the critical temperature in a magnetic field, the lines of induction are expelled from the material. This effect is called the Meissner effect. Figure 9.3(a) shows a superconductor in normal state, and the magnetic lines of force pass through it. But when the specimen is cooled below its transition temperature, as shown in Figure 9.3(b), the magnetic lines of force are expelled out the specimen.

Therefore, the Meissner effect is the phenomenon whereby magnetic lines of force are removed from a superconducting material when it is cooled below the transition temperature in a magnetic field.

Important Points

1. The Meissner effect can be reversed. The flux suddenly penetrates the specimen, and the material returns to its normal condition when the temperature is raised below T_c.

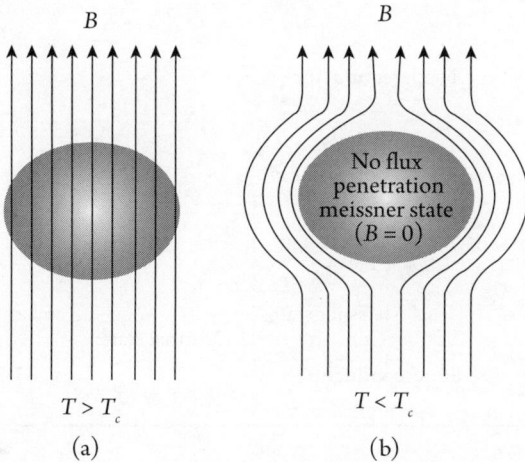

Fig. 9.3 Meissner effect: (a) flux penetration in a normal conductor; (b) flux expulsion from a superconductor.

2. A perfect diamagnetic is a superconductor. The explanation is that a superconductor has zero magnetic induction $\vec{B}$. Hence, in the case of superconductors,

$$B = \mu_0 (H + M) = 0,$$

or, $M = -H,$ \hfill (9.2)

where H is the magnetizing field intensity and M is magnetization.

The magnetic susceptibility χ_m is given by

$$\chi_m = \frac{M}{H}. \quad (9.3)$$

From equations (9.2) and (9.3), we get

$$\chi_m = \frac{-H}{H} = -1. \quad (9.4)$$

This is the maximum value of susceptibility of a diamagnetic material. In the sense, a superconductor is a perfect diamagnetic.

9.8 Type I and Type II Superconductors

The following two categories for superconductors are identified by their magnetic behavior.

1. Soft superconductors, also known as type I superconductors.
2. Hard superconductors, often known as type II superconductors.

9.8.1 Type I Superconductors

The dependence of magnetization of a superconductor of type I as a function of external field h is shown in Figure 9.4. It is obvious from the figure that up to the critical field strength H_c, the magnetization of a superconductor grows in proportional to the external field. As soon as the applied field H exceeds H_c the magnetization abruptly drops to zero.

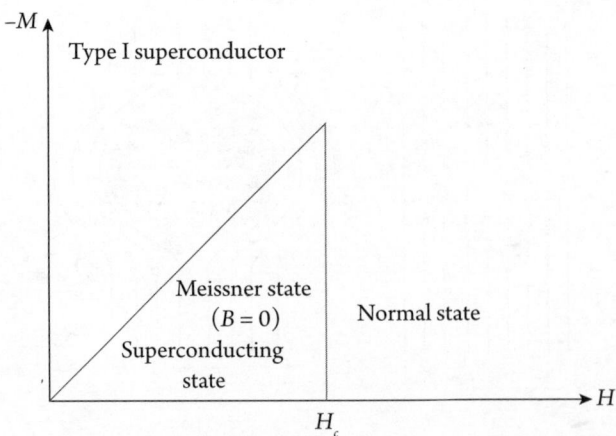

Fig. 9.4 Magnetization versus magnetic field for type I superconductors.

Superconducting Materials

So, *type I superconductor is one in which the transition from superconducting state to normal state in presence of magnetic field occurs sharply at the critical value H_c.*

When an external magnetic field is present, H is smaller than H_c. Perfectly diamagnetic type I superconductors exist in the superconducting state. The superconductor reaches the normal state, losing all of its diamagnetic properties, when H surpasses H_c. In this condition, the magnetic flux permeates the whole superconductor. For type I superconductors, the critical field value H_c is discovered to be extremely low. Superconductors of type I include aluminum, lead, and indium.

9.8.2 Type II Superconductors

The magnetization curve of type II superconductors is shown in Figure 9.5. A type II superconductor is characterized by two critical magnetic fields H_{c1} and H_{c2}.

The description of the curve is as follows:

1. The superconductor totally expels the magnetic field lines from its body at the field intensity below H_{c1} and functions as a perfect diamagnetic. The lower critical field is referred to as H_{c1}. AB is a representation of the curve.
2. Magnetic field lines begin to penetrate the material as the magnetic field strength grows from H_{c1}. Up to H_{c2}, the penetration increases. The higher critical field is referred to as the H_{c2}. At H_{c2}, the magnetization totally disappears, indicating that the external field has completely penetrated the superconductor and rendered it incapable of conducting electricity.
3. The material returns to its original form after H_{c2}.

Therefore, *type II superconductor is one that is characterized by two critical fields H_{c1} and H_{c2}, and the transition to the normal state takes place gradually as the magnetic field is increased from H_{c1} to H_{c2}.*

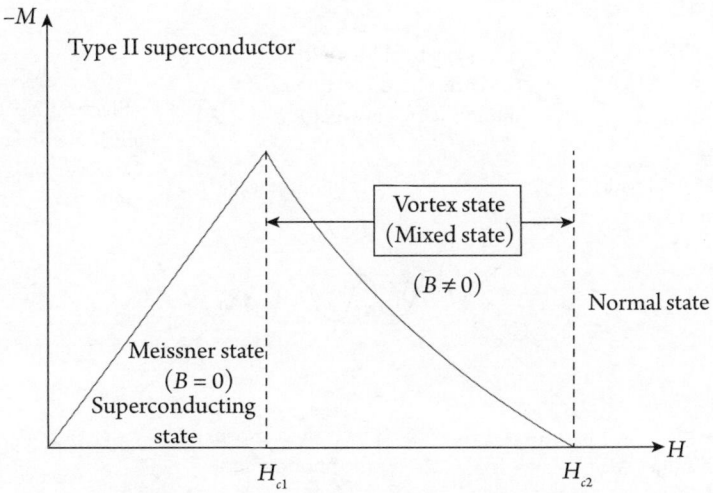

Fig. 9.5 Magnetization versus magnetic field for type II superconductors.

Now let us consider the state of superconductor between H_{c1} and H_{c2}. The state is called as mixed state or vertex state. In this state, though there is a flux penetration, yet the material retains its zero resistance property and it is still a superconductor.

The most important advantages of type II superconductors are that they have the value of critical field at H_{c2}, which is many more times (even 100 times) higher than the value of H_{c1} for Type I superconductor. So, the most important use of type II superconductors is to build up a device which can work in high magnetic fields, such as superconducting magnets.

9.9 BCS Theory (Explanation of Superconductivity)

In 1957, Bardeen, Cooper, and Schrieffer gave a theory to explain the phenomenon of superconductivity that is known as BCS theory. This theory based on the formation of cooper pairs.

In metals, the electrical resistance arises due to the collision of conduction electrons (free electrons) with the vibrating ions of the lattice. In normal state, the force between the electrons is repulsive. In superconducting state, the force between two electrons becomes attractive due to formation of cooper pair. The explanation is as follows:

When a current flows through a superconductor an electron (negative charge) comes near the positive ion core of the lattice, then the electron experiences an attractive force. Due to interaction between electron and ion core, the iron core is slightly displaced. This is known as lattice distortion. The distortion in the lattice then travels away as mechanical wave (phonons).

Now, suppose that another electron comes near the distorted lattice. The lattice vibration interacts with the second electron and hence there is a force of attraction between second electron and phonon. In this way, two electrons interact with each other through the lattice vibration. This process is called electron lattice electron interaction via phonon field (mechanical wave). This is shown in Figure 9.6.

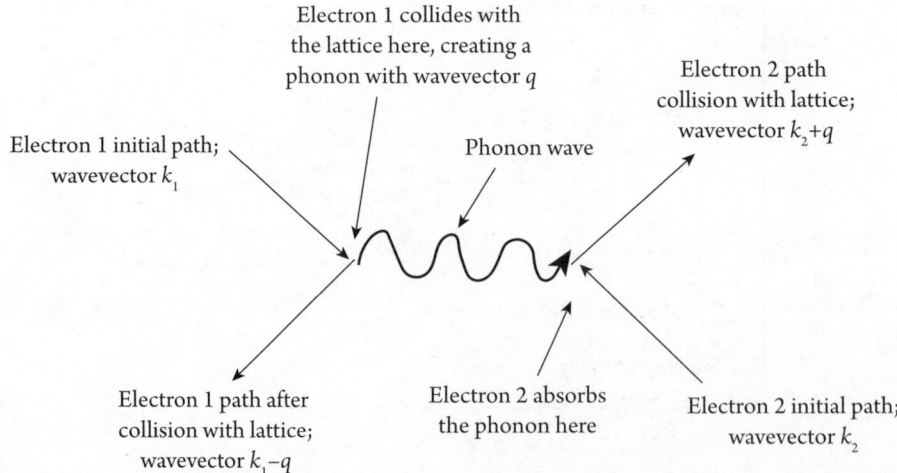

Fig. 9.6 A schematic diagram of Cooper pair formation.

Superconducting Materials

When an electron with vector k_1 distorts the lattice, then lattice gains momentum. As a result, the momentum of electron decreases. So, a phonon wave vector q is emitted. When another electron with the wave vector k_2 absorbs the energy from the phonon, it gains momentum. Therefore, due to interaction, there are two electrons with wave vector k_1-q and k_2+q. This pair of electrons is called a "Cooper pair." These Cooper pairs in a superconductor are bosons. The energy of cooper pair is lower than that of the individual electrons.

Therefore, Cooper pair is a bound pair of electrons by the interaction between the electrons in a phonon field. The two electrons that pair up have opposite momenta and spin.

The Cooper pair of electrons moves without suffering any deviation either by impurities or thermal vibrations. There is no exchange of energy between the pair of electrons and lattice ions. If an electric field is established inside the substance, the electrons gain additional kinetic energy and give rise to a current. The main important point is that the pair of electrons does not transfer any energy to the lattice. Therefore, these pair of electrons would not get slow down. As a consequence of this, the substance does not possess any electrical resistivity and conductivity is very large.

When the pair of electrons flows in the form of a Cooper pair, they do not encounter any scattering, and the resistance factor vanishes, meaning conductivity becomes infinite. This phenomenon is known as superconductivity.

The conduction of free electrons in a normal conductor can be compared to the behavior of people dancing to rock music on a crowded floor. The random motion leads to many collisions. In contrast, when current flows in a superconductor in the superconducting state, the pairs of electrons are like dancers who move in sync without collisions.

The maximum distance up to which the states of paired electrons are correlated to produce superconductivity is called coherence length ϵ_0. The properties of superconductor depend on the correlation of electrons within a volume of ϵ_0^3 called coherence volume. The ratio of London penetration depth to the coherence length is a number,

$$k = \frac{\lambda}{\epsilon_0}.$$

For type I superconductors $k < 1/\sqrt{2}$ and type II superconductors $k > 1/\sqrt{2}$. From the BCS theory, it is shown that intrinsic coherence length (ϵ_0) is related to the energy gap as

$$\epsilon_0 \approx \frac{\hbar v_F}{2\Delta},$$

where 2Δ is the energy gap and v_F is the Fermi velocity. It also has been observed that the energy gap decreases from a value of about $3.5\ k_B T_c$ at 0 K to zero at T_c.

9.10 HIGH-TEMPERATURE SUPERCONDUCTORS

In 1986, Bednorz and Alex Muller in Zurich, Switzerland, discovered the high temperature superconductivity in ceramics. They made a particular type of ceramic materials that remained a superconductor at a temperature as high as 30 K. The importance of the discovery was recognized immediately, and they were awarded the Nobel Prize in Physics in 1987.

The first high temperature superconductor material fabricated by Bednorz and Alex Muller, with critical temperature of 30 K, was lanthanum-barium cuprate (LBCO). After the discovery of LBCO, the second high temperature superconductors material discovered by Wu-Chu was yattrium-barium cuprate (YBCO), a material that superconducts at temperature above the temperature of liquid nitrogen 77 K. The figure shows the variation of resistivity will the temperature in YBCO.

Important Features of High-temperature Superconductors

1. All high temperature superconductors bear a particular type of crystal structure called the perovskite structures.
2. All high temperature superconductors are metal oxides (ceramics).
3. There are two types of high temperature superconductors: Cooperates and noncorporates

Comparison of High-temperature Superconductors with Low-temperature Superconductors

1. The critical temperatures for high temperature superconductors are higher than the critical temperature of classical superconductors.
2. High temperature superconductors are cuprate while classical superconductors are metals or alloys.
3. Classical superconductors can be either type I superconductors or type II superconductors, while high temperature superconductors are always type II superconductors.

9.11 AC Resistivity

A superconducting metal has no resistance, which implies that when current is conducted through it, there is no voltage drop along the metal, and no power is produced. This is only accurate for constant value direct current. If there is a change in the current, an electric field develops, and some power is lost. We must first quickly go over a few elements of conduction electrons in superconductors in order to see why this is the case.

Conduction electrons behave in two ways below the transition temperature: some behave as super electrons that can pass through the metal with no resistance, while the other group behaves as normal electrons that can scatter and thus behave similarly to conduction electrons in a typical metal. As the temperature rises toward the transition point, the proportion of super electrons seems to decrease. At absolute zero Kelvin, all conduction electrons act as super electrons, but as the temperature rises, some conduction electrons start to behave differently, and as the temperature rises even higher, the proportion of regular electrons rises. Once all the electrons have converted to regular electrons at the transition temperature, the metal loses its superconducting capabilities. Therefore, a superconductor seems to be penetrated by two electron flows, one of regular electrons fluid and the other of super electron fluid, below its transition temperature. The temperature affects the relative electron density between the two fluids. Thermodynamical considerations based on the findings of specific heat and related measurement superconductors indicate this two-fluid model.

Superconducting Materials

Both conventional and super electrons may often carry current in a superconducting metal. The superconducting electrons carry all of the current in the specific situation of a continuous direct current, though. The absence of an electric field in the metal is necessary for the current to remain constant; otherwise, the super electrons would be continually accelerated in the field, increasing the current forever. There won't be a normal current if there is no field since there won't be anything to move the regular electrons. Super electrons carry 100% of the current when the total current is constant. A superconducting metal is comparable to two parallel conductors, one with normal resistance and the other with zero resistance. The conventional electrons are damaged by the superconducting electrons. In this approach, the current tends to increase indefinitely when a material is exposed to a voltage source, like a battery, across a superconductor, but it is really constrained by the internal resistance of the source. The electrons must be accelerated while the current is changing by an electric field. Because electrons have a low inertial mass, the super current increases more slowly than electrons' rate of acceleration in the electric field. Due to the inertia of super electrons, if an alternating field is applied, the super current will lag behind the field. Because there is now an electric field present, some of the current will be carried by the regular electrons as a result of the super electrons' inductive impedance. The current is not carried entirely by the superconducting electrons as in the DC case. Since the normal electrons have an inertial mass by definition, the resistance caused by their dispersion in the metal entirely overpowers the inductive reactance that results. A perfect inductance running parallel to a resistance can be used to illustrate the general characteristics of a superconducting metal.

Power is dissipated in the typical manner by the portion of current that is redirected through the regular electrons. Since the mass of an electron is incredibly small, so is the inductance caused by their inertia. Because a typical superconductor's inductance in Henry is only about 10–12 of its normal resistance in ohm, at frequencies like 1000 Hz, for instance, only about 10–8 of the total current is carried by the normal electrons, and there is only a very small amount of power dissipation. However, in the DC instance, there is completely zero resistance.

A superconducting metal reacts in the same manner as regular metal if the frequency of an applied field is high enough. This is due to the fact that superconducting electrons have lower energies than regular electrons, but if the frequency of the applied field is high enough, the electromagnetic field's photons can excite superconducting electrons into higher energies where they behave like regular electrons. When the frequency exceeds 10^{11} Hz, something occurs. A superconductor behaves exactly like a regular metal at optical frequencies; hence, there is no difference in a superconductor's appearance when it is cooled below its transition temperature, for instance.

9.12 Entropy

In all superconductors, the entropy decreases on cooling below the critical temperature.

We know that entropy is a measure of disorder of system and hence the observed decrease in entropy between the normal state and the superconducting state tells us that the superconducting state is more ordered than the normal state.

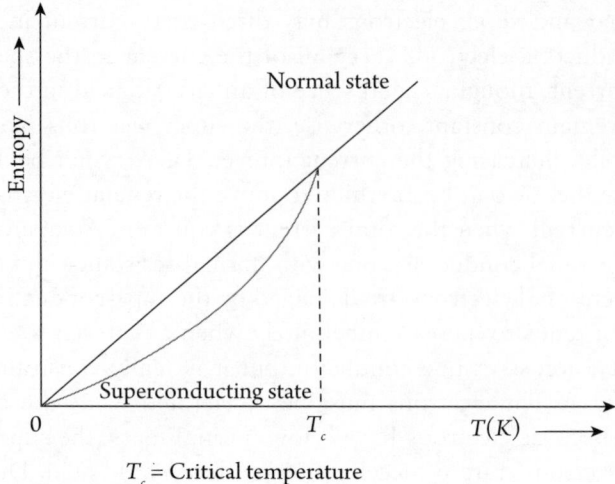

Fig. 9.7 Entropy of aluminum in the normal and superconducting states as a function of temperature.

For aluminum, the change in entropy is small and of the order of 10^{-14} K_B per atom. The decrease in entropy between the normal state and the superconducting state means that some or all of the electrons thermally excited in the normal state are ordered in the superconducting state. It has been predicted that in simple superconductors (type I), there is a spatial order that extends over a distance of the order of 10^{-6} m. This range is called coherence length. Entropy of aluminum in the normal and superconducting states as a function of temperature is plotted in the Figure 9.7.

9.13 SPECIFIC HEAT

The specific heat of the normal metal is seen to be of the form

$$C_n(T) = \gamma T + \beta T^3.$$

The term γT in the above equation is the specific heat of the electrons in the metal and the term βT^3 is the contribution of lattice vibrations at low temperature. The specific heat of the superconductor shows a jump at T_c. Since the superconductivity affects electrons mainly, it is natural to assume that the lattice vibration part remains unaffected, that is, it has the same value of βT^3 in the normal and superconducting states. On subtracting this, we notice that the electronic specific heat C_{es} is not linear with the temperature. It rather fits an exponential form,

$$C_{es} = Ae^{(-\Delta/k_B T)}.$$

This exponential form is an indication of the existence of a finite gap in the energy spectrum of electrons, separating the ground state from the lowest state (Figure 9.8).

The number of electrons thermally excited across the gap varies exponentially with the reciprocal of temperature. The energy gap, as shown in Figure 9.9, is believed to be a characteristic feature of the superconducting state that determines the thermal properties as well as high-frequency electromagnetic response of all superconductors.

Superconducting Materials

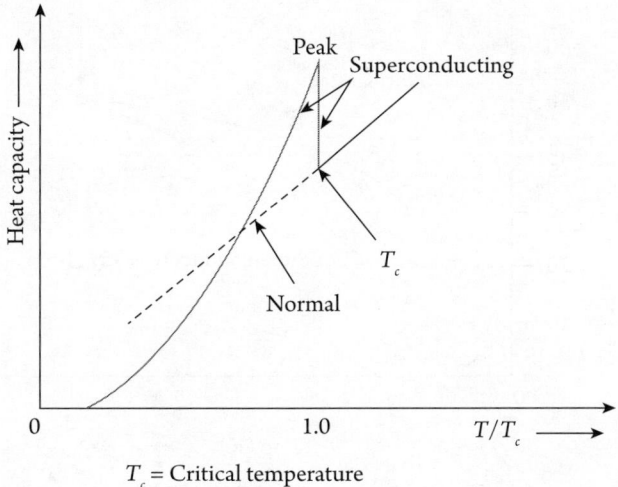

Fig. 9.8 Temperature dependence of specific heat in normal and superconducting states.

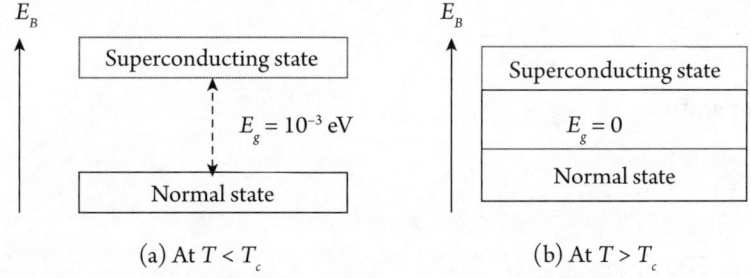

Fig. 9.9 Energy band gap for (a) superconductor and (b) normal conductor.

9.14 Thermal Conductivity

Professor Hulm and others discuss the result of thermal conductivity in superconductors. The thermal conductivity of superconductors undergoes a continuous change between the two faces and is usually lower in the superconducting phase, suggesting that the electronic contribution drops, the superconducting electrons possibly playing no part in the heat transfer. Thermal conductivity of tin at 2 K is 34 cm^{-1} K^{-1} for the normal phase and 16 cm^{-1} K^{-1} for superconducting phase. At 4 K, it is 50 W cm^{-1} K^{-1} (at 4 K, there is no superconducting phase for tin as T_c = 3.73 K), as shown in Figure 9.10.

9.15 Acoustic Attenuation

When a sound wave propagates through a metal, the microscopic electric field due to the displacement of ions can impart energy to electrons near the Fermi energy level, thereby removing energy form the wave. This is expressed by attenuation coefficient, α, of acoustic waves. The ratio of α for superconducting and normal state is given by

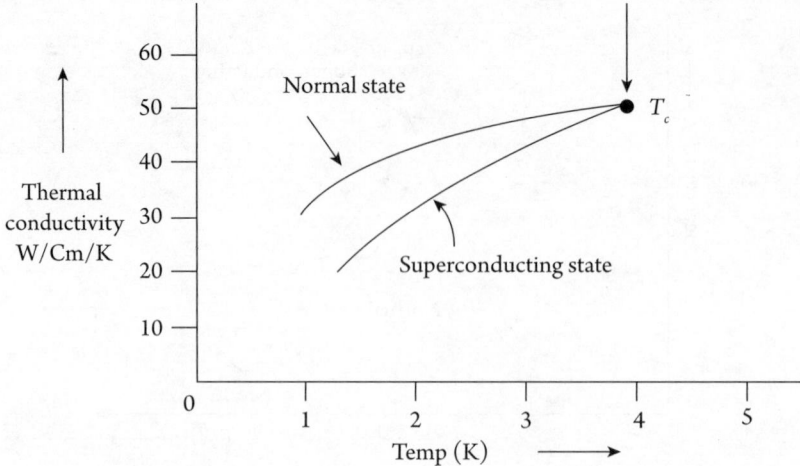

Fig. 9.10 Thermal conductivity of a specimen of tin in normal and superconducting states.

$$\frac{\alpha_s}{\alpha_n} = \frac{2}{1 + e^{(\Delta/k_B T)}}.$$

At low temperatures,

$$\frac{\alpha_s}{\alpha_n} = 2\left[e^{(-\Delta/k_B T)}\right].$$

9.16 THE ENERGY GAP

The heat capacity in the superconducting state varies with temperature in an exponential manner, as shown in Figure 9.11. It is of the form of $(-\Delta/k_B T)$ with $\Delta = bk_B T_c$, where b is the constant. This indicates, in accordance with the fact that the exponential form is compatible with the thermal excitation across a gap in the energy, that an energy gap may exist in the superconducting electron levels. The jump in the heat capacity at the critical temperature T_c supports this idea of the existence of the energy gap further. Superconductors' energy gap is fundamentally different from that of semiconductors and insulators. Since the energy gap in superconductors is bound to the Fermi gas in contrast to the energy gap in insulators, which is bound to the lattice, the former gap frequently has a completely different character. The gap in semiconductors blocks the passage of electrical current. Before current may flow, energy must be applied to elevate electrons from the valence band to the conduction band. On the other hand, current moves via a gap in a superconductor. The behavior of special electrons that carry current in a superconductor is unaffected by the energy gap. Superconductors contain normal electrons as well, and it is these electrons that are affected by the gap.

The existence of energy gap in superconductors has been confirmed by a number of experiments: electron tunneling observation across the superconducting junctions by Giaever being one of them. Some experiments have been employed for the experimental determination of its value. From theory and comparison with optical and other methods of determination of the gap, it is concluded that

Superconducting Materials

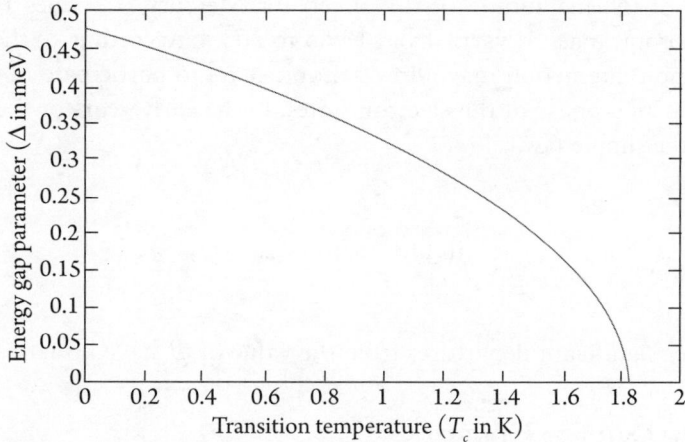

Fig. 9.11 A plot of temperature dependence of energy gap parameter.

$$E_g = 2\Delta = 2b(k_B T_c),$$

$$\left[\frac{E_g}{k_B T_c}\right] = 2b,$$

$2b$ is about 3.5, that is, the gap decreases from a value of about 3.5 $k_B T_c$ at 0 K to zero at the transition temperature. Values of energy gap of some selected superconductors are given in Table 9.2.

9.17 Isotope Effect

The critical temperature of superconductors has been shown to change with isotopic mass. The observation was first made by Maxwell and others, who used Mercury isotopes.

Table 9.2 Energy gap at 0 K for some superconductors

Element	$E_g(0)$ in 10^{-4} eV	T_c(K)	$\dfrac{E_g(0)}{k_B T_c}$
Niobium	30.5	9.5	3.7
Aluminum	3.4	1.2	3.3
Tantalum	14.0	4.48	3.6
Tin	11.6	3.72	3.6
Lead	27.3	7.18	4.4
Mercury	16.5	4.16	4.6
Zinc	2.4	0.9	3.1
Gallium	3.3	1.09	3.5

To give an idea of the magnitude of the effect, for Mercury, T_c varies from 4.185 K to 4.146 K as the isotopic mass M varies from 199.5 to 203.4. According to the reasoning, only the electron phonon interaction may allow isotopic mass to participate in the development of the superconducting phase of the electron states. In the early years of the development of the BCS theory, the simple law,

$$T_c \propto M^{-\beta}$$

with $\beta = +0.5$ was thought to be valid for materials.

Thus, $T_c M^\beta = Constant$.

However, later significant departures from the value of β have been observed.

9.18 Mechanical Effects

It has been demonstrated via experimentation that mechanical stress causes a superconductor's critical magnetic field and transition temperature to change without being audibly affected. The critical magnetic field strength is dependent on the difference between the two states' free energy, and many of the mechanical features of the superconducting and normal states are thermodynamically connected to these free energies. The bulk modulus of elasticity and the thermal expansion coefficient must also differ somewhat between the superconducting and normal states since there is a very minor change in volume when a normal material becomes superconducting. It is possible to derive expressions for these effects by straightforward thermodynamic manipulation, but the effects are extremely small and hence not considered here.

9.19 Characteristics of Superconductors

The crucial features of superconductors include the following:

1. Superconductors allow for long-term sustained current.
2. Superconductors are not permeable to magnetic fields.
3. A superconductor transforms into a regular metal over the critical magnetic field.
4. Superconductors are materials with high resistance.
5. Materials that are ferromagnetic or antiferromagnetic are not superconductors.

9.20 Organic Superconductor

An organic substance that is artificially created and demonstrates superconductivity at low temperatures is known as an organic superconductor.

The greatest critical temperature for an organic superconductor at standard pressure as of 2007 is 33 K (–240°C; –400°F), observed in the alkali-doped fullerene $RbCs_2C_{60}$.

Many substances fit the description of organic superconductors. These comprise the quasi-one-dimensional Bechgaard salts and Fabre salts, as well as quasi-two-dimensional materials such as alkali-doped fullerenes, k-BEDT-TTF2X charge-transfer complex, λ-BETS2X compounds, and Bechgaard and Fabre salts.

Scientists interested in room-temperature superconductivity and model systems that explain the origin of superconductivity are particularly interested in organic superconductors. In contrast to copper or osmium, organic compounds are primarily composed of carbon and hydrogen, two elements that are more prevalent on earth.

9.21 One-Dimensional Fabre and Bechgaard Salts

Fabre-salts are composed of tetramethyltetrathiafulvalene (TMTTF) and Bechgaard salts of tetramethyltetraselenafulvalene (TMTSF). These two organic molecules are similar except for the sulfur-atoms of TMTTF being replaced by selenium-atoms in TMTSF. The molecules are stacked in columns (with a tendency to dimerization) that are separated by anions. Typical anions are, for example, octahedral PF_6, AsF_6, or tetrahedral ClO_4 or ReO_4.

Both material classes are quasi-one-dimensional at room-temperature, only conducting along the molecule stacks, and share a very rich phase diagram containing antiferromagnetic ordering, charge order, spin–density wave state, dimensional crossover, and superconductivity.

Only one Bechgaard salt was found to be superconducting at ambient pressure, which is $(TMTTF)_2ClO_4$ with a transition temperature of $T_c = 1.4$ K. Several other salts become superconducting only under external pressure. The external pressure required to drive most Fabre-salts to superconductivity is so high that, under lab conditions, superconductivity was observed only in one compound. A selection of the transition temperature and corresponding external pressure of several one-dimensional organic superconductors is shown in Table 9.3.

9.22 Two-Dimensional $(BEDT-TTF)_2X$

BEDT-TTF is the short form of bisethylenedithio-tetrathiafulvalene, commonly abbreviated ET. Anions divide the planes that these molecules make. The arrangement of the molecules in the planes is not singular, but a variety of phases are developing based on the anion and the development circumstances. The alpha and theta phases, in which molecules arrange themselves in a fishbone pattern, as well as the beta and particularly k-phases, in which molecules arrange themselves in a checkerboard pattern and are dimerized in the k-phase,

Table 9.3 Some organic superconductors with transition temperature and external pressure

S. No.	Material	T_c(K)	p_{ext} (kbar)
1.	$(TMTSF)_2SbF_6$	0.36	10.5
2.	$(TMTSF)_2PF_6$	1.1	6.5
3.	$(TMTSF)_2AsF_6$	1.1	9.5
4.	$(TMTSF)_2ReO_4$	1.2	9.5
5.	$(TMTSF)_2TaF_6$	1.35	11
6.	$(TMTTF)_2Br$	0.8	26

are significant phases with regard to superconductivity. Because they are half-filled systems rather than quarter filled ones, the phases are unique because they enter superconductivity at greater temperatures than the other phases.

There are almost infinitely many potential anions that might separate two sheets of ET molecules. There are simple anions like I_3 and polymeric anions like the well-known $Cu[N(CN)_2]Br$, as well as anions that incorporate solvents like $Ag(CF_3)_4 \bullet 112DCBE$. The growth phase, the anion, and the amount of applied external pressure all affect the crystals' electrical characteristics. In comparison to Bechgaard salts, an ET salt with an insulating ground state may be driven to a superconducting one with a much less external pressure. For instance, it only requires a pressure of roughly 300 bar for $K-(ET)_2Cu[N(CN)_2]Cl$ to become superconducting, which may be accomplished by submerging a crystal in grease that has been frozen to 0°C (32°F) and then applying enough stress to cause the superconducting transition. The crystals are extremely sensitive, which can be viewed impressively when $-(ET)_2I_3$ is left out in the sun for many hours (or under more controlled conditions in an oven set to 40°C, 104°F). After receiving this treatment, one is left with superconducting tempered-$(ET)_2I_3$.

For all the ET-based salts, universal phase diagrams have only been presented, unlike the Fabre or Bechgaard salts. Such a phase diagram would be influenced by electrical correlations as well as temperature, pressure, and bandwidth (that is, bandwidth). In addition to their superconducting ground state, some materials exhibit charge order, antiferromagnetism, or remain metallic at extremely low temperatures. Even a spin liquid is anticipated for one substance. The phases with very comparable anions have the greatest transition temperatures both at atmospheric pressure and under external pressure. A pressure of 300 bar transforms deuterated $-(ET)_2Cu[N(CN)_2]Cl$ from an antiferromagnetic to a superconducting ground state with a transition temperature of T_c = 13.1 K. $-(ET)_2Cu[N(CN)_2]Br$ also undergoes a superconducting transition at T_c = 11.8 K at ambient pressure. Only a few representative superconductors of this class are included in the following Table 9.4.

Table 9.4 Some organic superconductors with transition temperature and external pressure

S. No.	Material	T_c(K)	p_{ext} (kbar)
1.	$\beta_H-(ET)_2I_3$	1.5	0
2.	$\theta-(ET)_2I_3$	3.6	0
3.	$k-(ET)_2I_3$	3.6	0
4.	$\alpha-(ET)_2KHg(SCN)_4$	0.3	0
5.	$\alpha-(ET)_2KHg(SCN)_4$	1.2	1.2
6.	$\beta'-(ET)_2SF_5CH_2CF_2SO_3$	5.3	0
7.	$\kappa-(ET)_2Cu[N(CN)_2]Cl$	12.8	0.3
8.	$\kappa-(ET)_2Cu[N(CN)_2]Cl$ deuterated	13.1	0.3
9.	$\kappa-(ET)_2Cu[N(CN)_2]Br$ deuterated	11.2	0
10.	$\kappa-(ET)_2Cu(NCS)_2$	10.4	0

Even more superconductors can be found by changing the ET-molecules slightly either by replacing the sulfur atoms by selenium (BEDT-TSF, BETS) or by oxygen (BEDO-TTF, BEDO). Some two-dimensional organic superconductors of the κ-$(ET)_2X$ and $\lambda(BETS)_2X$ families are candidates for the Fulde–Ferrell–Larkin–Ovchinnikov (FFLO) phase when superconductivity is suppressed by an external magnetic field.

9.23 Doped Fullerenes

Superconducting fullerenes based on C_{60} are fairly different from other organic superconductors. The building molecules are no longer manipulated hydrocarbons but pure carbon molecules. In addition, these molecules are no longer flat but bulky, which gives rise to a three-dimensional, isotropic superconductor. The pure C_{60} grows in an fcc-lattice and is an insulator. By placing alkali atoms in the interstitials, the crystal becomes metallic and eventually superconducting at low temperatures.

9.24 More Organic Superconductors

There are more organic systems that become superconducting at low temperatures or under pressure in addition to the three main types of organic superconductors. These are discussed in the following paragraphs.

Tetrathiafulvalene, or TTF, is the basis for the tetrathiapentalene (TTP)-based superconductors TMTTF and BEDT-TTF. TTP is used as the starting point to create a number of novel chemical compounds that act as cations in organic crystals. Some of them even possess superconductivity. Only lately has this class of superconductors been reported, and research is currently ongoing.

Superconductors of the Phenanthrene Type

The hydrocarbons picene and phenanthrene can now be used to create crystals in place of sulfated molecules or the relatively large Buckminster fullerenes.

Superconductivity may be achieved with transition temperatures as high as 18 K (255°C; 427°F) by doping crystals of picene and phenanthrene with alkali metals like potassium or rubidium. The superconductivity for axphenanthrene is unconventionally feasible. The term "phenanthrene-edge-type polycyclic aromatic hydrocarbon" refers to both phenanthrene and picene. Higher Tc is the result of benzene rings having more rings. Some organic superconductors with transition temperature and external pressure are listed in Tables 9.5 and 9.6.

Intercalated Graphite Superconductors

Even though neither the foreign molecule or atom nor the graphite layers are metallic, placing foreign molecules or atoms between hexagonal graphite sheets results in ordered structures and superconductivity. Several stoichiometries have been synthesized using mainly alkali atoms as anions.

Table 9.5 Some organic superconductors with transition temperature and external pressure

S. No.	Material	T_c(K)	p_{ext} (mbar)
1.	K_3C_{60}	18	0
2.	Rb_3C_{60}	30.7	0
3.	K_2CsC_{60}	24	0
4.	K_2RbC_{60}	21.5	0
5.	K_5C_{60}	8.4	0
6.	Sr_6C_{60}	6.8	0
7.	$(NH_3)_4Na_2CsC_{60}$	29.6	0
8.	$(NH_3)K_3C_{60}$	28	14.8

Table 9.6 Some organic superconductors with transition temperature

S. No.	Material	T_c(K)
1.	$(BDA-TTP)_2AsF_6$	5.8
2.	$(DTEDT)_3Au(CN)_2$	4
3.	$K_{3.3}$Picene	18
4.	$Rb_{3.1}$Picene	6.9
5.	K_3Phenanthrene	4.95
6.	Rb_3Phenanthrene	4.75
7.	CaC_5	11.5
8.	NaC_2	5
9.	KC_8	0.14

9.25 Applications of Superconductors

The applications of superconductor are as follows:

Superconductors are used to make high-field magnets known as super magnets and are essential for flux trapping and shielding. They play a crucial role in magnetic bearings, energy storage, and D.C. transformers, as well as in nuclear magnetic resonance, medical diagnostics, and spectroscopy. Additionally, superconductors are utilized in magnetic levitation, magnetic shielding, and large engineering machines, such as colliders, fusion confinement, and R.F. cavities. They contribute to the production of magnetic fusion, magnetohydrodynamics, magnetic energy storage, and electric power transmission. In transportation, superconductors are found in high-speed trains, including Maglev trains and bullet trains, as well as in ship drive systems. They are also employed in superconducting quantum interference devices (SQUIDs), Josephson devices (like square-law detectors, parametric amplifiers, and mixers), bolometers, and electromagnetic shielding. Furthermore, superconductors are integral to semiconductor–superconductor hybrids (A/D converters) and active superconducting elements (FETs), as well as in voltage standards, optoelectronics, and matched filters.

Superconducting Materials

Solved Problems

Ex. 1: The critical field for Aluminum is 1.2×10^4 A/m. Determine the critical current and current density that can flow through a long thin superconducting wire of aluminum of diameter 1 mm.

Solution:

Here, we have diameter = 1 mm = 10^{-3} m, $H_c = 1.2 \times 10^4$ A/m, radius $r = 0.5 \times 10^{-3}$ m.

Now, $I_c = 2\pi r H_c = 2 \times 3.14 \times 0.5 \times 10^{-3} \times 1.2 \times 10^4 = 37.68$ A.

Current density, $J_c = \dfrac{I_c}{A} = \dfrac{I_c}{\pi r^2} = \dfrac{37.68}{3.14 \times (0.5 \times 10^{-3})^2} \dfrac{A}{m^2} = 48 \times 10^8 \, A/m^2$.

Ex. 2: A superconducting material has a critical temperature of 3.7 K in zero magnetic fields and a critical field of 0.02 T at 0 K. Find the critical field of 3 K.

Solution:

Here, we have T_c = 3.7 K, $H_c(0)$ = 0.02 T, and T = 3 K. Now,

$$H_c(T) = H_c(0)\left[1 - \left(\dfrac{T}{T_c}\right)^2\right] = 0.02\left[1 - \left(\dfrac{3}{3.7}\right)^2\right] = 0.0069 \, T.$$

Ex. 3: The critical field for lead is 1.2×10^5 A/m at 8 K and 2.4×10^5 A/m at 0 K. Find the critical temperature of the material.

Solution:

Here, we have $H_c(T) = 1.2 \times 10^5$ A/m, $H_c(0) = 2.4 \times 10^5$ A/m, and T = 8 K. Now,

$$H_c(T) = H_c(0)\left[1 - \left(\dfrac{T}{T_c}\right)^2\right] \Rightarrow \dfrac{H_c(T)}{H_c(0)} = \left[1 - \left(\dfrac{T}{T_c}\right)^2\right] \Rightarrow \dfrac{T}{T_c} = \sqrt{1 - \dfrac{H_c(T)}{H_c(0)}}.$$

$$\Rightarrow T_c = \dfrac{T}{\sqrt{1 - \dfrac{H_c(T)}{H_c(0)}}} = \dfrac{8}{\sqrt{1 - \dfrac{1.2 \times 10^5}{2.4 \times 10^5}}} = 11.31 \, K.$$

Ex. 4: The critical fields for lead are 1.2×10^5 A/m and 3.6×10^5 A/m at 12 K and 10 K, respectively. Find its critical temperature and critical field at 0 K and 3.2 K.

Solution:

Here, we have $H_c(T) = 1.2 \times 10^5$ A/m at T = 12 K

$H_c(T) = 3.6 \times 10^5$ A/m at T = 10 K.

Now, $H_c(T) = H_c(0)\left[1 - \left(\dfrac{T}{T_c}\right)^2\right]$.

So, we have $\dfrac{H_c(T)}{H_c(0)} = \left[1-\left(\dfrac{T}{T_c}\right)^2\right] \Rightarrow \dfrac{1.2\times 10^5}{H_c(0)} = \left[1-\left(\dfrac{12}{T_c}\right)^2\right]$ (1)

and $\dfrac{H_c(T)}{H_c(0)} = \left[1-\left(\dfrac{T}{T_c}\right)^2\right] \Rightarrow \dfrac{3.6\times 10^5}{H_c(0)} = \left[1-\left(\dfrac{10}{T_c}\right)^2\right]$.   (2)

On dividing equation (1) by equation (2), we get

$1 - \dfrac{144}{T_c^2} = 0.33 - \dfrac{33}{T_c^2} \Rightarrow T_c^2 = 165.67 \Rightarrow T_c = 12.87\,K$.

Now, on substituting the value of T_c in equation (1), we get

$\dfrac{1.2\times 10^5}{H_c(0)} = \left[1-\left(\dfrac{12}{12.87}\right)^2\right] \Rightarrow H_c(0) = 9.12\times 10^5\,A/m$.

Critical field is given by

$H_c(T) = 9.12\times 10^5 \left[1-\left(\dfrac{3.2}{12.87}\right)^2\right] = 8.56\times 10^5\,A/m$.

Ex. 5: Determine the temperature at which the critical field becomes half of its value at 0 K if the critical temperature of a superconductor, when no magnetic field is present, is T_c.

Solution:

Here, we have $H_c(T) = \dfrac{H_c(0)}{2}$.

Now, using this formula in $H_c(T) = H_c(0)\left[1-\left(\dfrac{T}{T_c}\right)^2\right] \Rightarrow \dfrac{H_c(0)}{2} = H_c(0)\left[1-\left(\dfrac{T}{T_c}\right)^2\right]$

$\Rightarrow T = \dfrac{T_c}{\sqrt{2}} = 0.71\,T_c$.

Previous Year Questions (University Examination)

1. How would you define superconductivity? Discuss the superconductor's sensitivity to temperature.
2. What are superconductors, exactly? Describe how superconductors are affected by magnetic fields.
3. Explain Meissner's effect in superconductivity. Show that a superconductor is perfectly diamagnetic.
4. What is critical current, and why? Explain the superconducting Silsbee effect.
5. What do you understand by critical current density? Give some applications of superconducting materials.
6. What do you mean by organic superconductors? Give some examples of organic superconductors with their applications.
7. Write short notes on type I and type II superconductors.
8. What are superconductors? Discuss applications of superconductors.
9. What is BCS theory? Explain in detail.

Superconducting Materials

10. Explain the behavior of superconductor in magnetic field.
11. Write short notes on high temperature superconductors. Discuss the characteristics of superconductors.

Multiple Choice Questions

1. The resistance of a metal increases with temperature
    (a) Decreases
    (b) Become zero
    (c) Increases
    (d) Remains constant
2. At 0 K, the average kinetic energy of a free electron gas is
    (a) $\frac{5}{3}E_F$
    (b) $\frac{3}{5}E_F$
    (c) $\frac{2}{5}E_F$
    (d) E_F
3. At low temperature, the resistivity of a metal is proportional to
    (a) T^5
    (b) T^4
    (c) T
    (d) T^2
4. Superconductors are generally
    (a) Ferromagnetic and antiferromagnetic metals
    (b) Monovalent metals
    (c) Amorphous thin films of Be and Bi
    (d) Thin films of barium titanate
5. A superconductor on being subjected to the critical field changes to
    (a) Normal conductor
    (b) Remains unaffected
    (c) Nano-material
    (d) None of the above
6. The width of energy gap of a superconductor is maximum at
    (a) 0 K
    (b) Room temperature
    (c) Transition temperature
    (d) None of these
7. The temperature at which conductor becomes superconductor is known as
    (a) Room temperature
    (b) Critical temperature
    (c) Curies temperature
    (d) None of these
8. The resistance of conductor in superconducting state
    (a) Infinite
    (b) Zero
    (c) Residual resistant
    (d) None of these
9. Mercury's transition temperature is
    (a) 1 K
    (b) 1.14 K
    (c) 4.12 K
    (d) 9.22 K
10. In superconducting state, the entropy and thermal conductivity
    (a) Increases
    (b) Decreases
    (c) Remains constant
    (d) None of the above

11. Hard superconductors observe
    (a) Breakdown of Silsbee's rule
    (b) Incomplete Meissner effect
    (c) High critical field and transition temperature
    (d) All of the above

12. Soft superconductors observe
    (a) Meissner effect   (b) Silsbee's rule
    (c) Both (a) and (b)   (d) None of the above

13. At $T = T_c$, the specific heat of a superconducting material changes abruptly, jumping to a high value for
    (a) $T < T_c$ (b) $T > T_c$
    (c) $T = T_c$ (d) None of the above

14. Superconductivity is the outcome of
    (a) Crystal structure with ∞ atomic vibrations at 0 K
    (b) Crystal structure with 0 atomic vibrations at 0 K
    (c) Conductivity only
    (d) None of the above

15. The majority of superconducting materials' transition temperatures fall between
    (a) Zero to 10 K   (b) 10 K to 20 K
    (c) 20 K to 50 K   (d) Above 50 K

16. When exposed to a magnetic field, a superconducting material will
    (a) Bring the magnetic field to the center of it.
    (b) Reject any magnetic force lines that are going through it.
    (c) Draw the magnetic field while concentrating it in one area.
    (d) Not have an impact on the magnetic field

17. The critical current density can be determined by
    (a) Temperature and Magnetic field strength
    (b) Electric field
    (c) Silsbee effect
    (d) Permeability

18. At 0 K, a superconductor's energy gap is about
    (a) 0 J   (b) $3.5\,K_B T_c$
    (c) $K_B T_c$ (d) $300\,K_B T_c$

19. Cooper pairs are
    (a) Fermions   (b) Bosons
    (c) Classical particles   (d) None of these

20. The paired electrons' coherence length is
    (a) 0.25 nm   (b) 250 nm
    (c) 0.01 nm   (d) 0.001 nm

BIBLIOGRAPHY

1. Cardwell, D. A. and Ginley, D. S., eds. (2003). *Handbook of Superconducting Materials*, vol. 1. CRC Press.
2. Wesche, R. (2013). *High-temperature Superconductors: Materials, Properties, and Applications*, vol. 6. Springer Science & Business Media.
3. Wang, N. L., Hosono, H., and Dai, P., eds. (2012). *Iron-based Superconductors: Materials, Properties and Mechanisms*. CRC Press.
4. Phillips, J. (2012). *Physics of High-Tc Superconductors*. Elsevier.
5. Gabovich, A., ed. (2012). *Superconductors: Materials, Properties and Applications*. BoD–Books on Demand.
6. Shi, D., ed. (1995). *High-temperature Superconducting Materials Science and Engineering: New Concepts and Technology*. Elsevier.
7. Rey, C., ed. (2015). *Superconductors in the Power Grid: Materials and Applications*. Elsevier.
8. Roberts, B. W. (1969). *Superconductive Materials and Some of Their Properties*, vol. 482. US Department of Commerce, National Bureau of Standards.
9. Seidel, P., ed. (2015). *Applied Superconductivity: Handbook on Devices and Applications*. John Wiley & Sons.
10. Roberts, B. W. (1972). *Properties of Selected Superconductive Materials*. vol. 724. US Department of Commerce, National Bureau of Standards.

Keys

1. (c) 2. (c) 3. (c) 4. (c) 5. (a) 6. (a) 7. (b) 8. (b) 9. (c) 10. (b)
11. (d) 12. (c) 13. (a) 14. (b) 15. (a) 16. (b) 17. (a) 18. (b) 19. (b) 20. (d)

CHAPTER 10

Crystal Structure

10.1 CRYSTALLOGRAPHY

The science that deals with the geometrical structure and physical properties of crystalline solids is called crystallography.

Solids are classified into two categories:

1. Crystalline solids
2. Amorphous solids

10.2 CRYSTALLINE SOLIDS

Crystalline solids are those that contain the regular repeated pattern of atoms or molecules, as shown in Figure 10.1. The physical properties of crystalline solids are different in different directions. Therefore, crystalline solids are anisotropic. Examples are rock salt, quartz, calcite, sugar, and so on.

10.3 AMORPHOUS SOLIDS

Figure 10.2 illustrates an amorphous material, which lacks the regular arrangement of atoms or molecules. The amorphous solid's physical characteristics are uniform throughout. As a result, amorphous solids are isotropic. Examples are glass, rubber, polymers, and so on.

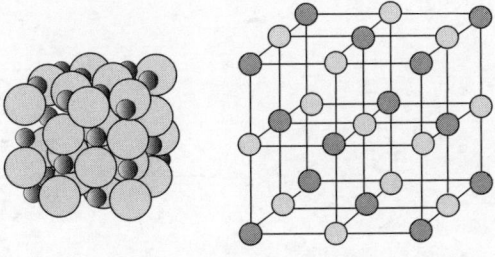

Fig. 10.1 Crystalline solids.

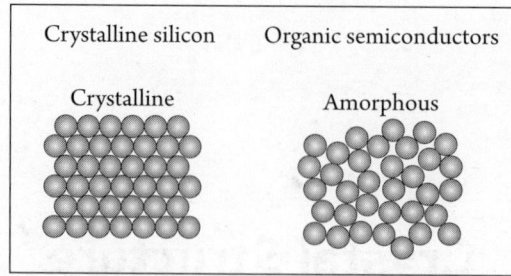

Fig. 10.2 Crystalline solids and amorphous solids.

10.4 SPACE LATTICES

A crystal is made up of identical structural units (atoms, molecules, or ions) that are infinitely repeated in space; each unit can be replaced by a geometrical point. The outcome is a pattern of dots with crystal-like geometrical characteristics. The crystal lattice or space lattice is this geometric arrangement. Lattice points are the name given to the geometrical points.

The regular pattern of points that describes the three-dimensional arrangement of points (atoms, molecules, or ions) in the crystal structure is called the crystal lattice or space lattice.

10.5 BASIS

The unit assembly of atoms, molecules, or ions identical in the composition, arrangement, and orientation is called basis. If we add the basis to every lattice point, then it forms a crystal structure, as shown in Figure 10.3.

Space lattice + Basis = Crystal structure.

10.6 TRANSLATIONAL VECTORS

Let us consider a three-dimensional space lattice with the translational vectors a, b, and c along x, y, and z directions. If we consider an origin "O" at any point, then the position vector from this origin in any direction is given by

$$\vec{T} = n_1\vec{a} + n_2\vec{b} + n_3\vec{c},$$

where n_1, n_2, and n_3 are integers.

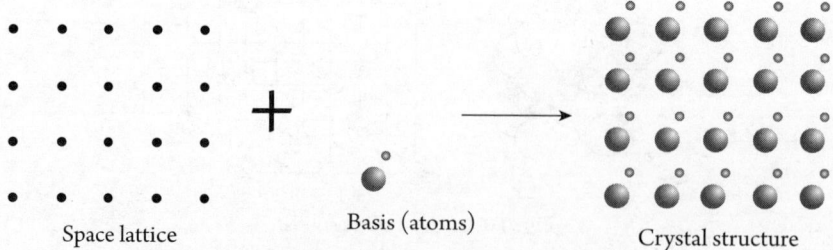

Fig. 10.3 Space lattice and basis together form crystal structure.

Crystal Structure

The position vector from the origin for a two-dimensional space lattice is given by

$\vec{T} = n_1 \vec{a} + n_2 \vec{b}.$

10.7 Unit Cell and Lattice Parameter

The unit cell is a basic structural unit or building block of the crystal, and its atomic arrangement, when repeated in the three dimensions, gives the total structure of the crystal.

The majority of crystals have parallelopiped or cubes with three sets of parallel faces as their unit cells. A unit cell of the space lattice is a parallelopiped generated in space by the three basis vectors a, b, and c as linearly independent edges. The three lattice vectors (a, b, c) and interfacial angles (α, β, γ) constitute the lattice parameter of the unit cell, as shown in Figure 10.4.

10.8 Primitive Cell

A primitive cell is a unit cell having lattice points at the corners only, as shown in Figure 10.5. It is a minimum volume unit cell containing one lattice point, each at the corner only. Thus, every primitive cell is a unit cell, but every unit cell is not a primitive cell, as shown in Figure 10.6. The volume of the primitive cell is given by $a \cdot (b \times c)$.

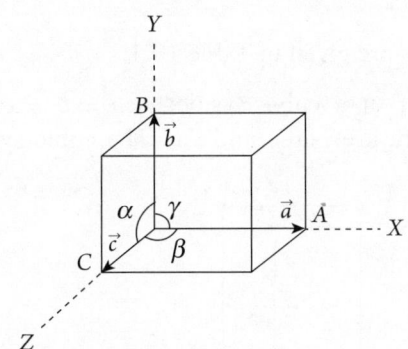

Fig. 10.4 Unit cell with lattice parameter.

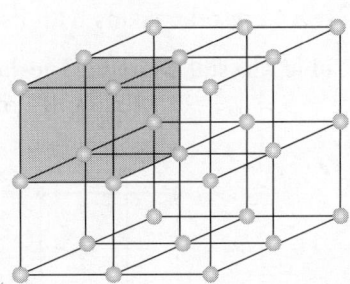

Representation of unit cell in cristal lattice

Fig. 10.5 Primitive cell.

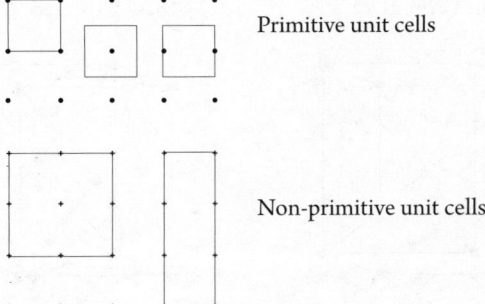

Fig. 10.6 Primitive cell and non-primitive cell.

10.9 Types of Unit Cell

Arrangement of lattice points in the unit cell:

1. Eight lattice points situated at eight corners (*P*).
2. Eight lattice points situated at eight corners, and one lattice point situated at the body center (*I*).
3. Eight lattice points situated at eight corners, and six lattice points situated at the center of six faces (*F*).
4. Eight lattice points situated at eight corners, and two lattice points situated at the centers of the opposite face (*A/B/C*), as shown in Figure 10.7.

10.10 Seven Crystal Systems and Fourteen Bravais Lattices

There are seven crystal systems on the basis of the shape of the unit cell. The classification of the crystal system has been made on the basis of the length of the unit cells and the angles included between them (i.e., lattice parameters). Bravais combined the crystal lattice system with various possible lattice centering and showed that there are fourteen different types of lattices under the seven crystal systems. The fourteen unique lattices in the three dimensions are known as Bravais lattices.

The seven crystal systems with its Bravais lattices are given in Table 10.1:

1. **Cubic Crystal System:** The lattice parameter of a cubic system is $a = b = c$ and $\alpha = \beta = \gamma = 90°$. Examples of a cubic system are NaCl and Cu. The cubic system

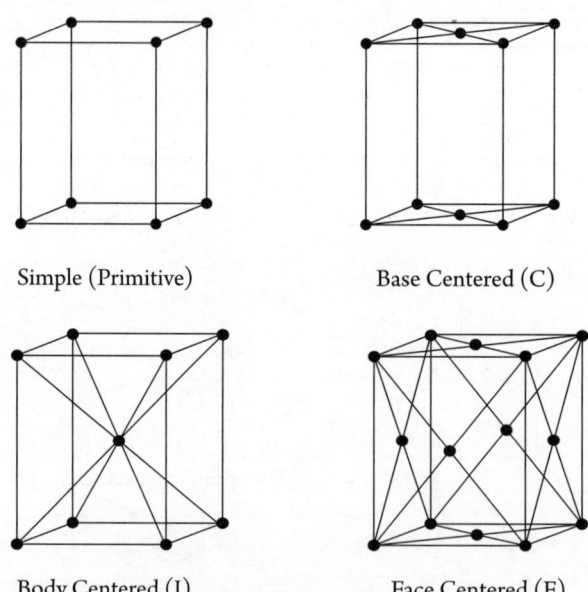

Simple (Primitive)   Base Centered (C)

Body Centered (I)   Face Centered (F)

Fig. 10.7 Types of unit cell.

Crystal Structure

Table 10.1 Seven crystal systems with fourteen Bravais lattices

S. No.	Crystal System	Lattice Parameter	Bravais Lattice
1.	Cubic	$a = b = c$, $\alpha = \beta = \gamma = 90°$	P, I, F
2.	Tetragonal	$a = b \neq c$, $\alpha = \beta = \gamma = 90°$	P, I
3.	Orthorhombic	$a \neq b \neq c$, $\alpha = \beta = \gamma = 90°$	P, I, F, C
4.	Rhombohedral	$a = b = c$, $\alpha = \beta = \gamma \neq 90°$	P
5.	Hexagonal	$a = b \neq c$, $\alpha = \beta = 90°$, $\gamma = 120°$	P
6.	Monoclinic	$a \neq b \neq c$, $\alpha = \gamma = 90° \neq \beta$	P, C
7.	Triclinic	$a \neq b \neq c$, $\alpha \neq \beta \neq \gamma$	P

consists of three Bravais lattices, that is, primitive (*P*), body-centered cubic (*I*), and face-centered cubic (*F*). Figure 10.8 shows all three types of cubic systems.

2. **Tetragonal Crystal System:** The lattice parameter of a tetragonal crystal system is $a = b \neq c$ and $\alpha = \beta = \gamma = 90°$. Examples of this tetragonal crystal system are $NiSo_4$ and TiO_2. The cubic system consists of two Bravais lattices, that is, primitive (*P*) and body-centered cubic (*I*). Figure 10.9 shows two types of tetragonal crystal systems.

3. **Orthorhombic Crystal System:** The lattice parameter of an orthorhombic crystal system is $a \neq b \neq c$ and $\alpha = \beta = \gamma = 90°$. An example of this orthorhombic crystal system is KNO_3. An orthorhombic crystal system consists of four Bravais lattices, that is, primitive (*P*), body-centered cubic (*I*), face-centered cubic (*F*), and base-centered (*C*). Figure 10.10 shows two types of orthorhombic crystal systems.

4. **Rhombohedral Crystal System:** The lattice parameter of a rhombohedral crystal system is $a = b = c$ and $\alpha = \beta = \gamma \neq 90°$. Examples for this rhombohedral crystal system

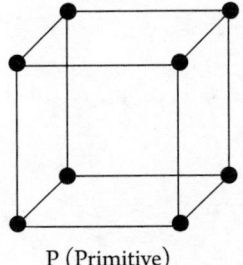
P (Primitive)

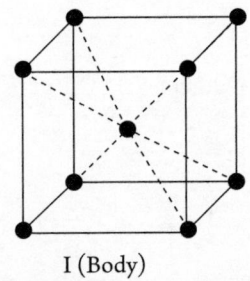
I (Body)

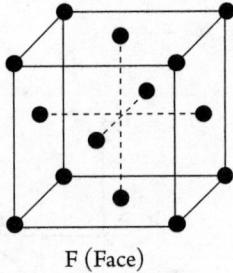
F (Face)

Fig. 10.8 Cubic crystal system.

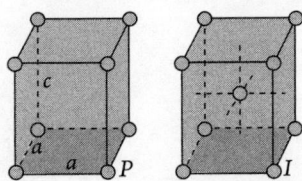

Fig. 10.9 Tetragonal crystal system.

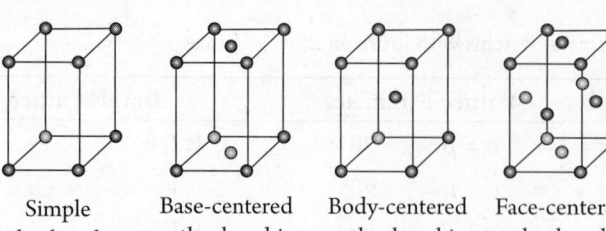

| Simple orthorhombic | Base-centered orthorhombic | Body-centered orthorhombic | Face-centered orthorhombic |

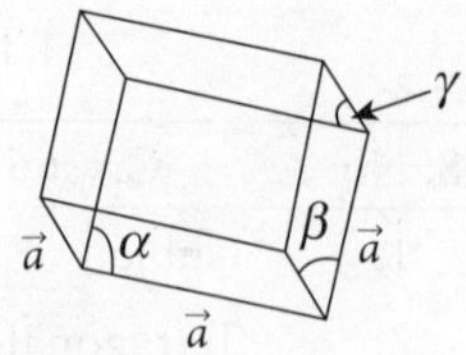

Fig. 10.11 Rhombohedral crystal system.

Fig. 10.10 Orthorhombic crystal system.

are Sb and Bi. The rhombohedral crystal system consists of one Bravais lattice, that is, primitive (P). Figure 10.11 shows a rhombohedral crystal system.

5. **Hexagonal Crystal System:** The lattice parameter of a hexagonal crystal system is $a = b \neq c$, $\alpha = \beta = 90°$ and $\gamma = 120°$. An example for this hexagonal crystal system is Mg. The hexagonal crystal system consists of one Bravais lattice, that is, primitive (P). Figure 10.12 shows a hexagonal crystal system.

6. **Monoclinic Crystal System:** The lattice parameter of a monoclinic crystal system is $a \neq b \neq c$ and $\alpha = \gamma = 90° \neq \beta$. An example for this monoclinic crystal system is $FeSo_4$. The monoclinic crystal system consists of two Bravais lattices, that is, primitive (P) and base-centered (C). Figure 10.13 shows two types of monoclinic crystal systems.

7. **Triclinic Crystal System:** The lattice parameter of a triclinic crystal system is $a \neq b \neq c$ and $\alpha \neq \beta \neq \gamma$. An example of this triclinic crystal system is $K_2Cr_2O_7$. The triclinic crystal system consists of one Bravais lattice, that is, primitive (P). Figure 10.14 shows a triclinic crystal system.

Some Important Definitions

The number of lattice points per unit cell is given by

$$N = N_i + \frac{N_f}{2} + \frac{N_c}{8},$$

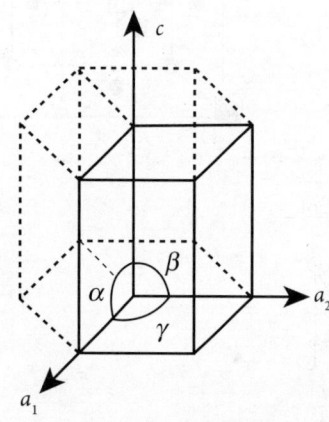

Fig. 10.12 Hexagonal crystal system.

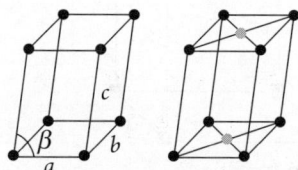

Fig. 10.13 Monoclinic crystal system.

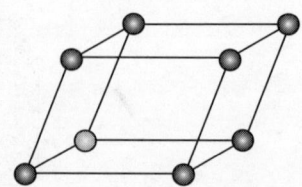

Fig. 10.14 Triclinic crystal system.

Crystal Structure

where N_i = Number of lattice points inside the unit cell.

N_f = Number of lattice points at the faces of unit cell.

N_c = Number of lattice points at the corners of the unit cell.

Coordination Number: In a crystal, the number of nearest-neighboring atoms to an atom in the structure is called coordination number (CN). The CN gives an idea of clones of packing of an atom.

The Distance of Neighboring Atom and Atomic Radius

The distance of the neighboring atom is measured from the center of an atom. The atomic radius is the half of the distance of the neighboring atom, as shown in Figure 10.15.

Packing Factor: The packing factor (PF) is the ratio of a unit cell's volume to the volume of atoms that make up that unit cell.

$$\text{Atomic Packing Factor (APF/PF)} = \frac{\text{Volume of atoms in unit cell}}{\text{Volume of unit cell}},$$

or $\text{APF} = \dfrac{\text{No. of atoms in unit cell} \times \text{Volume of atom}}{\text{Volume of unit cell}}.$

10.11 CRYSTAL SYSTEM STRUCTURE

1. **Simple Cubic Structure (SCC)**

    The diagram of simple cubic structure is shown in Figure 10.16.

    a. **Number of Atoms per Unit Cell:**

        Each atom is placed at eight corners of a unit cell, which is shared by eight unit cells, as shown in Figure 10.17. Therefore, the number of atoms per unit cell is given by

        $N = 8 \times \dfrac{1}{8} = 1 \text{ atom}.$

    b. **Atomic Radius:** In a simple cubic cell, the atoms are in contact along the edges of the cube. The length of the size of the unit cell, known as the lattice constant, is denoted by "a," as shown in Figure 10.18. The atoms touch along the cube edges; therefore, the nearest neighbor distance is

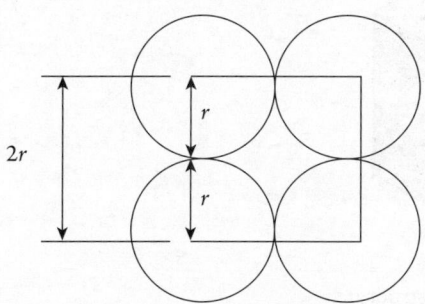

Fig. 10.15 Top view of simple cubic unit cell.

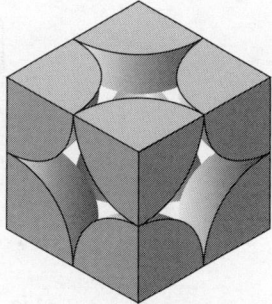

Fig. 10.16 Simple cubic structure.

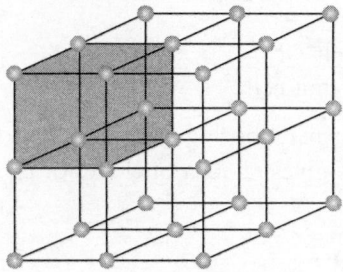

Fig. 10.17 Simple cubic unit cell in three-dimensional visualization.

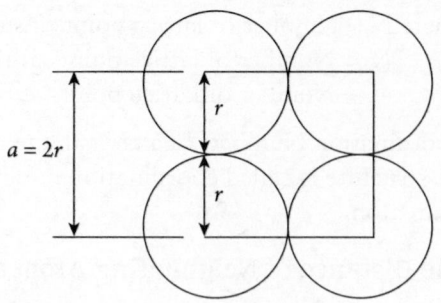

Fig. 10.18 Atomic radius view for simple cubic cell.

$a = 2r,$

$r = \dfrac{a}{2}.$

c. **Coordination Number:** Each corner atom interacts with two atoms vertically above and below it, as well as four atoms at the corners in horizontal planes. Therefore, as shown in Figure 10.17, there are six nearest neighbors that are equally spaced apart, and each is located at a distance from that neighbor of "a." Therefore, the CN is six.

d. **Packing Factor:** The PF is the ratio of the volume of the atoms occupied by a unit cell to the volume of the unit cell.

$$PF = \dfrac{\text{No. of atoms in unit cell} \times \text{Volume of atom}}{\text{Volume of unit cell}}.$$

$$PF = \dfrac{1 \times \dfrac{4}{3}\pi r^3}{a^3} = \dfrac{\pi}{6} = 0.52 \quad [\because a = 2r],$$

or the percentage of PF = 52%.

Therefore, SCC is a loosely packed structure. Polonium is the only element that exhibits this structure.

2. **Body-Centered Cubic (BCC) Structure:**

The diagram of a body-centered cubic structure is shown in Figure 10.19.

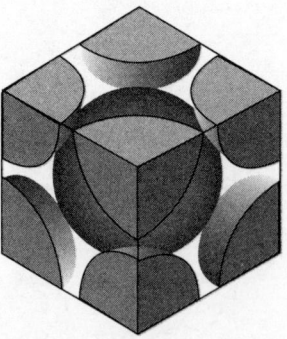

Fig. 10.19 Body-centered structure.

Crystal Structure

a. **Number of Atoms per Unit Cell:**
   Each atom is placed at eight corners of a unit cell, which is shared by eight unit cells. One atom is placed at the center of the unit cell, which is shared by the unit cell only, as shown in Figure 10.21. Therefore, the number of atoms per unit cell is given by
   $$N = 8 \times \frac{1}{8} + 1 = 2 \text{ atoms.}$$

b. **Atomic Radius:** In a BCC structure, the body-centered atom touches eight corner atoms; then, from Figure 10.20,
   $$(AF)^2 = (AC)^2 + (FC)^2$$
   $$(AF)^2 = (AB)^2 + (BC)^2 + (FC)^2$$
   $$(4r)^2 = (a)^2 + (a)^2 + (a)^2 = 3a^2$$
   $$a = \frac{4}{\sqrt{3}} r$$
   $$r = \frac{\sqrt{3}}{4} a.$$

c. **Coordination Number:** Its CN is the quantity of nearby atoms around the center atom. A single atom from the body center and one from each of the unit cell's eight corners are present in the unit cell. It is the body-centered atom alone that is closest to the corner atom. Accordingly, Figure 10.20 depicts of the nearest neighbors of the CN, showing that there are eight of them.

d. **Packing Factor:** The PF is the ratio of the volume of the atoms occupied by a unit cell to the volume of the unit cell.
   $$PF = \frac{\text{No. of atoms in unit cell} \times \text{Volume of atom}}{\text{Volume of unit cell}}.$$
   $$PF = \frac{2 \times \frac{4}{3} \pi r^3}{a^3} = 0.68 \qquad \left[\because a = \frac{4}{\sqrt{3}} r \right]$$
   or the percentage of PF = 68%.

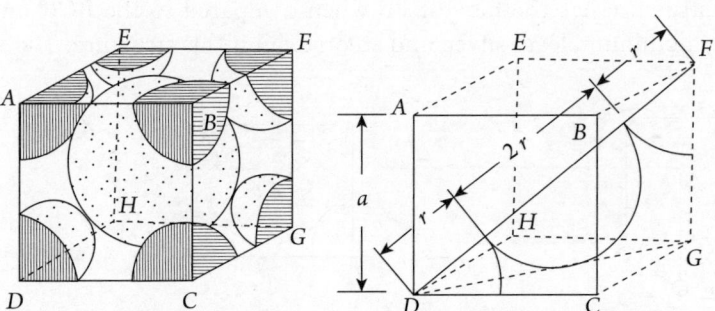

Fig. 10.20 Body-centered cubic cell and its atomic radius view.

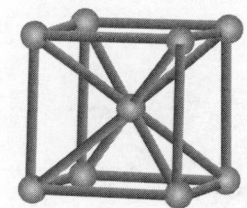

Fig. 10.21 Atomic location in body-centered cubic unit cell.

Therefore, the PF of BCC > SCC. Thus, BCC is a not a loosely packed structure. Tungsten, sodium, iron and chromium exhibit this structure.

3. **Face-Centered Cubic Structure:**

   The diagram of a face-centered cubic structure structure is shown in Figure 10.22.

   a. **Number of Atoms per Unit Cell:**
      Each atom is placed at eight corners of a unit cell, which is shared by eight unit cells. Six atoms are face-centered shared by the two unit cell. Therefore, the number of atoms per unit cell is given by

      $$N = 8 \times \frac{1}{8} + 6 \times \frac{1}{2} = 4 \text{ atoms.}$$

   b. **Atomic Radius:** In an FCC structure, atoms are in contact along the face diagonals, as shown in Figure 10.23.

      $$(AC)^2 = (AB)^2 + (BC)^2$$
      $$(4r)^2 = (a)^2 + (a)^2 = 2a^2$$
      $$a = 2\sqrt{2}r$$
      $$r = \frac{a}{2\sqrt{2}}.$$

   c. **Coordination Number:** Any corner atom's closest neighbors are the unit cell's face-centered atoms. Any corner atom has four of these atoms in its own plane, four more in the plane above it, and four more in the plane below it. Thus, the number of nearest neighbors is 12, as shown in Figure 10.23—that is, the CN is 12.

   d. **Packing Factor:** The PF is the ratio of the volume of the atoms occupied by a unit cell to the volume of the unit cell.

      $$PF = \frac{\text{No. of atoms in unit cell} \times \text{Volume of atom}}{\text{Volume of unit cell}}.$$

      $$PF = \frac{4 \times \frac{4}{3}\pi r^3}{a^3} = 0.74 \qquad \left[\because a = 2\sqrt{2}r\right],$$

      or the percentage of PF = 74%.

      Therefore, the FCC structure has the highest PF when compared to the BCC and SCC structure. Copper, aluminum, lead, silver, and so on exhibit this structure.

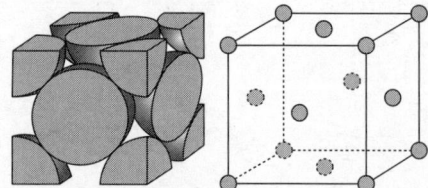

Fig. 10.22 Face-centered structure.

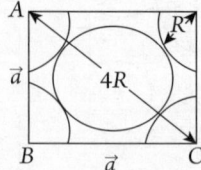

Fig. 10.23 Atomic radius view for face-centered unit cell.

10.12 SOME IMPORTANT CRYSTAL STRUCTURE

1. **NaCl Crystal Structure:** The NaCl crystal has an FCC structure, with Na⁺ and Cl⁻ ions as the basis, as shown in Figure 10.24. A unit cell of the NaCl lattice is shown in Figure 10.25. Both the cube's center and its four corners are occupied by Na⁺ ions. In terms of relative displacement along each axis, the chlorine ions are spaced halfway from the unit cell's border. Therefore, the FCC Na⁺ and Cl⁻ sublattices can be regarded as the constituents of the NaCl crystal. Since there are eight corners on each cell, each corner is shared by eight cells.

    Thus, an ion at the corner of the cell is shared by eight cells. So, only one-eighth ion belongs to any one cell. Similarly, an ion at the center of the faces is shared by two cells. So, only half of an ion belongs to any one cell. Therefore, it has $\left(8 \times \dfrac{1}{8} + 6 \times \dfrac{1}{2}\right) = 4$ ions of one kind (Na⁺) and similarly four ions of the other kind (Cl⁻). Thus, there are four Na⁺–Cl⁻ ions pair per unit cell. There are four molecules of NaCl in each unit cube, with ions in the positions

    Na: $(0,0,0); \left(\dfrac{1}{2},\dfrac{1}{2},0\right); \left(\dfrac{1}{2},0,\dfrac{1}{2}\right); \left(0,\dfrac{1}{2},\dfrac{1}{2}\right).$

    Cl: $\left(\dfrac{1}{2},\dfrac{1}{2},\dfrac{1}{2}\right); \left(0, 0, \dfrac{1}{2}\right); \left(0,\dfrac{1}{2},0\right); \left(\dfrac{1}{2},0,0\right).$

Coordination Number: Each Na ion has six chlorine ions as its nearest neighbors. Similarly, each Cl ion has six Na ions as its nearest neighbors, as shown in Figure 10.25. Hence, the CN of NaCl for the opposite kind of ions is six and the nearest neighbor distance is a/2. There will be no change in the crystal structure if the positions of the Na and Cl atoms are interchanged.

Calculation of Distance between Na and Cl Atoms

The molecular weight of NaCl = 35.5 + 23 = 58.5 gm.

The mass of each molecule = $58.5/N_A = 58.5/6.023 \times 10^{23}$.

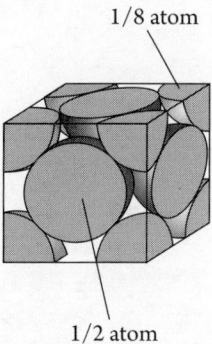

Fig. 10.24 NaCl crystal structure.

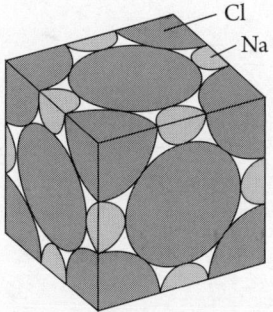

Fig. 10.25 Atomic location in NaCl unit cell.

Mass of four molecules = $4 \times \dfrac{5.85}{6.023 \times 10^{23}}$.

Let "a" be the length of the unit cell. Then, the volume of the unit cell is a^3. The mass of each unit cell of NaCl can be calculated by making use of the density, which is 2.18 gm/cm³.

So, the mass of each unit cell = $a^3 \times 2.18$ gm.

Thus, we can write

$$4 \times \dfrac{5.85}{6.023 \times 10^{23}} = a^3 \times 2.18.$$

$$a = 5.63 \times 10^{-8} \text{ cm} = 5.63 \text{Å}.$$

This is the distance between two atoms of the same kind. Thus, the distance between Na and Cl will be

$$\dfrac{a}{2} = 2.815 \text{Å}.$$

1. **Diamond:** The diamond structure is a combination of two interpenetrating FCC sub-lattices. One sub-lattice (say x) has its origin at $(0, 0, 0)$. The other sub-lattices (say y) have their origin quarter of the way along the body diagonal, that is, at the point $\left(\dfrac{a}{4}, \dfrac{a}{4}, \dfrac{a}{4}\right)$.

    The diamond crystal structure is loosely packed, since each atom has only four nearest neighbors, as shown in Figure 10.26.

Number of Atoms per Unit Cell

In the unit cell of diamond, there are eight atoms at the corner of the unit cell. Furthermore, there are six face-centered atoms and four atoms located inside the unit cell. Therefore, the total number of atoms per unit cell is given by

$$N = \left(8 \times \dfrac{1}{8} + 6 \times \dfrac{1}{2}\right) + 4 = 8 \text{ atoms}.$$

Atomic Radius

From Figure 10.27, we can obtain

$$(XZ)^2 = \left(\dfrac{a}{4}\right)^2 + \left(\dfrac{a}{4}\right)^2 = 2\left(\dfrac{a}{4}\right)^2$$

$$XZ = \dfrac{a}{4}\sqrt{2}$$

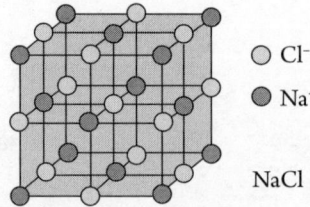

Fig. 10.26 Diamond structure.

Crystal Structure

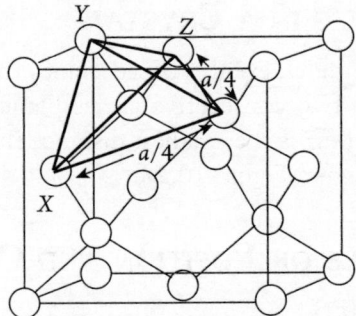

Fig. 10.27 Atomic radius view for face-centered cubic cell.

$$\therefore (XY)^2 = 2\left(\frac{a}{4}\right)^2 + \left(\frac{a}{4}\right)^2 = 3\left(\frac{a}{4}\right)^2$$

$$XY = \frac{a}{4}\sqrt{3}.$$

However, $XY = 2r$, where "r" is the atomic radius and "a" is the lattice constant.

Therefore, $r = \frac{\sqrt{3}}{8}a \Rightarrow a = \frac{8}{\sqrt{3}}r.$

Coordination Number: The directed covalent bonds that connect the carbon atoms in the diamond crystal structure contribute to the coordination number. In a tetrahedral structure, each carbon atom associates with four other carbon atoms at each of the four corners of the cube. Each atom makes covalent connections with its four closest neighbors in a diamond lattice. As a result, the diamond crystal's CN is 4.

Packing Factor: The PF is the ratio of the volume of the atoms occupied by a unit cell to the volume of the unit cell.

$$PF = \frac{\text{No. of atoms in unit cell} \times \text{Volume of 1 atom}}{\text{Volume of unit cell}}.$$

$$PF = \frac{8}{a^3} \times \frac{4}{3}\pi r^3 = \frac{\sqrt{3}}{16}\pi = 0.34.$$

The percentage of PF = 34%.

10.13 Calculation of Lattice Constant

Let us consider the lattice constant of unit cell is "a." Then, the volume of the unit cell is a^3. If ρ is the density of the material, then the mass of the unit cell is ρa^3. If M is the molecular weight of the molecules, N_A is Avogadro's number, and "n" is the number of molecules (lattice points) per unit cell, then the mass of the molecules in the unit cell is $\left(\frac{nM}{N_A}\right)$.

Thus, the lattice constant is given by

$$a = \left(\frac{nM}{N_A \rho}\right)^{\frac{1}{3}}.$$

10.14 LATTICE PLANES IN A CRYSTAL

A stack of atoms flowing on parallel, evenly spaced planes through lattice points makes up a crystal. The spacing between two consecutive lattice planes is known as the interplanar spacing, indicated by the letter "d," is the smallest distance that may exist between any two adjacent planes in a set, as shown in Figure 10.28.

10.15 MILLER INDICES OR POSITION AND ORIENTATION OF LATTICE PLANES

The position and orientation of lattice planes in a crystal are specified by a system of indices. These indices of lattice planes are known as Miller indices and denoted by h, k, l or as (hkl).

The three smallest numbers in a crystal miller matrix that have the same ratios to one another as the reciprocals of the plane's intercepts on the three crystal axes are the plane's Miller indices. The three numbers h, k, and l are referred to as the Miller indices of that plane or any other plane that is parallel to it. The plane is specified as (hkl), as shown in Figure 10.30.

To determine the Miller indices, the procedure is as follows:

1. Choose a plane that does not pass through the origin at $(0, 0, 0)$.
2. Find the intercepts of the plane on the three crystal axes (OX, OY, OZ). Let it be pa, qb, and rc, where a, b, and c are corresponding lengths of primitives and p, q, and r may be integer or fraction.
3. Take reciprocals of these intercepts, that is, the reciprocal of the number p, q, r as $\frac{1}{p}, \frac{1}{q}, \frac{1}{r}$.
4. Reduce the reciprocals to the smallest integers having the same ratio. The smallest possible integer is given by
$$h : k : l = \frac{1}{p} : \frac{1}{q} : \frac{1}{r}.$$

For example: Let the intercepts are $2a$, $2b$, and $3c$ on X, Y, and Z axes, respectively, as shown in Figure 10.29:

$\therefore p = 2, q = 2, l = 3.$

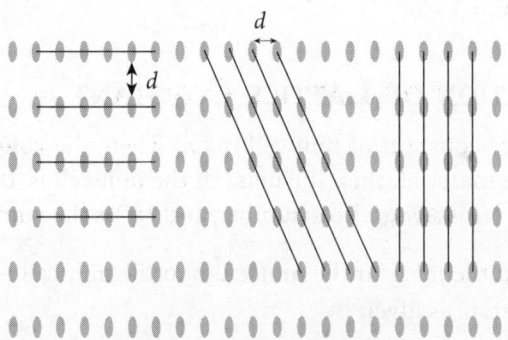

Fig. 10.28 Lattice planes.

Crystal Structure

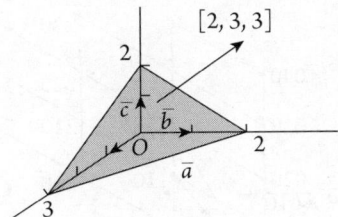

Fig. 10.29 Example intercepts.

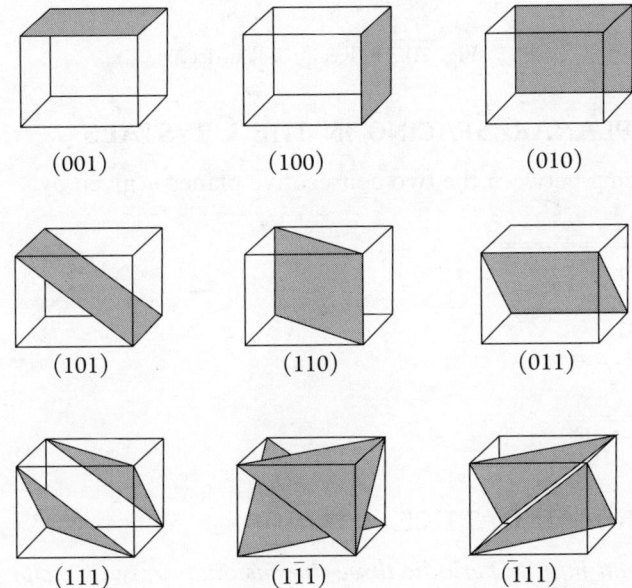

Fig. 10.30 Different lattice planes with Miller indices.

$$\therefore h:k:l = \frac{1}{2}:\frac{1}{2}:\frac{1}{3} = 3:3:2.$$

$$\therefore h = 3, k = 3, l = 2.$$

$$(hkl) = (332).$$

Some Important Features of Miller Indices

1. For an intercept at infinity, the corresponding index is zero.

   $$\frac{1}{p}:\frac{1}{q}:\frac{1}{r} = h:k:l = 1:0:0.$$

2. If the plane cuts axes on the negative side of the origin, the corresponding index is negative.

3. The indices do not define a particular plane but a set of parallel planes.

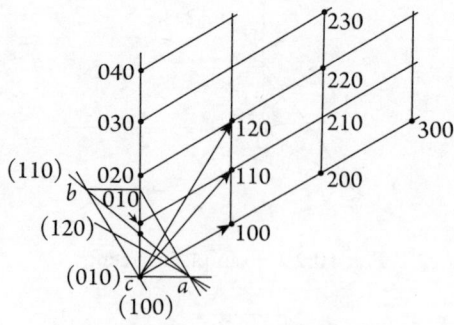

Fig. 10.31 Reciprocal lattice.

10.16 Interplanar Spacing in the Crystals

The interplanar spacing between the two consecutive planes is given by

$$d_{hkl} = \frac{1}{\sqrt{\left(\frac{h}{a}\right)^2 + \left(\frac{k}{b}\right)^2 + \left(\frac{l}{c}\right)^2}}.$$

For cubic crystals, $a = b = c$,

$$\therefore d_{hkl} = \frac{1}{\sqrt{h^2 + k^2 + l^2}}.$$

10.17 Reciprocal Lattice

Reciprocal lattice is an infinite periodic three-dimensional array of reciprocal lattice points whose spacing varies inversely as the distance between the planes in the direct lattice of the crystal. Reciprocal lattice is shown in Figure 10.31.

Properties of Reciprocal Lattice

1. The reciprocal of the reciprocal lattice is the direct lattice.
2. The primitive vectors of a direct lattice have dimensional length, while the primitive vectors in the reciprocal lattice have the dimensions of the inverse of length.
3. Diffraction pattern of a crystal is a map of the reciprocal lattice of the crystal, while the microscopic image is the map of direct lattice.

Solved Problems

Ex. 1: Calculate the distance between the two nearest copper atoms in the FCC structure. Given the density and atomic weight of copper are 8.96 gm/cm³ and 63.5 respectively. The Avogadro constant is 6.02×10^{23} /gm-mole.

Crystal Structure

Solution:

The lattice constant a is given as

$$a = \left(\frac{nM}{N\rho}\right)^{1/3} = \left(\frac{4 \times 63.5}{6.02 \times 10^{26} \times 8960}\right)^{1/3} = 3.41 \times 10^{-8} \text{ cm} = 3.41 \text{Å}.$$

The distance between the two nearest copper atoms in the FCC structure is $a/\sqrt{2}$. Therefore,

$$d = \frac{a}{\sqrt{2}} = \frac{3.61}{\sqrt{2}} = 2.55 \text{Å}$$

Ex. 2: Calculate the lattice parameter of NaCl crystal of density 2189 Kgm^{-3} and the Avogadro constant is 6.02×10^{23} /gm-mole.

Solution:

The lattice parameter for NaCl crystal is given as

$$a = \left(\frac{nM}{N\rho}\right)^{1/3} = \left(\frac{4 \times 58.5}{6.02 \times 10^{26} \times 2189}\right)^{1/3} = 5.61 \times 10^{-10} \text{ m} = 5.61 \text{Å}.$$

Ex. 3: Calculate the lattice parameter of iron if it crystallizes in an FCC structure. The density and atomic weight of iron are 7870 Kgm^{-3} and 55.8 respectively.

Solution:

The FCC lattice has $n = 4$; therefore,

$$a = \left(\frac{nM}{N\rho}\right)^{1/3} = \left(\frac{4 \times 55.8}{6.02 \times 10^{26} \times 7870}\right)^{1/3} = 2.86 \times 10^{-10} \text{ m} = 2.86 \text{Å}.$$

Ex. 4: In a crystal, a lattice plane cuts intercepts of a, $2b$, and $3c$ along the three axes, where $\vec{a}$, $\vec{b}$, and $\vec{c}$ are the primitive vectors of the unit cell. Determine the Miller indices of the given plane.

Solution:

According to the law of rational indices, we have

$$a : 2b : 3c = \frac{a}{h} : \frac{b}{k} : \frac{c}{l},$$

where h, k, l are the Miller indices. Therefore,

$$\frac{1}{h} : \frac{1}{k} : \frac{1}{l} = 1 : 2 : 3$$

$$h : k : l = \frac{1}{1} : \frac{1}{2} : \frac{1}{3} = 6 : 3 : 2.$$

Therefore, the Miller indices of the plane are (6 3 2).

Ex. 5: Calculate the Miller indices of a plane in an orthorhombic crystal which cuts the intercepts at $3a, -2b, 3c/2$ along the three axes.

Solution:

According to the law of rational indices, we have

$$a : 2b : 3c = \frac{a}{h} : \frac{b}{k} : \frac{c}{l},$$

where h, k, l are the Miller indices. Therefore,

$$3a : -2b : 3c/2 = \frac{a}{h} : \frac{b}{k} : \frac{c}{l}$$

$$3 : -2 : 3/2 = \frac{1}{h} : \frac{1}{k} : \frac{1}{l}$$

$$h : k : l = \frac{1}{3} : -\frac{1}{2} : \frac{2}{3}$$

$$h : k : l = 2 : -3 : 4.$$

Therefore, the Miller indices of the plane are $(2\bar{3}2)$.

Ex. 6: Determine the Miller indices of the plane that intercepts the three axes in the ratio $a:3b:4c$, where a, b, c are primitive lattice vectors.

Solution:

According to the law of rational indices, we have

$$a : 2b : 3c = \frac{a}{h} : \frac{b}{k} : \frac{c}{l}.$$

On comparing with the given ratio of intercepts, we get

$$a : 3b : 4c = \frac{a}{h} : \frac{b}{k} : \frac{c}{l}$$

$$1 : 3 : 4 = \frac{1}{h} : \frac{1}{k} : \frac{1}{l}$$

$$h : k : l = \frac{1}{1} : \frac{1}{3} : \frac{1}{4}$$

$$h : k : l = 12 : 4 : 3$$

Therefore, the Miller indices of the plane are $(12\ 4\ 3)$.

Ex. 7: Calculate the interplanar spacing of a lattice plane in a simple cubic lattice with edge 2Å that cuts the axis in the intercept ratio 3:4:5.

Solution:

The relationship between interplanar spacing and the Miller indices is given by

$$d = \frac{1}{\sqrt{\left(\frac{h}{a}\right)^2 + \left(\frac{k}{b}\right)^2 + \left(\frac{l}{c}\right)^2}}$$

For cubic lattice structure, $a = b = c = 2$; therefore,

$$d = \frac{2}{\sqrt{h^2 + k^2 + l^2}}$$

The Miller indices can be obtained as

$$3 : 4 : 5 = \frac{a}{h} : \frac{b}{k} : \frac{c}{l}$$

$$3 : 4 : 5 = \frac{2}{h} : \frac{2}{k} : \frac{2}{l}$$

Crystal Structure

$$h:k:l = \frac{2}{3}:\frac{2}{4}:\frac{2}{5}$$

$$h:k:l = 40:30:24$$

Now, the interplanar spacing of a lattice plane is given by

$$d = \frac{2}{\sqrt{40^2 + 30^2 + 24^2}} = \frac{2}{\sqrt{2976}} \text{ Å}.$$

Ex. 8: Determine the glancing angle of a plane (1 1 0) of a simple cubic crystal with lattice constant 2.81Å corresponding to the second order diffraction maximum for X-rays of wavelength 0.71Å.

Solution:
The distance between successive lattice planes is given by

$$d_{hkl} = \frac{a}{\sqrt{h^2 + k^2 + l^2}} \Rightarrow \frac{2.81}{\sqrt{1^2 + 1^2 + 0^2}} \Rightarrow \frac{2.81}{\sqrt{2}} \text{ Å}.$$

According to Bragg's equation for X-ray diffraction, we have

$$2d_{hkl} \sin\theta = n\lambda$$

$$\sin\theta = \frac{n\lambda}{d_{110}} = \frac{2 \times 0.71}{2 \times 1.99} = 0.357 \Rightarrow \theta = \sin^{-1}(0.357) = 21°.$$

Thus, the glancing angle of a plane (1 1 0) of a simple cubic crystal is $21°$.

Previous Year Questions (University Examination)

1. Define (a) space lattice, (b) basis, and (c) crystal structure.
2. Define primitive and unit cell of a space lattice.
3. Write the names of seven crystal systems.
4. Discuss four different types of possible unit cells in a lattice system.
5. Define simple cubic, BCC, and FCC cell.
6. What is Bravais space lattice?
7. What do you mean by atomic radii in a crystal?
8. Define coordination number.
9. What do you understand by APF in a lattice?
10. Show that APF for an FCC lattice is $\pi(2)^{1/2}/6$.
11. What are Miller indices?
12. Write some important features of Miller indices.
13. Define crystal planes.
14. Describe inter-planar spacing in a crystal structure.
15. Derive a relation between interplanar distance and cube edge.
16. Derive a relation between lattice constant and density of a crystal material.

17. What is reciprocal lattice?
18. Give two properties of reciprocal lattice.
19. What are soft and hard X-rays?
20. Distinguish between continuous and characteristic X-rays.
21. Briefly describe diffraction of X-rays by crystal planes.
22. What is Bragg's law?
23. What are the outcomes of Laue's X-rays diffraction experiment?
24. What are the practical applications of X-rays?
25. What was the Bragg's explanation about formation of Laue's spots in X-ray diffraction.
26. Give some important properties of X-rays.

Previous Year Questions (University Examination): Long Questions

1. What do you mean by unit cell? Find the number of atoms per unit cell in SC, BCC, and FCC lattices with suitable diagrams.
2. What exactly does "space lattice" mean? Describe the seven crystal systems. List the several kinds of lattices used in the cubic system, and explain each one them with examples.
3. What is crystal structure? Explain its types.
4. Define coordination number. Compute the same for simple cubic, BCC, and FCC lattices. Also, find the distance between nearest neighbors in these three.
5. Define atomic packing density or PF. Calculate PFs for SC, BCC, and FCC.
6. Describe the crystal structure of NaCl crystal. Give its main features. Explain with diagram how the lattice is FCC while its coordination number is that of simple cubic lattice.
7. Describe the crystal structure of diamond, and calculate the number of carbon atoms per unit cell.
8. What are Miller indices? How are they obtained?
9. What are lattice planes of a crystal? How are they represented in terms of Miller indices? Sketch the plane (100), (010), (110), and (111) planes in a cubic crystal.
10. What do you mean by interplanar spacing or distance in a crystal structure? Derive a relation between interplanar distance and cube edge.
11. Deduce formula for the distance between two adjacent planes of a simple cubic lattice. Show that in simple cubic lattice the spacing between successive lattice planes (100), (110), and (111) is in the ratio of 1:0.71:0.58.
12. Show that for a cubic lattice, the lattice constant "a" is given by

    $a = (nM/N\rho)^{1/3}$,

    where n is the no. of molecules per unit cell, M is the molecular weight, N is Avogadro's number, and ρ is the density.
13. What is a reciprocal lattice? Explain its importance and give some of its important properties.

Crystal Structure

Previous Year Questions (University Examination): Numerical Questions

1. Iron at 20°C is a BCC with atoms of atomic radius 0.124 nm. Calculate the value of lattice constant "a" for the cube edge of an iron unit cell. Also, calculate the volume of the unit cell.
2. (a) Calculate the next nearest distance in the simple cubic lattice.

    (b) In a simple cubic lattice, find the ratio of nearest neighbors distance to the next neighbor distance.
3. Show that the maximum radius of the sphere that can adjust into the void at the body center of an FCC structure coordinated by the facial atoms is $0.414r$, where r is the radius of the atom.
4. Iron changes from BCC to FCC at 910°C, and correspondingly the atomic radius varies from 1.26 Å to 1.29 Å. Calculate the percent volume change during this structure change.
5. Assuming the atoms to be hard sphere, calculate the atomic PF for the BCC unit cell.
6. Prove that the packing fraction of simple cubic, BCC, and FCC crystals are, respectively, 52%, 68%, and 74%.
7. A substance with FCC lattice has density 6,250 kg/m^3 and molecular weight 60.2. Calculate the lattice constant "a."
8. The density of α iron is 7870 kg/m^3, and its atomic weight is 55.8. Given that iron crystallizes in BCC space lattice, deduce its lattice constant.
9. Copper has a density of 8.96 gm/cc and an atomic weight of 63.5. Calculate the distance between the two nearest copper atoms in the FCC structure. The Avogadro's number is 6.02×10^{23} per gram mole.
10. Copper has an FCC structure, and the radius of copper atom is 1.278 Å. Calculate its density.
11. Find the Miller indices of a group of parallel planes that intersect the X and Y axes in the ratio $3a:4b$ and are parallel to the Z axis, where a, b, and c are the lattice primitives.
12. Deduce the Miller indices of a plane that cuts off intercepts in the ratio $1a:3b:-2c$ along the three coordinate axes, where a, b, and c are the primitives.
13. The lattice constants of a crystal are 1.21 Å, 1.84 Å, and 1.97 Å. A plane with Miller indices $(1,-2,-3)$ intercepts the X axis at 2.42 Å. What are the intercepts along the Y and Z axes?
14. Find the Miller indices of the cubic crystal plane that intersects the position coordinates (1, 1/4, 0), (1, 1, 1/2), (3/4, 1, 1/4) and all coordinate axes.
15. Determine interplanar spacing of a lattice plane in a simple cubic lattice with edge 2 Å, which cuts the axis in the ratio 3:4:5.

Multiple Choice Questions

1. Atoms arranged regularly in a substance are referred to as
    - (a) Crystalline
    - (b) Amorphous
    - (c) Dielectric
    - (d) All of the above
2. The dimension of the crystal is
    - (a) 10^{-20} cm
    - (b) 10^{-4} cm
    - (c) 10^{-10} cm
    - (d) 10^{-23} m

3. The study of crystal structures can be performed with
   (a) 1–100 Å
   (b) 1000–6000 Å
   (c) 6000–9000 Å
   (d) All of the above

4. Three-dimensional crystals have an array of lattice points that is known as
   (a) Base
   (b) Space lattice
   (c) Crystal unit
   (d) Unit prize

5. Crystallization includes
   (a) Base and unit prize
   (b) Unit prize and lattice unit
   (c) Basis and lattice point
   (d) All of the above

6. The crystal's foundation is the
   (a) Periodicity and orientations of atoms
   (b) Irregular atoms
   (c) Atoms
   (d) Short-range atoms

7. A material that is crystalline has a base made up of more than two atoms. Such a base is known as
   (a) Quadatomic
   (b) Quantatomic
   (c) Multi-atomic
   (d) Hexa-atomic

8. In the case of a simple cubic lattice, the coordination number is
   (a) 21
   (b) 9
   (c) 6
   (d) 2

9. A cubic crystal does not contain which of the following Bravais lattices?
   (a) Primitive cubic
   (b) Body-centered
   (c) Face-centered
   (d) Base-centered

10. A diamond's packing percentage in its crystal structure is
    (a) 34.0%
    (b) 57.48%
    (c) 61.2%
    (d) 73.58%

Bibliography

1. Giacovazzo, C., et al. (2011). *Fundamentals of Crystallography*. Oxford University Press.
2. Sands, D. E. (1993). *Introduction to Crystallography*. Courier Corporation.
3. Hammond, C. (2015). *The Basics of Crystallography and Diffraction*, vol. 21. International Union of Crystal.
4. Mc Kie, D., and Mc Kie, C. (1986). *Essentials of crystallography*. OSTI.
5. Woolfson, M. M. (1997). *An Introduction to X-ray Crystallography*. Cambridge University Press.
6. Borchardt-Ott, W. (2011). *Crystallography: An Introduction*. Springer Science & Business Media.
7. Matthewman, J. C., Thompson, P., and Brown, P. J. (1982). "The Cambridge Crystallography Subroutine Library." *Journal of Applied Crystallography* 15, no. 2: 167–173.

Keys

1. (a) 2. (b) 3. (a) 4. (b) 5. (c) 6. (a) 7. (c) 8. (c) 9. (d) 10. (a)

CHAPTER 11

Wave Optics
Interference

Wave optics is the branch of modern physics in which the nature of light and its propagation are studied.

11.1 Interference

When two waves of the same frequency, having a constant phase difference between them, and traveling in the same medium are allowed to superimpose each other, there is a modification in the intensity pattern. This phenomenon is known as interference of light.

When the resultant amplitude at certain points is the sum of the amplitudes of the two waves, this interference is known as constructive interference.

When the resultant amplitude at certain points is the difference of the amplitudes of the two waves, this interference is known as destructive interference, as shown in Figure 11.1.

11.2 Coherent Sources

Two sources are said to be coherent if the waves emitted from them have a constant phase difference with time.

11.3 Theory of Interference

Let us consider two coherent sources S_1 and S_2 that are equidistant from source S. Let a_1 and a_2 be the amplitudes of the waves originated from source S_1 and S_2, respectively, as shown in Figure 11.2. Then the displacement y_1 from the source S is given by

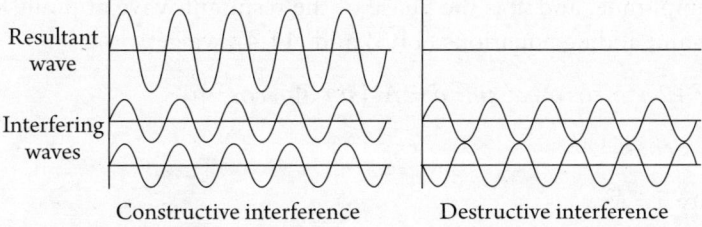

Fig. 11.1 Interference phenomena due to two individual waves.

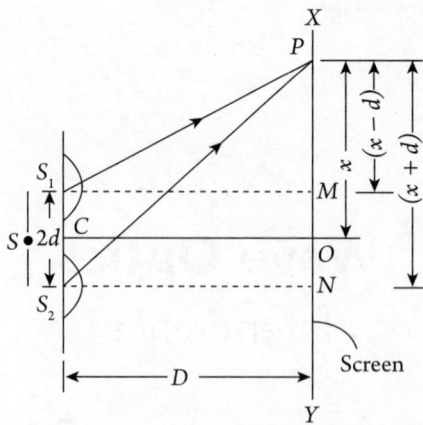

Fig. 11.2 Interference phenomenon by two coherent sources.

$y_1 = a_1 \sin \omega t.$ (11.1)

Similarly, y_2 will be

$y_2 = a_2 \sin(\omega t + \delta),$ (11.2)

where δ is the phase difference between the two waves.
Now, according to the law of superposition, the resultant wave is given by

$y = y_1 + y_2$

$y = a_1 \sin \omega t + a_2 \sin(\omega t + \delta)$

$y = a_1 \sin \omega t + a_2 \sin \omega t \cos \delta + a_2 \cos \omega t \sin \delta$

$y = (a_1 + a_2 \cos \delta) \sin \omega t + (a_2 \sin \delta) \cos \omega t.$

Let $a_1 + a_2 \cos \delta = A \cos \Phi$ (11.3)

$a_2 \sin \delta = A \sin \Phi.$ (11.4)

On substituting these values, the resultant wave is given by

$y = A \cos \Phi \sin \omega t + A \sin \Phi \cos \omega t$

$y = A \sin(\omega t + \Phi),$ (11.5)

where A is the amplitude, and Φ is the phase of the resultant wave at point P.
On squaring and adding equations (11.3) and (11.4), we get

$a_1^2 + a_2^2 \cos^2 \delta + 2 a_1 a_2 \cos \delta + a_2^2 \sin^2 \delta = A^2 (\sin^2 \Phi + \cos^2 \Phi)$

$a_1^2 + a_2^2 + 2 a_1 a_2 \cos \delta = A^2.$ (11.6)

Since intensity $I = A^2$,

$I = a_1^2 + a_2^2 + 2 a_1 a_2 \cos \delta.$ (11.7)

Wave Optics

On dividing equation (11.4) by equation (11.3), we get

$$\tan\phi = \frac{a_2 \sin\delta}{a_1 + a_2 \cos\delta}. \tag{11.8}$$

Conditions for Maximum and Minimum Intensities:

Case 1: When the phase difference is $\delta = 0, 2\pi, 4\pi, \ldots\ldots 2n\pi$ or even a multiple of π, the intensity is maximum because $\cos\delta = 1$ in equation (11.7). Therefore, the maximum intensity is given by

$$I_{max} = a_1^2 + a_2^2 + 2a_1 a_2$$

$$I_{max} = (a_1 + a_2)^2. \tag{11.9}$$

Since we know that path difference $= \dfrac{\lambda}{2\pi} \times$ phase difference,

$$\text{path difference} = \frac{\lambda}{2\pi} \times 2n\pi = (2n)\frac{\lambda}{2} = n\lambda \quad [\because \delta = 2n\pi]. \tag{11.10}$$

Where $n = 0, 1, 2,\ldots$. Thus, the intensity is maximum when the path difference between interfering beams is even multiple of half wavelength.

Case 2: When the phase difference is $\delta = \pi, 3\pi, \ldots\ldots(2n+1)\pi$ or odd multiple of π, the intensity is minimum because $\cos\delta = -1$ in equation (11.7). Therefore, the minimum intensity is given by

$$I_{min} = a_1^2 + a_2^2 - 2a_1 a_2$$

$$I_{min} = (a_1 - a_2)^2. \tag{11.11}$$

Since path difference $= \dfrac{\lambda}{2\pi} \times$ phase difference,

$$\text{path difference} = \frac{\lambda}{2\pi} \times (2n+1)\pi = (2n+1)\frac{\lambda}{2} \quad (12) \, [\because \delta = (2n+1)\pi].$$

Where $n = 0, 1, 2,\ldots$. Thus, the intensity is minimum when the path difference between the interfering beams is an odd multiple of half wavelength.

11.4 Energy Distribution and Conservation of Energy

The average energy at any point in the region of superposition is equal to the sum of intensity of individual waves. Mathematically, it can be given as

$$I_{avg.} = \frac{\int_0^{2\pi} I \, d\delta}{\int_0^{2\pi} d\delta}$$

$$I_{avg.} = \frac{1}{2\pi}\int_0^{2\pi} \left(a_1^2 + a_2^2 + 2a_1 a_2 \cos\delta\right) d\delta$$

$$I_{avg.} = \frac{1}{2\pi}[(a_1^2 \delta)_0^{2\pi} + (a_2^2 \delta)_0^{2\pi} + (2a_1 a_2 \sin\delta)_0^{2\pi}]$$

$$I_{avg.} = \frac{2\pi a_1^2 + 2\pi a_2^2}{2\pi} = a_1^2 + a_2^2$$

$$I_{avg.} = I_1 + I_2.$$

This means that the energy of the maxima is at the expense of energy of minima. Thus, the energy is conserved in the phenomena of interference.

11.5 Stokes' Treatment

When a ray is reflected from the denser medium, it suffers a phase change of π or a path difference of $\left(\dfrac{\lambda}{2}\right)$. However, when reflection takes place from a rarer medium, no phase change takes place.

11.6 Interference in Thin Film

When a thin film of a transparent material, such as oil drop, is spread over the surface of water and exposed to an extended light source, it appeared colored. A soap bubble appeared colored in sunlight. This is due to the phenomenon of interference in the thin film.

Case 1: Interference in Thin Film of Uniform Thickness

A. In Reflected Light

Let us consider a thin film of refractive index μ and thickness t. A ray AB of monochromatic light of wavelength λ is incident on the thin film. It is partly reflected from the upper surface of the film and partly refracted from the lower surface of the film, such that the two rays BC and EF interfere with each other in reflected light, as shown in Figure 11.3.

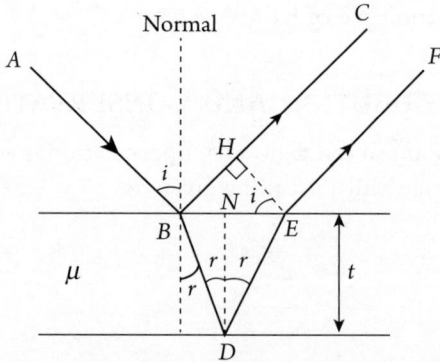

Fig. 11.3 Interference in thin film of uniform thickness in reflected light.

Wave Optics

The path difference between the two interfering beams is given by

Path difference = $(BD + DE)$ in film – BH in air

Path difference = $[(BD + DE) \times \mu] - [BH \times 1]$

Path difference = $(BD + DE)\mu - BH.$ \hfill (11.12)

Since from Figure 11.3, the $\angle BDN$ and $\angle NDE$ are equal, BD = DE. Thus, equation (11.12) becomes

Path difference = $2\mu BD - BH.$ \hfill (11.13)

From the right-angled triangle $\triangle BND$, we have

$$\frac{DN}{BD} = \cos r$$

$$\frac{t}{BD} = \cos r$$

$$BD = \frac{t}{\cos r}. \hfill (11.14)$$

From the right-angled triangle $\triangle BHE$, we have

$$\frac{BH}{BE} = \sin i.$$

$BH = BE \sin i = (BN + NE)\sin i$

Since $BN = NE$, we get

$BH = 2BN \sin i.$ \hfill (11.15)

Again, from the right-angled triangle $\triangle BND$, we have

$$\frac{BN}{DN} = \tan r$$

$BN = DN \tan r = t \tan r.$

On substituting the value of BN in equation (11.15), we get

$BH = 2t \tan r \sin i.$ \hfill (11.16)

On substituting the value of BD and BH in equation (11.13), we have

$$\text{Path difference} = 2\mu \frac{t}{\cos r} - 2t \tan r \sin i$$

$$\text{Path difference} = 2\mu \frac{t}{\cos r} - 2t \frac{\sin r}{\cos r} \sin i \times \frac{\sin r}{\sin r}$$

$$\text{Path difference} = 2\mu \frac{t}{\cos r} - 2\mu t \frac{\sin^2 r}{\cos r} \quad \left[\because \mu = \frac{\sin i}{\sin r}\right]$$

$$\text{Path difference} = 2\mu t \frac{(1 - \sin^2 r)}{\cos r}$$

Path difference = $2\mu t \cos r.$

According to Stokes' treatment, when a ray is reflected from the denser medium, it suffers a phase change of π or a path difference of $\left(\dfrac{\lambda}{2}\right)$. Thus, the total path difference is given by

$$\text{Path difference} = 2\mu t \cos r + \dfrac{\lambda}{2}.$$

Condition for Maxima and Minima
i. The condition for maxima of interference in the thin film is given by

$$2\mu t \cos r + \dfrac{\lambda}{2} = n\lambda$$

$$2\mu t \cos r = (2n-1)\dfrac{\lambda}{2},$$

where $n = 1, 2, 3,\ldots$. For least thickness of the film, $n = 1$. The wavelength satisfies the above equation and causes maxima at that point on the thin film.

ii. The condition for minima of interference in the thin film is given by

$$2\mu t \cos r + \dfrac{\lambda}{2} = (2n+1)\dfrac{\lambda}{2}$$

$$2\mu t \cos r = n\lambda,$$

where $n = 1, 2, 3,\ldots$. For least thickness of the film, $n = 1$. The wavelength satisfies the above equation and causes minima at that point on the thin film.

B. In Transmitted Light

Let us consider a thin film of refractive index μ and thickness t. A ray AB of monochromatic light of wavelength λ is incident on the thin film. It is partly reflected from the upper surface of the film and partly refracted from the lower surface of the film, such that the two rays DR and MS interfere with each other in the transmitted light. Figure 11.4 illustrates the phenomena of interference in transmitted light.

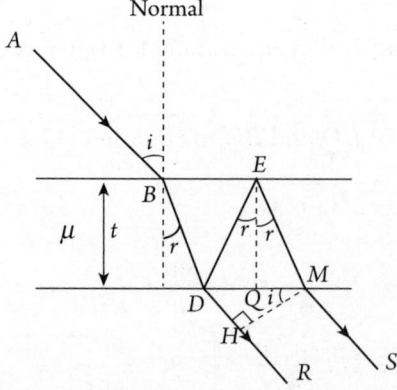

Fig. 11.4 Interference in thin film of uniform thickness in transmitted light.

Wave Optics

The path difference between the two interfering beams is given by

Path difference = (DE + EM) in film − DH in air

Path difference = $[(DE + EM) \times \mu] - [DH \times 1]$

Path difference = $(DE + EM)\mu - DH.$ (11.17)

Since, from Figure 1.4, the $\angle BDN$ and $\angle NDE$ are equal, DE = EM. Thus, equation (11.17) becomes

Path difference = $2\mu DE - DH.$ (11.18)

From right-angled triangle ΔDEQ, we have

$$\frac{EQ}{DE} = \cos r$$

$$\frac{t}{DE} = \cos r$$

$$DE = \frac{t}{\cos r}.$$ (11.19)

From the right-angled triangle ΔDMH, we have

$$\frac{DH}{DM} = \sin i$$

$DH = DM \sin i = (DQ + QM)\sin i.$

Since BN = NE, we get

$DH = 2DQ \sin i.$ (11.20)

Again from the right-angled triangle ΔDEQ, we have

$$\frac{DQ}{QE} = \tan r$$

$DQ = QE \tan r = t \tan r.$

On substituting the value of DQ in equation (11.20), we get

$DH = 2t \tan r \sin i.$ (11.21)

On substituting the value of DE and DH in equation (11.18), we have

Path difference = $2\mu \dfrac{t}{\cos r} - 2t \tan r \sin i$

Path difference = $2\mu \dfrac{t}{\cos r} - 2t \dfrac{\sin r}{\cos r} \sin i \times \dfrac{\sin r}{\sin r}$

Path difference = $2\mu \dfrac{t}{\cos r} - 2\mu t \dfrac{\sin^2 r}{\cos r}$ $\left[\because \mu = \dfrac{\sin i}{\sin r}\right]$

Path difference = $2\mu t \dfrac{(1 - \sin^2 r)}{\cos r}$

Path difference = $2\mu t \cos r.$ (11.22)

Condition for Maxima and Minima

i. The condition for maxima of interference in a thin film is given by

$$2\mu t \cos r + \frac{\lambda}{2} = (2n+1)\frac{\lambda}{2}$$

$$2\mu t \cos r = n\lambda,$$

where $n = 1, 2, 3, \ldots$. For least thickness of the film, $n = 1$. The wavelength satisfies the above equation causes maxima at that point on the thin film.

ii. The condition for minima of interference in the thin film is given by

$$2\mu t \cos r + \frac{\lambda}{2} = n\lambda$$

$$2\mu t \cos r = (2n-1)\frac{\lambda}{2},$$

where $n = 1, 2, 3, \ldots$. For least thickness of the film, $n = 1$. The wavelength satisfies the above equation and causes minima at that point on the thin film.

It is obvious that the conditions for transmitted light are opposite to those obtained with reflected light. Therefore, the interference pattern in the thin film is reflected, and transmitted light is complementary.

Case 2: Interference Due to Wedge-shaped Thin Film

A. In Reflected Light

Let us consider θ be the angle of wedge of a thin film with refractive index μ bounded by two surfaces OX and OX′, as shown in Figure 11.5. When a monochromatic light ray is incident on this wedge-shaped thin film, some part of light intensity is reflected from the upper surface of the film and partly reflected from the lower surface of the film such that

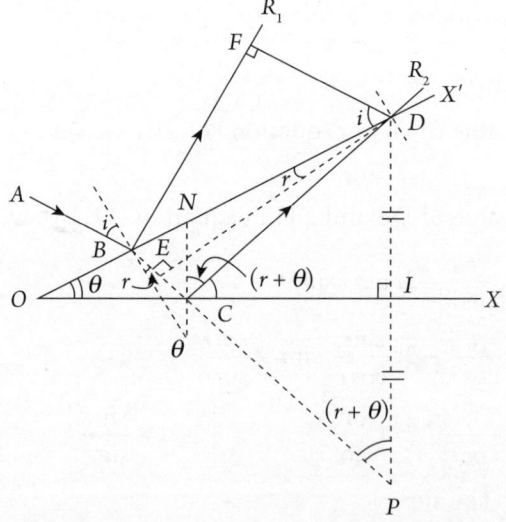

Fig. 11.5 Interference in a wedge-shaped thin film in reflected light.

Wave Optics

the two reflected beams from the film interfere to produce an interference pattern. The path difference between the two interfering beams is given by

Path difference = (BC + CD) in film − BF in air.

On solving the geometry, the path difference will be

Path difference = $2\mu t \cos(r + \theta)$.

According to Stokes' treatment, the total path of reflected light is given by

Path difference = $2\mu t \cos(r + \theta) + \dfrac{\lambda}{2}$.

Condition for Maxima and Minima

i. The condition for maxima of interference in a thin film is given by

$$2\mu t \cos(r + \theta) + \dfrac{\lambda}{2} = n\lambda.$$

$$2\mu t \cos(r + \theta) = (2n - 1)\dfrac{\lambda}{2},$$

where $n = 1, 2, 3, \ldots$. For least thickness of the film, $n = 1$. The wavelength satisfies the above equation and causes maxima at that point on the thin film.

ii. The condition for minima of interference in a thin film is given by

$$2\mu t \cos(r + \theta) + \dfrac{\lambda}{2} = (2n + 1)\dfrac{\lambda}{2}$$

$$2\mu t \cos(r + \theta) = n\lambda,$$

where $n = 1, 2, 3, \ldots$. For least thickness of the film, $n = 1$. The wavelength satisfies the above equation and causes minima at that point on the thin film.

B. In Transmitted Light

i. The condition for maxima of interference in a thin film is given by

$$2\mu t \cos(r + \theta) + \dfrac{\lambda}{2} = (2n + 1)\dfrac{\lambda}{2}$$

$$2\mu t \cos(r + \theta) = n\lambda,$$

where $n = 1, 2, 3, \ldots$. For least thickness of the film, $n = 1$. The wavelength satisfies the above equation and causes maxima at that point on the thin film.

ii. The condition for minima of interference in a thin film is given by

$$2\mu t \cos(r + \theta) + \dfrac{\lambda}{2} = n\lambda$$

$$2\mu t \cos(r + \theta) = (2n - 1)\dfrac{\lambda}{2},$$

where $n = 1, 2, 3, \ldots$. For least thickness of the film, $n = 1$. The wavelength satisfies the above equation and causes minima at that point on the thin film.

11.7 Fringe Width

The separation or distance between the two consecutive dark or bright fringes is called fringe width (w). The fringe width is given by

$$\omega = \frac{D\lambda}{2d}.$$

The condition for maxima in a wedge-shaped thin film in reflected light is given by

$$2\mu t \cos(r + \theta) = (2n - 1)\frac{\lambda}{2}. \tag{11.23}$$

From the figure,

$$\tan\theta = \frac{t}{x_n}.$$

Since θ is very small, $\tan\theta \approx \theta$. Then, we have

$$\theta = \frac{t}{x_n} \Rightarrow t = x_n \theta. \tag{11.24}$$

On substituting the value of t in equation (11.23), we have

$$2\mu x_n \tan\theta \; \cos(r + \theta) = (2n - 1)\frac{\lambda}{2}. \tag{11.25}$$

Similarly, the condition for $(n+1)^{th}$ fringe is given by

$$2\mu x_{n+1} \tan\theta \; \cos(r + \theta) = (2n + 1)\frac{\lambda}{2}. \tag{11.26}$$

On subtracting equation (11.26) from equation (11.25), we have

$$2\mu(x_{n+1} - x_n)\tan\theta \; \cos(r + \theta) = \lambda$$

$$\omega = (x_{n+1} - x_n) = \frac{\lambda}{2\mu \tan\theta \cos(r + \theta)}. \tag{11.27}$$

This is an expression of fringe width in a wedge-shaped thin film with an angle of wedge θ.

For normal incidence with very small angle of wedge, $\theta \approx 0$. Then, fringe width is given by

$$w = \frac{\lambda}{2\mu\theta}. \qquad [\because \tan\theta \approx 0 \text{ and } i = r = 0] \tag{11.28}$$

11.8 Newton's Ring

The formation of a Newton's ring is a special case of interference in a thin air film of variable thickness. When a plano-convex lens of large focal length is placed on a glass plate, a thin film of air is formed between the lower surface of the plano-convex lens and the upper surface of the glass plate. When a monochromatic light is allowed to fall normally on the film, circular fringes are observed. In reflected light, the center of the circular fringe is dark, followed by alternative bright and dark circular fringes, and vice versa, in transmitted light. **The fringes are circular because the thickness of the developed air film between the lower**

Wave Optics

surface of the plano-convex lens and the upper surface of the glass plate is symmetrical about the point of contact.

These circular fringes were first investigated by Newton and hence called "Newton's rings."

Experimental Arrangements
In Reflected Light
The Newton's ring experimental setup in reflected light is shown in Figure 11.6. If the extended monochromatic light source S is placed at the focus of convex lens L, then the rays become parallel, and the incidence on a plane glass plate G is inclined at an angle of 45^0. The rays are partly reflected from the plane glass plate and fall normally on a plano-convex lens of large focal length L_1 placed over a plane glass plate N. The air film is formed between the plane glass plate N and L_1 around the point of contact O. The interference takes place between the rays reflected from the lower surface of the plano-convex lens and the upper surface of the plane glass plate, which is viewed by microscope M.

Theory
The Newton's ring is the special case of a wedge-shaped thin film of air. Therefore, the total path difference between the two interfering beams is given by

$$\text{Path difference} = 2\mu t \cos(r+\theta) + \frac{\lambda}{2}.$$

Since the angle of wedge is very small, that is, $\theta \approx 0$, and for normal incidence $i = r = 0$, the total path difference is given by

$$\text{Path difference} = 2\mu t + \frac{\lambda}{2}.$$

Thus, the condition for bright ring or constructive interference is given by

$$2\mu t + \frac{\lambda}{2} = n\lambda$$

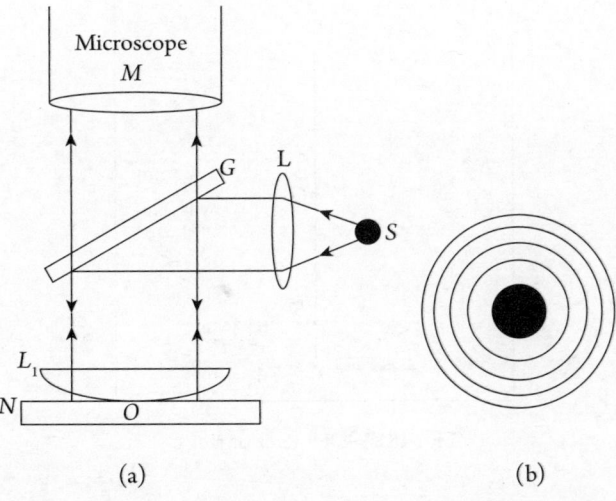

(a) (b)

Fig. 11.6 (a) Experimental setup for Newton's ring. (b) Newton's ring.

or $2\mu t = (2n-1)\dfrac{\lambda}{2}$, (11.29)

where $n = 1, 2, 3,...$ and so on.
Similarly, the condition for dark ring or destructive interference is given by

$2\mu t + \dfrac{\lambda}{2} = (2n+1)\dfrac{\lambda}{2}$

$2\mu t = n\lambda$, (11.30)

where $n = 1, 2, 3,...$ and so on.
At point of contact ($t = 0$), the path difference is equal to $\dfrac{\lambda}{2}$, which is the condition of minimum intensity; thus, at point of contact, a dark fringe will appear.

Diameter of Rings

Consider a plano-convex lens DAE is placed on a plane glass plate at point of contact A of radius of curvature R, as shown in Figure 11.7.

Hence, by the property of the circle,

$DB \times BE = BA \times CB$. (11.31)

Since $DB = BE = r$ (radius of the ring) equation (11.31) becomes

$r \times r = BA \times CB \quad [\because CB = 2R - t \text{ and } BA = t]$

$t(2R - t) = r^2$

$2Rt - t^2 = r^2$.

Since $t^2 \ll 2Rt$ (the radius of curvature is very large as compared to the thickness of the film), t^2 is neglected.

$2Rt = r^2$

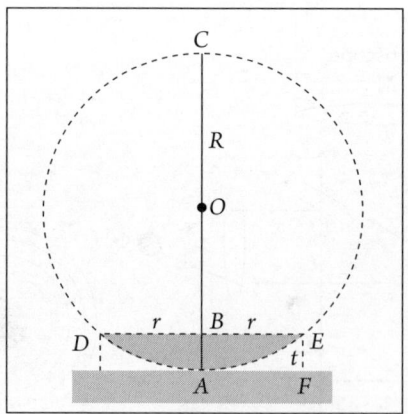

Fig. 11.7 Diameter of rings.

Wave Optics

$$t = \frac{r^2}{2R}. \tag{11.32}$$

This is a relation between the thickness of the film and radius of the ring.

a. Diameter of Bright Rings

The condition of a bright ring is given by

$$2\mu t = (2n-1)\frac{\lambda}{2}.$$

On substituting the value of t from equation (11.32),

$$2\mu \frac{r^2}{2R} = (2n-1)\frac{\lambda}{2}$$

$$r^2 = (2n-1)\frac{\lambda R}{2\mu}$$

$$\frac{D^2}{4} = (2n-1)\frac{\lambda R}{2} \quad \left[\because r^2 = \frac{D^2}{4}, \mu = 1\right]$$

$$D_n = \sqrt{2(2n-1)\lambda R}. \tag{11.33}$$

Hence, from equation (11.33), it is clear that the diameter of the bright ring is proportional to the square root of odd natural numbers.

b. Diameter of Dark Rings

The condition of dark ring is given by

$$2\mu t = n\lambda$$

$$2\mu \frac{r^2}{2R} = n\lambda$$

$$r^2 = \frac{n\lambda R}{\mu}$$

$$\frac{D^2}{4} = n\lambda R \quad \left[\because r^2 = \frac{D^2}{4}, \mu = 1\right]$$

$$D_n = \sqrt{4n\lambda R}. \tag{11.34}$$

Hence, from equation (11.34), it is clear that the diameter of the dark ring is proportional to the square root of natural numbers.

The diameter of $(n+1)^{th}$ dark ring is given by

$$D_{n+1} = \sqrt{4(n+1)\lambda R}.$$

So, the fringe width is given by

$$D_{n+1} - D_n = \sqrt{4(n+1)\lambda R} - \sqrt{4n\lambda R}$$

$$D_{n+1} - D_n = \sqrt{4\lambda R}\left[\sqrt{(n+1)} - \sqrt{n}\right]$$

Thus, for different values of n, we have:

For $n = 1$, $D_2 - D_1 = \sqrt{4\lambda R}\left[\sqrt{2} - \sqrt{1}\right]$.

For $n = 2$, $D_3 - D_2 = \sqrt{4\lambda R}\left[\sqrt{3} - \sqrt{2}\right]$.

For $n = 3$, $D_4 - D_3 = \sqrt{4\lambda R}\left[\sqrt{4} - \sqrt{3}\right]$.

For $n = 4$, $D_5 - D_4 = \sqrt{4\lambda R}\left[\sqrt{5} - \sqrt{4}\right]$.

Hence, from the above equations, it is clear that the spacing between the rings decreases with the increasing order of the rings, as shown in Figure 11.8.

11.9 Determination of Wavelength of Sodium Light Using Newton's Ring

The diameter for n^{th} dark ring is given by

$$D_n = \sqrt{4n\lambda R}. \tag{11.35}$$

The diameter for $(n+p)^{th}$ dark ring is given by

$$D_{n+p} = \sqrt{4(n+p)\lambda R}, \tag{11.36}$$

where p is an integer.

From, equation (11.35) and equation (11.36), we can write,

$$D_{n+p}^2 - D_n^2 = 4(n+p)\lambda R - 4n\lambda R$$

$$D_{n+p}^2 - D_n^2 = 4p\lambda R$$

$$\lambda = \frac{D_{n+p}^2 - D_n^2}{4pR}.$$

This is an expression for determining a monochromatic wavelength of light source using Newton's ring.

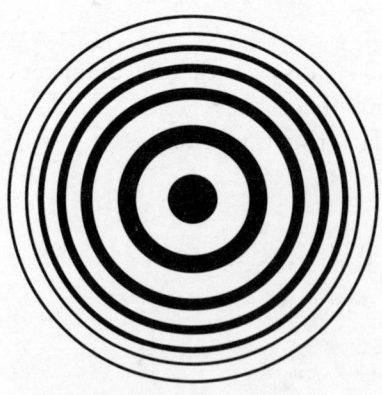

Fig. 11.8 Newton's ring pattern.

Wave Optics

11.10 Determination of Refractive Index of an Unknown Liquid

Let D_n and D_{n+p} be the corresponding diameters of nth and $(n+p)^{th}$ dark ring. Then for air film,

$$(D_{n+p}^2 - D_n^2)_{air} = 4p\lambda R. \tag{11.37}$$

Now, the above equation for an unknown liquid of refractive index μ is given by

$$(D_{n+p}^2 - D_n^2)_{liquid} = \frac{4p\lambda R}{\mu}. \tag{11.38}$$

On dividing equation (11.37) by equation (11.38), we get

$$\mu = \frac{(D_{n+p}^2 - D_n^2)_{air}}{(D_{n+p}^2 - D_n^2)_{liquid}}. \tag{11.39}$$

Hence, an unknown refractive index of a liquid can be determined by measuring the diameters of Newton's ring in air and in liquid.

11.11 Newton's Ring in Transmitted Light

The Newton's ring is the special case of a wedge-shaped thin film of air. Therefore, the total path difference between the two interfering beams in transmitted light is given by

Path difference = $2\mu t \cos(r + \theta)$.

Since the angle of wedge is very small, that is, $\theta \approx 0$, and for normal incidence $i = r = 0$, the total path difference is given by

Path difference = $2\mu t$.

Thus, the condition for bright ring or constructive interference is given by

$$2\mu t = n\lambda, \tag{11.40}$$

where $n = 1, 2, 3,...$ and so on.

Similarly, the condition for dark ring or destructive interference is given by

$$2\mu t = (2n-1)\frac{\lambda}{2}, \tag{11.41}$$

where $n = 1, 2, 3,...$ and so on.

At point of contact ($t = 0$), the path difference is equal to zero, which is the condition of maximum intensity; thus, at point of contact, a bright fringe will appear in transmitted light.

Case 1: When a Lower Surface of Plano-convex Lens Is Convex

The total thickness of air film at any point, as shown in Figure 11.9, is given by

$$t = t_1 + t_2. \tag{11.42}$$

If R_1 and R_2 be the radii of the curvature, then

$$t = \frac{r_n^2}{2R_1} + \frac{r_n^2}{2R_2}. \tag{11.43}$$

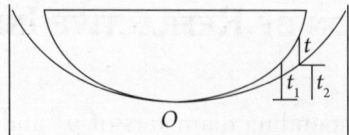

Fig. 11.9 Plano-convex lens placed on concave lens.

The condition for dark ring is given by

$2\mu t = n\lambda$

$2t = n\lambda \quad [\because \mu = 1].$

On substituting the value of t from equation (11.43), the condition for dark ring can be written as

$$r_n^2\left(\frac{1}{R_1}+\frac{1}{R_2}\right) = n\lambda \quad \left[\because r^2 = \frac{D^2}{4}\right]$$

$$\frac{D_n^2}{4}\left(\frac{1}{R_1}+\frac{1}{R_2}\right) = n\lambda$$

$$D_n^2 = \frac{4n\lambda}{\left(\dfrac{1}{R_1}+\dfrac{1}{R_2}\right)}. \tag{11.44}$$

The condition for bright ring is given by

$2\mu t = (2n-1)\dfrac{\lambda}{2}$

$2t = (2n-1)\dfrac{\lambda}{2}[\because \mu = 1].$

On substituting the value of t from equation (11.43), the condition for bright ring can be written as

$$r_n^2\left(\frac{1}{R_1}+\frac{1}{R_2}\right) = (2n-1)\frac{\lambda}{2}$$

$$\frac{D^2}{4}\left(\frac{1}{R_1}+\frac{1}{R_2}\right) = (2n-1)\frac{\lambda}{2} \quad \left[\because r^2 = \frac{D^2}{4}\right]$$

$$D_n^2 = \frac{2(2n-1)\lambda}{\left(\dfrac{1}{R_1}+\dfrac{1}{R_2}\right)}. \tag{11.45}$$

Case 2: When a Lower Surface of Plano-convex Lens Is Concave

The total thickness of air film is given by

$t = t_1 - t_2.$ \tag{11.46}

Wave Optics

If R_1 and R_2 be the radii of the curvature, then

$$t = \frac{r_n^2}{2R_1} - \frac{r_n^2}{2R_2}. \tag{11.47}$$

The condition for dark ring is given by

$2\mu t = n\lambda$

$2t = n\lambda \quad [\because \mu = 1]$.

On substituting the value of t from equation (11.47), the condition for dark ring can be written as

$$r_n^2\left(\frac{1}{R_1} - \frac{1}{R_2}\right) = n\lambda \quad \left[\because r^2 = \frac{D^2}{4}\right]$$

$$\frac{D_n^2}{4}\left(\frac{1}{R_1} - \frac{1}{R_2}\right) = n\lambda$$

$$D_n^2 = \frac{4n\lambda}{\left(\frac{1}{R_1} - \frac{1}{R_2}\right)}. \tag{11.48}$$

The condition for bright ring is given by

$2\mu t = (2n-1)\dfrac{\lambda}{2}$

$2t = (2n-1)\dfrac{\lambda}{2} [\because \mu = 1]$.

On substituting the value of t from equation (11.47), the condition for bright ring can be written as

$$r_n^2\left(\frac{1}{R_1} - \frac{1}{R_2}\right) = (2n-1)\frac{\lambda}{2}$$

$$\frac{D^2}{4}\left(\frac{1}{R_1} - \frac{1}{R_2}\right) = (2n-1)\frac{\lambda}{2} \quad \left[\because r^2 = \frac{D^2}{4}\right]$$

$$D_n^2 = \frac{2(2n-1)\lambda}{\left(\frac{1}{R_1} - \frac{1}{R_2}\right)}. \tag{11.49}$$

In general case, the diameter for dark ring in both cases can be combined to write

$$D_n^2 = \frac{4n\lambda}{\left(\frac{1}{R_1} \mp \frac{1}{R_2}\right)}. \tag{11.50}$$

In general case, the diameter for bright ring in both cases can be combined to write

$$D_n^2 = \frac{2(2n-1)\lambda}{\left(\frac{1}{R_1} \mp \frac{1}{R_2}\right)}. \tag{11.51}$$

11.12 Effect of Increasing the Distance between Lens and Plate or Lifting up the Lens from the Flat Surface

The order of the ring increases as the distance between the lens and the plate increases or as the lens is gradually raised off the flat surface. As a result, the rings become closer and closer together until they can no longer be seen individually.

11.13 Effect of Placing the Lens on a Silver Glass Plate or Mirror

If the top of the glass plate is fully polished with silver, the ring on the reflected system would disappear and a uniform illumination is observed. The reason for this is that there wouldn't be any ray transmission; however, any rays that were transmitted would also be reflected at the silvered surface. As a result, the two complementary ring systems superimpose on one another and provide uniform illumination.

11.14 Newton's Rings Are Circular, but Air-wedge Fringes Are Straight

Each fringe is the location of points of equal film thickness in both the Newton's ring and the air-wedge fringe configurations. In Newton's ring arrangement, the locus of points of equal thickness of air film lies on a circle, with the point of contact of the plano-convex lens and the glass plate as a center. Hence, the fringes are circular and concentric in Newton's ring. In case of wedge-shaped air film, the loci of points of equal thickness are straight lines parallel to the edge of the wedge. Hence, the fringes are straight and parallel.

11.15 The Effect of Placing the Concave Surface of the Plano-concave Lens Toward the Plane Glass Plate

In this situation, the fringes are still circular. The lens and the plate will be in contact along the circumference of a circle where the thickness of the air film is zero. The thickness of the air film increases as we move from either side toward the center. Therefore, in this case, the order of the ring is reverse, that is, the order of the maximum is at the center and zero at the periphery of the lens. Hence, the spacing between two consecutive rings goes on decreasing as we move toward the center of the lens.

11.16 Effects of Using a Lens of Small Radius of Curvature

The expressions for diameters of bright and dark rings $D_n = \sqrt{2(2n-1)\lambda R}$ and $D_n = \sqrt{4n\lambda R}$ clearly indicate that the diameters of bright and dark rings are directly proportional to the square root of radius of curvature R of the lens. Hence, if we use a lens of small radius of curvature, the diameter of the rings will be small.

Wave Optics

Solved Problems

Ex. 1: A soap film of refractive index 1.43 is illuminated by white light incident at an angle of 30°. The refracted light is examined by a spectroscope in which a dark band corresponding to the wavelength 6×10^{-7} m is observed. Calculate the thickness of the film.

Solution:

The condition for minima of interference in a thin film is given by

$2\mu t \cos r = n\lambda$.

Given $i = 30°$, $\mu = 1.43$, $n = 1$, $\lambda = 6 \times 10^{-7}$ m.

From Snell's law, $\mu = \dfrac{\sin i}{\sin r} \Rightarrow \sin r = \dfrac{\sin i}{\mu} = \dfrac{\sin 30°}{1.43} = 0.38$

$\cos r = \sqrt{1 - (\sin r)^2} = \sqrt{1 - (0.38)^2} = 0.92$

$t = \dfrac{n\lambda}{2\mu \cos r} = \dfrac{1 \times 6 \times 10^{-7}}{2 \times 1.43 \times 0.92} = 2.28 \times 10^{-7}$ m.

Ex. 2: Calculate the thickness of a soap bubble film (refractive index: 1.46) that will result in constructive interference in the reflected light if the film is illuminated with light whose wavelength in free space is 6000 Å.

Solution:

The condition for maxima of interference in a soap bubble thin film is given by

$2\mu t \cos r = (2n-1)\dfrac{\lambda}{2}$.

For normal incidence, $r = 0 \Rightarrow \cos r = 1$. Given $\lambda = 6000$ Å and $\mu = 1.46$ for least thickness $n = 1$, we have

$t = \dfrac{(2n-1)\lambda}{4\mu \cos r} = \dfrac{6000}{4 \times 1.46 \times 1} = 1027.4$ Å.

Ex. 3: Light of wavelength 5893 Å is reflected at nearly normal incidence from a soap film of refractive index $\mu = 1.42$. Calculate the least thickness of the film that will appear (i) dark and (ii) bright.

Solution:

(i) The condition for minima of interference in a thin film is given by

$2\mu t \cos r = n\lambda$.

For normal incidence, $r = 0 \Rightarrow \cos r = 1$. Given $\mu = 1.42$, $n = 1$, $\lambda = 5893$ Å.

$t = \dfrac{n\lambda}{2\mu \cos r} = \dfrac{1 \times 5893}{2 \times 1.42 \times 1} = 2075$ Å.

(ii) The condition for maxima of interference in a soap bubble thin film is given by

$2\mu t \cos r = (2n-1)\dfrac{\lambda}{2}$.

For normal incidence, $r = 0 \Rightarrow \cos r = 1$. Given $\mu = 1.42$ and $\lambda = 5893$ Å for least thickness $n = 1$, we have

$$t = \frac{(2n-1)\lambda}{4\mu \cos r} = \frac{5893}{4 \times 1.42 \times 1} = 1037.5 \text{ Å}.$$

Ex. 4: White light is incident on a soap film at an angle $\sin^{-1}(4/5)$ and the reflected light is observed with a spectroscope. It is found that two consecutive dark bands correspond to wavelengths 6.1×10^{-5} cm and 6×10^{-5} cm. If the refractive index of the film is $4/3$, calculate the thickness of the film.

Solution:
The condition for minima of interference in a thin film is given by,

$2 \mu t \cos r = n\lambda$.

If n and $(n + 1)$ are the orders of consecutive dark bands for wavelengths λ_1 and λ_2 respectively, then, $2 \mu t \cos r = n\lambda_1$ and $2 \mu t \cos r = (n + 1)\lambda_2$

$$n\lambda_1 = (n+1)\lambda_2 \Rightarrow n = \frac{\lambda_2}{\lambda_2 - \lambda_1}.$$

Therefore, the condition for minima of interference in a thin film becomes

$$2\mu t \cos r = n\lambda_1 \Rightarrow 2\mu t \cos r = \frac{\lambda_1 \lambda_2}{\lambda_2 - \lambda_1} \Rightarrow t = \frac{\lambda_1 \lambda_2}{(\lambda_2 - \lambda_1) 2\mu \cos r}.$$

From Snell's law, $\mu = \frac{\sin i}{\sin r} \Rightarrow \sin r = \frac{\sin i}{\mu} = \frac{\frac{4}{5}}{\frac{4}{3}} = 4/3$

$\cos r = \sqrt{1 - (\sin r)^2} = \sqrt{1 - (4/3)^2} = 4/5$.

Given $\lambda_1 = 6.0 \times 10^{-5}$ cm and $\lambda_2 = 6.1 \times 10^{-5}$ cm.

$$t = \frac{6.1 \times 10^{-5} \times 6.1 \times 10^{-5}}{(6.1 \times 10^{-5} - 6.1 \times 10^{-5}) \times 2 \times 4/3 \times 4/5} = 0.0017 \text{ cm}.$$

Ex. 5: Light of wavelength 6000 Å falls normally on a thin, wedge-shaped film of refractive index 1.4, forming fringes that are 2 mm apart. Calculate the angle of the wedge.

Solution:
For normal incidence with a very small angle of wedge $\theta \approx 0$, the fringe width is given by

$$\omega = \frac{\lambda}{2\mu\theta} \Rightarrow \theta = \frac{\lambda}{2\mu\omega}.$$

Given $\lambda = 6000 \times 10^{-10}$ m, $\mu = 1.4$, $\omega = 2 \times 10^{-3}$ m.

$$\theta = \frac{6000 \times 10^{-10}}{2 \times 1.4 \times 2 \times 10^{-3}} = 1.07 \times 10^{-4} \text{ radians}.$$

Ex. 6: Newton's rings are observed normally in reflected light of wavelength 6000 Å. The diameter of the 10th dark ring is 0.5 cm. Calculate the radius of curvature of the lens and the thickness of the film.

Wave Optics

Solution:

The diameter of the nth dark ring is given by

$D_n^2 = 4n\lambda R.$

Given $D_n = 0.5$ cm, $n = 10$, and $\lambda = 6000 \times 10^{-8}$ cm. Then

$R = \dfrac{D_n^2}{4n\lambda} = \dfrac{0.5 \times 0.5}{4 \times 10 \times 6000 \times 10^{-8}} = 106$ cm.

Ex. 7: In Newton's rings experiment, the diameters of the 4th and 12th dark rings are 0.4 cm and 0.7 cm, respectively. Calculate the diameter of the 20th dark ring.

Solution:

The difference between the $(n + p)$th and nth dark rings is given by

$D_{n+p}^2 - D_n^2 = 4p\lambda R.$

Here, $(n + p) = 12$, $n = 4$, $D_{12} = 0.7$ cm, and $D_4 = 0.4$ cm.

$D_{12}^2 - D_4^2 = 4 \times 8 \times \lambda R$ (1)

$D_{20}^2 - D_4^2 = 4 \times 16 \times \lambda R$ (2)

On dividing equation (2) by equation (1), we have

$\dfrac{D_{20}^2 - D_4^2}{D_{12}^2 - D_4^2} = \dfrac{4 \times 16 \times \lambda R}{4 \times 8 \times \lambda R} = 2$

$D_{20}^2 - D_4^2 = 2\left(D_{12}^2 - D_4^2\right)$

$D_{20}^2 = 2D_{12}^2 - D_4^2 \Rightarrow D_{20}^2 = 2(0.7)^2 - (0.4)^2 = 0.82$

$D_{20} = \sqrt{0.82} = 0.9$ cm.

Ex. 8: Newton's rings are observed by keeping a spherical surface of 100 cm radius on a plane glass plate. If the diameter of the 15th bright ring is 0.59 cm and the diameter of the 5th ring is 0.336 cm, calculate the wavelength of light used.

Solution:

The wavelength of monochromatic light is given by,

$\lambda = \dfrac{D_{n+p}^2 - D_n^2}{4pR}.$

Given $D_{15} = 0.59 \times 10^{-2}$ m, $D_5 = 0.336 \times 10^{-2}$ m, $p = (15 - 5) = 10$, $R = 100$ cm $= 1$ m,

$\lambda = \dfrac{D_{15}^2 - D_5^2}{4pR} = \dfrac{0.59 \times 10^{-2} - 0.336 \times 10^{-2}}{4 \times 10 \times 1} = 5880 \times 10^{-10}$ m.

Ex. 9: Newton's rings are formed in reflected light of wavelength 6000 Å with a liquid between the plane and curved surfaces. If the diameter of the 6th bright ring is 3.1 mm and the radius of curvature of the curved surface is 100 cm, calculate the refractive index of the liquid.

Solution:

Given, $D_n = 3.1 \times 10^{-3}$ m, $\lambda = 6000 \times 10^{-10}$ m, $n = 6$, $R = 1$ m.

The diameter of nth bright ring is given by

$$D_n^2 = \frac{2(2n-1)\lambda R}{\mu} \Rightarrow \mu = \frac{2(2n-1)\lambda R}{D_n^2}.$$

$$\mu = \frac{2(2 \times 6 - 1) 6000 \times 10^{-10} \times 1}{(3.1 \times 10^{-3})^2} = 1.374.$$

Previous Year Questions (University Examination): Short Questions

1. What do you mean by monochromatic light?
2. What do you mean by interference of light?
3. What are coherent sources? How are they obtained in practice?
4. What is a wavefront?
5. What is the main condition to produce interference?
6. What are the conditions for sustained interference?
7. What do you mean by incoherent sources?
8. How will you locate zero-order fringes in a biprism experiment?
9. Two independent sources cannot be coherent. Why?
10. What do you mean by fringe width?
11. What is Fresnel's biprism?
12. What do you mean by displacement of fringes?
13. What is the deviation method for the determination of distance between two virtual sources in a biprism experiment?
14. Sketch the wavefront that corresponds to a beam of light (i) coming from a very far-away source and (ii) diverging radially from a point source.
15. Explain three methods for obtaining coherent sources.
16. Describe how velocity, wavelength, and frequency of a wave changes on entering a medium of refractive index μ.
17. What is the phase difference corresponding to a path difference of λ?
18. Oil floating on water looks colored due to interference of light. What should be the approximate thickness of the film for such effects to be visible?
19. What is the ratio of intensities at two points x and y on a screen in Young's double-slit experiment, where waves from S_1 and S_2 have path difference of 0 and $\lambda/4$?
20. How does the interference pattern by reflection in thin films differ from that of a transmitted system?
21. Is it necessary to have two waves of equal intensity to study interference pattern?

Wave Optics

22. Give the shape of interference fringes observed in (a) Young's double-slit experiment, (b) in the air wedge experiment, and (c) from a thin air film formed by placing a convex lens on top of a flat glass plate.
23. How will you obtain Newton's rings with bright center due to reflected light?

Previous Year Questions (University Examination): Long Questions

1. Discuss why two independent sources of light of the same wavelength cannot produce interference fringes,
2. Can an interference pattern be produced with white light?
3. If Young's double-slit experiment were performed under water, how will the observed interference pattern change?
4. If white light is used in Young's double-slit experiment, rather than monochromatic light, how does the interference pattern change?
5. Why don't we see interference patterns if we look at a glass window?
6. What is the effect on the interference fringes in Young's double-slit experiment if the separation between the two slits is increased?
7. In Young's double-slit experiment, a thin plate of some transparent material is introduced in the path of one of the interfering beams. What would happen to the fringe pattern?
8. Explain why an excessively thin film seen in reflected white light appears perfectly black.
9. Why does a compact disc show a rainbow of colors?
10. Why do we prefer a point source of light in Young's double-slit experiment? How does the extended source affect the interference pattern?
11. In the Young's double-slit experiment, the source gives out white light, but one slit is covered with a red filter and the other with a blue filter. What will be the nature of the interference pattern?
12. Explain what happens when the width of the slit in the biprism experiment is increased.
13. Why are the refracting angles of the two prisms made so small?
14. The width of the interference fringes for red light is double that of violet. Why?
15. Explain the changes in the interference pattern when the distance between coherent sources increases.
16. What is the effect of changing the distance between the slit and biprism on the fringe width?
17. Can we obtain interference pattern if the two coherent sources are separated by less than the wavelength of light?
18. Why broad (extended) source of light is needed to observe colors in the thin films?
19. Explain why a thick film seen by reflected light shows no colors but appears white?
20. The interference pattern on the surface of a soap bubble changes continuously. Why?
21. Why can't the entire thin film be viewed at a glance with a point source?
22. Why Newton's rings are circular in shape?

23. In Newton's ring experiment, what happens when (i) light is not monochromatic and (ii) the plane glass plate is replaced by a plane mirror?
24. Why is the central fringe dark in the Newton's ring experiment when we view the interference pattern by reflection?
25. Why do the rings get closer as the order of the rings increases in Newton's rings?
26. What will happen if we use a plano-concave lens in place of plano-convex lens in Newton's rings?
27. What will happen if few drops of a transparent liquid are introduced between the lens and glass plate?
28. Thin films like soap bubble or thin layer of oil on water show beautiful colors when illuminated by white light. Explain the observation.
29. What is interference of light? Describe and explain Young's experiment demonstrating interference of light.
30. State the essential condition for observing the phenomenon of interference of light.
31. What are the coherent sources of light? Is it possible to obtain coherent sources from two separate sources? If not, why?
32. What are coherent sources? How two coherent sources are produced?
33. Discuss three methods of producing coherent sources.
34. Calculate the separation between two consecutive bright and dark fringes in the interference pattern. What will happen if white light is replaced with sodium light?
35. There is a path difference between two coherent interfering photons of identical intensity. The resultant intensity as a function of should be expressed. Create a curve that depicts the change in consequent intensity I versus δ.
36. Discuss why two independent sources of light of the same wavelength cannot show interference?
37. Show that, in the phenomenon of interference, the energy is conserved.
38. Explain the conditions of interference of optical waves and differentiate between interference due to the division of wavefront and division of amplitude, giving one example for each.
39. Explain the formation of interference fringes by means of biprism using monochromatic source of light. How is wavelength measured by biprism experiment? Derive the expression for fringe width.
40. Explain the formation of interference fringes by means of biprism using monochromatic source of light. How is wavelength measured by biprism experiment?
41. Fresnel's biprism is used to deduce the formula for the fringe width and to explain how interference fringes originate.
42. Explain the phenomenon of interference in thin films due to reflected light.
43. In a biprism experiment, what happens if you place a thin plate of mica in the path of one of the interfering beams? Determine an expression for the fringe displacement.
44. Discuss the phenomenon of interference of light due to thin films of uniform thickness in reflected and transmitted light, and find the conditions of maxima and minima. The interference patterns of monochromatic light that are reflected and transmitted in thin films must be shown to be complementary.
45. Explain the colors when a thin film illuminated by white light is observed in reflected light.

46. When seen by reflected light, a soap coating on a wire loop suspended in air looks black in its thinnest part. On the other hand, when seen similarly from the air above, a thin oil layer floating on water looks brilliant at its thinnest region. Why do these things happen?

47. Discuss the formation of interference fringes due to a wedge-shaped thin film seen by normally reflected sodium light and obtain an expression for the fringe width.

48. What are Newton's rings? Explain the circular nature of Newton's rings. Establish the relationship between the widths of the brilliant rings in the reflected light and the square root of odd natural numbers. Show that dark ring diameters are related to the square roots of natural integers.

49. Give a description and an explanation of how Newton's rings arise in monochromatic light that is reflected. Show that the widths of the black circles in the reflected light are related to the square root of the natural numbers.

50. Describe the Newton's ring method to determine the wavelength of sodium light. What will happen to fringes if air film between the plano-convex lens and glass plate is filled with a liquid of refractive index μ. Explain?

51. Describe how monochromatic light reflected from Newton's rings experiment results in the development of brilliant and dark circular fringes. Describe how these rings may be utilized to determine the light's wavelength. What will happen to the rings that were seen in the following scenarios?
    1. If a drop of water is placed between the lens and the glass plate, and 2) if the lens is slowly raised off a flat surface?

52. Show that the diameter D_n of the nth Newton's ring, when two surfaces with radii R_1 and R_2 are placed in contact, is given by the relation
$$\frac{1}{R_1} \pm \frac{1}{R_2} = \frac{4n\lambda}{D_n^2}.$$

53. Describe how Newton's ring experiment can be used to determine the refractive index of a liquid.

Previous Year Questions (University Examination): Numerical Questions

1. The path difference between the two interfering rays at a point on the screen is 1/8th of a wavelength. Find the ratio of intensity at this point to that at the center of a bright fringe.

2. Two coherent sources whose intensity ratio is 100:1 produce interference fringes. Find the ratio of maximum intensity to minimum intensity in the interference pattern.

3. In an interference pattern, the amplitude of intensity variation is found to be 5% of the average intensity. Calculate the relative intensities of the interfering source

4. On a screen held at a distance of 1 m from two coherent sources of monochromatic light of wavelength 6000, an interference pattern appears. The distance between two consecutive bright fringes on the screen is 0.5 mm. Find the distance between the two coherent sources.

5. In a two-slit interference pattern at a point, we observe the 10th order maximum for a wavelength of 7000 Å. What order will be visible here if the source of light is changed to a wavelength 5000 Å?

6. In Young's double-slit experiment, the slits are 0.5 mm apart, and interference is observed on a screen placed at a distance of 100 cm from the slits. It is found that the 9th bright fringe is at a distance of 8.835 mm from the second dark fringe from the center of the pattern. Find the wavelength of light used.

7. A beam of light consisting of two wavelengths 6500 Å and 5200 Å is used to obtain interference fringes in a Young's double-slit experiment. The distance between the slits is 2.0 mm, and the distance between the plane of slits and the screen is 120 cm. (a) Find the distance of the third bright fringe on the screen from the central maximum for the wavelength 6500 Å. (b) What is the least distance from the central maximum where the bright fringes due to both wavelengths coincide?

8. A biprism is placed 5 cm away from a slit illuminated by sodium light ($\lambda = 5890$ Å). The width of the fringes obtained on a screen placed at a distance of 75 cm from the biprism is $9.424*10^{-2}$ cm. What is the distance between the two coherent sources?

9. Each of the distances between the slit and the biprism and between the biprism and the screen is 50 cm. The biprism's refractive index is 1.5, and it has an obtuse angle of 179°. Determine the light's wavelength if the fringes' width is 0.0135 cm.

10. In a Fresnel biprism experiment, the fringe width observed is 0.087 mm. What will it become if the slit to biprism distance is reduced to 3/4 the original distance (all else remaining unchanged)?

11. In a biprism experiment, the distance between the slit and the screen is 160.0 cm. The biprism is 40 cm away from the slit, and its refractive index is 1.52. When a source of wavelength 5893 Å is used, the fringe width is found to be 0.01 cm. Find the angle of prism.

12. A two-slit Young's interference experiment is done with monochromatic light of wavelength 6000 Å. The slits are 2 mm apart, and the fringes are observed on a screen placed 10 cm away from the slits, and it is found that the interference pattern shifts by 5 mm when a transparent plate of thickness 0.5 mm is introduced in the path of one of the slits. What is the refracting index of the transparent plate?

13. Interference fringes are produced by biprism in the focal plane of an eye-piece 200 cm away from the slit. The two images of the slit formed for each of the two positions of a convex lens placed between the biprism and eye-piece are found to be separated by 5.1 mm and 3.14 mm, respectively. If the width of interference fringes be 0.45 mm, find the wavelength of light used.

14. On placing a thin sheet of mica of thickness $1.2*10^{-4}$ cm in the path of one of the interfering beams in a biprism experiment using light of wavelength 5890 Å, the central fringe shifts to a position originally occupied by 12th bright fringe. Calculate the refractive index of the sheet.

15. On introducing a thin sheet of mica of thickness $1.2*10^{-4}$ cm in the path of one of the interfering beams in a biprism experiment, the central fringe is shifted through a distance equal to the spacing between successive bright fringes. Calculate the refractive index of the mica sheet. (Given: $\lambda = 6*10^{-7}$ m.)

16. In a biprism experiment with a source of light of wavelength 5890 Å, a thin mica sheet of refractive index 1.6 is placed in the path of one of the interfering beams and the central bright fringe is shifted to a position of third bright fringe from the center. Calculate the thickness of the mica sheet.

17. A monochromatic light of $\lambda = 5000$ Å is incident on two slits separated by a distance of $5*10^{-4}$ m. The interference pattern is seen on a screen placed at a distance of 1 m from the slits. A thin glass plate of thickness $1.5*10^{-6}$ m and refractive index $\mu = 1.5$ is placed between one of the slits and the screen. Find the intensity at the center of the screen. Also, find the lateral shift of the central maximum.

Wave Optics

18. A man whose eyes are 150 cm above the oil film on water surface observes greenish color at a distance of 100 cm from his feet. Calculate the probable thickness of the film.

19. A soap film of refractive index 1.43 is illuminated by white light incident at an angle of 30°. The refracted light is examined by a spectroscope in which a dark band corresponding to the wavelength $6*10^{-7}$ m is observed. Calculate the thickness of the film.

20. Calculate the thickness of soap bubble film (refractive index: 1.46) that will result in constructive interference in the reflected light if the film is illuminated with light whose wavelength in free space is 6000 Å.

21. Light of wavelength 5893 Å is reflected at nearly normal incidence from a soap film of refractive index $\mu = 1.42$. What is the least thickness of the film that will appear (i) dark and (ii) bright?

22. White light is incident on a soap film at an angle $\sin^{-1}(4/5)$ and the reflected light is observed with a spectroscope. It is found that two consecutive dark bands correspond to wavelengths $6.1*10^{-5}$ cm and $6*10^{-5}$ cm. If the refractive index of the film is 4/3, calculate the thickness.

23. White light falls normally on a film of soapy water whose thickness is $1.5*10^{-5}$ cm and refractive index 1.33. Which wavelength in the visible region will be reflected most strongly?

24. A film of refractive index μ is illuminated by white light at an angle of incidence i. In reflected light, two consecutive bright fringes of wavelengths λ_1 and λ_2 are found overlapping. Obtain an expression for the thickness of the film.

25. Light of wavelength 6000 Å falls normally on a thin wedge-shaped film of refractive index 1.4, forming fringes that are 2 mm apart. Find the angle of the wedge in seconds.

26. Two plane glass surfaces in contact along one edge are separated at the opposite edge by a thin wire. If 20 fringes are observed between these edges in sodium light for normal incidence, what is the thickness of the wire?

27. A square piece of cellophane film with index of refraction 1.5 has a wedge-shaped section so that its thickness at two opposite sides is t_1 and t_2. If the number of fringes appearing with wavelength $\lambda = 6000$ Å is 10, calculate the difference (t_2-t_1).

28. Newton's rings are observed normally in reflected light of wavelength 6000 Å. The diameter of the 10th dark ring is 0.5 cm. Find the radius of curvature of the lens and the thickness of the film.

29. Newton's rings are observed by keeping a spherical surface of 100 cm radius on a plane glass plate. If the diameter of the 15th bright ring is 0.59 cm and the diameter of the 5th ring is 0.336 cm, what is the wavelength of light used?

30. In Newton's ring experiment the diameter of the 4th and 12th dark rings are 0.4 cm and 0.7 cm, respectively. Deduce the diameter of the 20th dark ring.

31. Newton's rings are formed in reflected light of wavelength 6000 Å with a liquid between the plane and curved surfaces. If the diameter of the 6th bright ring is 3.1 mm and the radius of curvature of the curved surface is 100 cm, calculate the refractive index of the liquid.

32. Two plano-convex lenses, each of radius of curvature 100 cm, are placed with their curved surfaces in contact with each other. Newton's rings are formed by using a light of wavelength $6*10^{-5}$ cm. Find the distance between the 10th and 20th rings.

33. The convex surface of radius 300 cm of a plano-convex lens rests on the concave spherical surface of radius 400 cm. If Newton's rings are viewed with reflected light of wavelength $6*10^{-5}$ cm, calculate the radius of the 12th dark and 13th bright ring.

Multiple Choice Questions

1. The interference of light phenomenon is explained by
   (a) Wave theory
   (b) Particle theory
   (c) F. D. Statistics
   (d) None of the above

2. In interference, the energy is only transferred from the points of
   (a) Maximum to the minimum displacement
   (b) Minimum to the maximum displacement
   (c) Both (a) and (b)
   (d) None of the above

3. Which of the following device is an example of division of wavefront?
   (a) Newton's rings
   (b) Fresnel's mirrors
   (c) Michelson interferometer
   (d) None of the above

4. Which of the following device is an example of division of amplitude?
   (a) Newton's rings
   (b) Interference in thin film
   (c) Michelson interferometer
   (d) All of the above

5. The central fringe in Newton's ring experiment is
   (a) Bright
   (b) Dark
   (c) First dark then bright
   (d) None of these

6. The wedge-shaped fringe pattern will always begin with
   (a) Dark fringe
   (b) Bright fringe
   (c) Maximum intensity
   (d) None of these

7. The actual shape of interference fringes in Young's double-slit experiment is
   (a) Elliptical
   (b) Parabolic
   (c) Hyperbolic
   (d) Circle

Bibliography

1. Lesurf, J. C. G. (2017). *Millimetre-wave Optics, Devices and Systems*. CRC Press.
2. Poon, T-C., and Liu, J. P. (1973) "Fundamentals of Optics." *Optical and Digital Image Processing*. 1–23.
3. Dallas, W. J. (2007). "Wave Optics." *The Optics Encyclopedia*. 1–37.
4. Germain, C. (2005). *Introduction to Optics*. Springer.
5. Ghatak, A. (2012). *Contemporary Optics*. Springer Science & Business Media.

Keys

1. (a) 2. (b) 3. (b) 4. (d) 5. (b) 6. (a) 7. (c)

CHAPTER 12

Wave Optics
Diffraction

12.1 Diffraction

Diffraction is a phenomenon in which a light beam bends around the corner of an obstacle and spreads into the geometric shadow of that obstacle.

12.2 Fresnel and Fraunhofer Diffraction

Diffraction can be classified into two categories:

1. Fresnel diffraction
2. Fraunhofer diffraction

The distinction between these two categories is as follows:

a. In Fresnel diffraction, the screen and source are at a finite distance from an obstacle. The distances are important in this class. In Fraunhofer diffraction, the source and screen are at an infinite distance from an obstacle. Therefore, inclination is important.
b. The incident wavefront in Fresnel diffraction is either spherical or cylindrical, whereas the incident wavefront in Fraunhofer diffraction is planar.
c. In Fresnel diffraction, the central point of the screen is either bright or dark depending on the number of zones, whereas in Fraunhofer diffraction, the central point of the screen is always bright.

12.3 Fraunhofer Diffraction due to Single Slit

Let us consider a monochromatic light source of wavelength λ placed at the focus of convex lens L_1. The collimated rays of plane wavefront are incident on a single-slit AB of width "e." The un-deviated rays from the slit reaches at point O, and the rays diffracted by an angle θ reach at P on the screen, as shown in Figure 12.1.

The path difference between the two extreme rays at point P from right-angled triangle ΔANB is given by

Path difference = BN

Path difference = $AB \sin \theta$

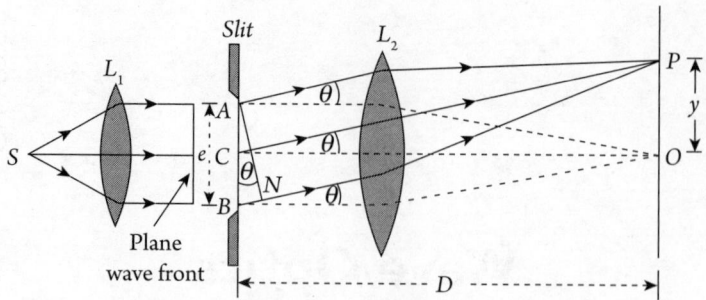

Fig. 12.1 Fraunhofer diffraction due to single slit.

Path difference = $e \sin \theta$. (12.1)

Phase difference = $\dfrac{2\pi}{\lambda} e \sin \theta$. (12.2)

Let AB be divided into n equal parts; then, the phase difference between two consecutive parts is given by

$$\frac{1}{n} \frac{2\pi}{\lambda} e \sin \theta = \delta \, (say).$$

The resultant amplitude at point P by the theory of n vibrations having amplitude "a" and a common phase difference δ between two successive waves is given by

$$R = a \frac{\sin\left(\dfrac{n\delta}{2}\right)}{\sin\left(\dfrac{\delta}{2}\right)}$$

$$R = a \frac{\sin\left(\dfrac{\pi}{\lambda} e \sin \theta\right)}{\sin\left(\dfrac{\pi}{\lambda n} e \sin \theta\right)}.$$

Let $\dfrac{\pi}{\lambda} e \sin \theta = \alpha$, then

$$R = a \frac{\sin \alpha}{\sin\left(\dfrac{\alpha}{n}\right)}.$$

Since $\left(\dfrac{\alpha}{n}\right)$ is very small, $\sin\left(\dfrac{\alpha}{n}\right) \approx \left(\dfrac{\alpha}{n}\right)$. Thus, we have

$$R = a \frac{\sin \alpha}{\left(\dfrac{\alpha}{n}\right)} = na \frac{\sin \alpha}{\alpha}$$

$$R = A \frac{\sin \alpha}{\alpha}, \qquad (12.3)$$

Wave Optics

where $A = na$. Thus, this is an expression of resultant amplitude at point P, while the intensity at point P is given by

$$I = a^2 = R^2$$

$$I = A^2 \frac{\sin^2 \alpha}{\alpha^2}. \tag{12.4}$$

Conditions for Maxima and Minima

The expression of resultant amplitude at point P from equation (12.3) is

$$R = A \frac{\sin \alpha}{\alpha} = \frac{A}{\alpha}\left[\alpha - \frac{\alpha^3}{3!} + \frac{\alpha^5}{5!} - \frac{\alpha^7}{7!} + \ldots\right]$$

$$R = A\left[1 - \frac{\alpha^2}{3!} + \frac{\alpha^4}{5!} - \frac{\alpha^6}{7!} + \ldots\right].$$

When $\alpha = 0 \Rightarrow \frac{\pi}{\lambda} e \sin \theta = 0 \Rightarrow \sin \theta = 0$ or $\theta = 0$, the resultant amplitude will be maximum; that is, $R = A$. Thus, the resultant intensity $I = A^2$ at $\theta = 0$ is called "principal maxima."

Condition for Minima

The expression of resultant amplitude at point P from equation (12.3) is

$$R = A \frac{\sin \alpha}{\alpha}.$$

The intensity at point P will be minimum when $\sin \alpha = 0$ but $\alpha \neq 0$,

$$\alpha = \pm m\pi, \text{ where } m = 1, 2, 3, \ldots.$$

$$\frac{\pi}{\lambda} e \sin \theta = \pm m\pi$$

$$e \sin \theta = \pm m\lambda.$$

This is the condition of minima, where $m = 1, 2, 3, \ldots$ gives the direction of first minima, second minima, third minima, and so on.

Conditions for Secondary Maxima

In order to obtain the secondary maximum intensities, differentiate equation (12.4); thus, we have

$$\frac{dI}{dt} = \frac{d}{d\alpha}\left[A^2 \frac{\sin^2 \alpha}{\alpha^2}\right] = 0$$

$$\frac{dI}{dt} = A^2 \frac{2\sin \alpha}{\alpha}\left[\frac{\alpha \cos \alpha - \sin \alpha}{\alpha^2}\right] = 0.$$

From the above equation, we have

Either $\frac{\sin \alpha}{\alpha} = 0$ or $\left[\frac{\alpha \cos \alpha - \sin \alpha}{\alpha^2}\right] = 0$.

Since $\frac{\sin \alpha}{\alpha} = 0$ for the condition of minima,

$$\left[\frac{\alpha\cos\alpha - \sin\alpha}{\alpha^2}\right] = 0$$

$\alpha = \tan\alpha$.

From the above equation, we can plot a graph, as shown in Figure 12.2.

The other values of α can be obtained by the intersection of these two curves $Y = \alpha$ and $Y = \tan\alpha$, which is given as

$$\alpha = 0, \pm\frac{3\pi}{2}, \pm\frac{5\pi}{2}, \pm\frac{7\pi}{2}, \ldots$$

Since $\alpha = 0$ is the condition for the principal maxima, the values of α for secondary maxima will be

$$\alpha = \pm\frac{3\pi}{2}, \pm\frac{5\pi}{2}, \pm\frac{7\pi}{2}, \ldots$$

The above values give the first, second, third, and so on secondary maxima. The intensity of the first secondary maxima is given by

$$I_1 = A^2 \left[\frac{\sin\frac{3\pi}{2}}{\frac{3\pi}{2}}\right]^2$$

$$I_1 = \frac{4A^2}{9\pi^2}.$$

Similarly, the intensity of the second secondary maxima is given by

$$I_2 = A^2 \left[\frac{\sin\frac{5\pi}{2}}{\frac{5\pi}{2}}\right]^2$$

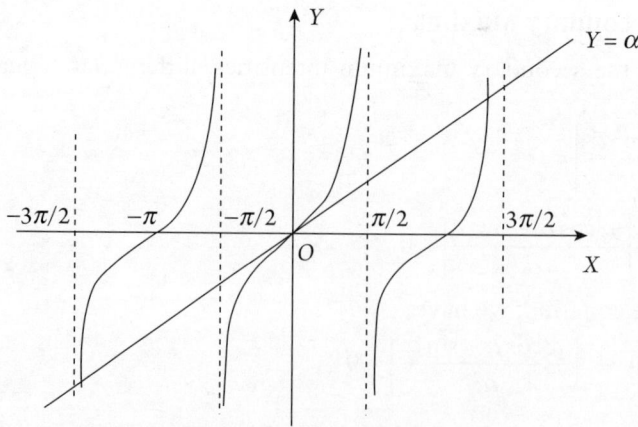

Fig. 12.2 Plot of $y = \alpha$, $y = \tan\alpha$.

$$I_2 = \frac{4A^2}{25\pi^2}.$$

Hence, the relative intensities of successive maxima are given by

$$I_0 : I_1 : I_2 : \ldots = 1 : \frac{4}{9\pi^2} : \frac{4}{25\pi^2} : \ldots$$

From the expression of the intensities of the secondary maxima, it is clear that most of the light is concentrated on the principal maximum. The intensity of the first secondary maxima on either side of the principal maximum is about 4.5% of the light of the principal maximum. Thus, the intensity of the secondary maxima decreases very rapidly. The minima are equally spaced, and their intensity is zero. The diffraction patterns of relative intensities of successive maxima are shown in Figure 12.3.

Spread of Central Diffraction Maximum

The direction of first minima is given by

$$e \sin\theta = \pm\lambda \quad (\because n = 1)$$

$$\sin\theta = \pm\frac{\lambda}{e} \Rightarrow \theta = \sin^{-1}\left(\pm\frac{\lambda}{e}\right). \tag{12.5}$$

Therefore, the central maximum extends between $\theta = \sin^{-1}\left(\frac{\lambda}{e}\right)$ and $\theta = \sin^{-1}\left(-\frac{\lambda}{e}\right)$. It means θ is the angular half width of the central maximum.

If the lens toward the screen, that is, L_2, is very near to the slit AB or the screen is far away from the lens L_2, shown in Figure 12.1, then we have

$$\sin\theta = \pm\frac{r}{f}, \tag{12.6}$$

where r is the linear half width of the central maximum, and f is the focal length of lens L_2.

Comparing equations (12.5) and (12.6), we get

$$\frac{r}{f} = \frac{\lambda}{e} \Rightarrow r = \frac{\lambda f}{e}. \tag{12.7}$$

Hence, the width of the central maximum is $2r = \frac{2\lambda f}{e}$.

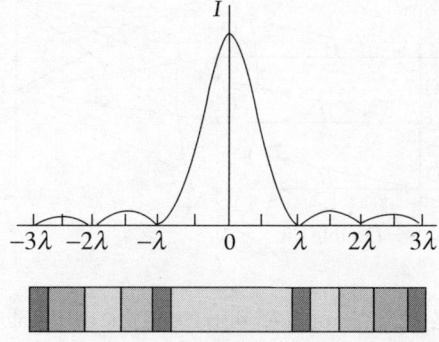

Fig. 12.3 Diffraction pattern.

The width of the central maxima is obviously exactly proportional to λ, illustrated by equation (12.7). Red light has a wider dispersion from the central maxima than violet light because red light has a longer wavelength than violet light.

12.4 Effect of Making Slit Narrower

The first minima on either side of the central maximum occur in the direction θ given by
$$e \sin\theta = \pm\lambda.$$

When the slit is narrowed by reducing e, the angle of diffraction θ increases, which means that the central maximum becomes wider. When the width of the slit is made equal to the wavelength of light, that is, $e = \lambda$, the minimum occurs at $\theta = 90°\left(\sin\theta = \dfrac{\lambda}{\lambda}\right)$; it means that the central maximum occupies the whole space.

Distinction between Single-slit Diffraction Pattern and Double-Slit Diffraction Pattern:
  a. The diffraction pattern due to single slit of width "e" differs in many respects with the interference pattern due to double slit of separation "e."
  b. In the diffraction pattern, the principal maximum has the highest and brightest intensity and has non-symmetrical weak secondary maxima on either side.
  c. In the interference pattern, the maxima and minima are equidistant and equally wide, and all the maxima have the same intensity.

12.5 Fraunhofer Diffraction due to Double (Two) Slits

Let AB and CD be two parallel slits of equal width "e", separated by an opaque distance "d." A plane wavefront is incident normally upon the slits. The light diffracted from these slits is focused by lens L_2 on the screen XY, as shown in Figure 12.4.

The path difference between two extreme rays is given by
$$S_2 N = S_1 S_2 \sin\theta$$
$$S_2 N = (e+d)\sin\theta. \tag{12.8}$$

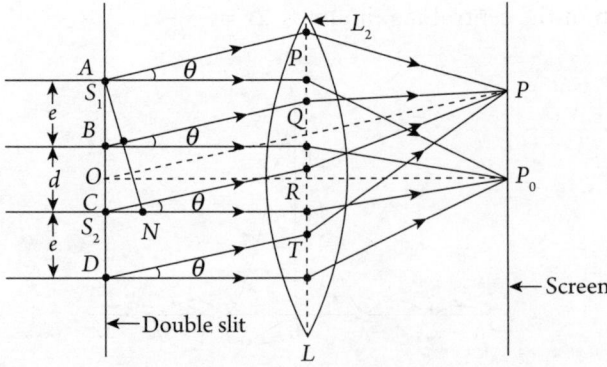

Fig. 12.4 Fraunhofer diffraction due to double slit.

Wave Optics

Phase difference $= \dfrac{2\pi}{\lambda}(e+d)\sin\theta = (2\beta)$. (12.9)

The resultant intensity at point P' due to double-slit S_1 and S_2 is given by

$R'^2 = R^2 + R^2 + 2R^2 \cos(2\beta)$

$R'^2 = 2R^2 [1 + \cos(2\beta)]$

$R'^2 = 2R^2 \left[1 + 2\cos^2\left(\dfrac{2\beta}{2}\right) - 1\right]$

$R'^2 = 4R^2 \cos^2\beta \qquad \left[\because R^2 = A^2 \dfrac{\sin^2\alpha}{\alpha^2}\right].$

Thus, the intensity at point P is given by

$I = R'^2 = 4A^2 \dfrac{\sin^2\alpha}{\alpha^2} \cos^2\beta.$ (12.10)

Hence, the resultant intensity at any point depends on two variable factors: diffraction factor $\left(A^2 \dfrac{\sin^2\alpha}{\alpha^2}\right)$ and interference factor $(\cos^2\beta)$, which are reflected in Figure 12.5.

Conditions for Maxima and Minima

A principal maximum is obtained at $\alpha = 0$, $\theta = 0$.

The minima condition is given by

$e\sin\theta = \pm m\lambda$.

This is the condition of minima, where $m = 1, 2, 3, \ldots$ gives the direction of first minima, second minima, third minima, and so on.

The secondary maxima can be obtained at the values of α as

$\alpha = \pm\dfrac{3\pi}{2}, \pm\dfrac{5\pi}{2}, \pm\dfrac{7\pi}{2}, \ldots$

Conditions for Maxima in Interference Pattern

For maxima, $\cos^2\beta = 1$, that is, $\beta = \pm n\pi$. So,

$(e+d)\sin\theta = \pm n\lambda.$

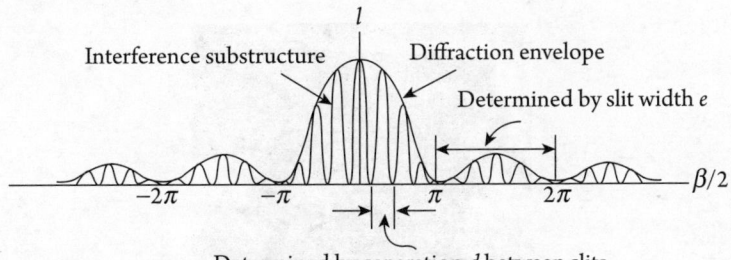

Fig. 12.5 Diffraction pattern due to double slit.

Here, $n = 0, 1, 2, 3, \ldots$. When $n = 0$, $\theta = 0$, the central maxima is obtained.

For minima, $\cos^2 \beta = 0$, that is, $\beta = \pm \dfrac{(2n-1)\pi}{2}$. So,
$(e+d)\sin\theta = \pm \dfrac{(2n-1)\lambda}{2}$.

Now, if e is kept constant and d is varied, the position of maxima and minima due to diffraction remains unchanged while those due to interference undergo change.

Absent Spectra or Missing Spectra in Double Slit Diffraction

The condition for diffraction minima is given by
$$e\sin\theta = \pm m\lambda. \tag{12.11}$$
The condition for interference maxima is given by
$$(e+d)\sin\theta = \pm n\lambda. \tag{12.12}$$
The condition for diffraction minima coincides with interference maxima at the same angle θ. Therefore, the interference fringe will be absent.

On dividing equation (12.12) by equation (12.11), we get
$$\dfrac{e+d}{e} = \dfrac{n}{m}$$
$$n = \dfrac{(e+d)}{e} m. \tag{12.13}$$
where $m = 1, 2, 3, \ldots$.

12.6 Fraunhofer Diffraction due to Plane Diffraction Grating

A diffraction grating is made up of several parallel slits that are of the same width and are spaced equally apart by opaque gaps. A grating is created by the diamond creating a succession of incredibly thin, evenly spaced, parallel lines, with the gaps between the lines being transparent to light. Such a grating has 15000 lines per inch, which causes visible light to be diffracted. The term "diffraction element" or "grating" refers to the distance $(e + d)$ between consecutive slits, as shown in Figure 12.6. If the lines are drawn on a silvered mirror's surface, the light is reflected from the mirror's locations between any two lines, creating a concave or flat reflection grating.

Fig. 12.6 Diffraction grating

Wave Optics

Theory

Let us consider a monochromatic light source of wavelength λ placed on the focus of convex lens L_1; then, the collimated rays of plane wavefront is incident on a single-slit AB of width "e." The undeviated rays from the slit reaches at point C, and the rays diffracted by an angle θ reach at P on the screen, as shown in Figure 12.7.

The path difference between the two extreme rays at point P is given by

$$S_2N = S_1S_2 \sin\theta$$

$$S_2N = (e+d)\sin\theta. \tag{12.14}$$

The phase difference is given by

$$\text{Phase difference} = \frac{2\pi}{\lambda}(e+d)\sin\theta = (2\beta). \tag{12.15}$$

According to the theory of n vibrations, the resultant amplitude at point P is given by

$$R' = R\frac{\left(\sin\dfrac{2N\beta}{2}\right)}{\left(\sin\dfrac{2\beta}{2}\right)}$$

$$R' = R\frac{(\sin N\beta)}{(\sin \beta)} \quad \left[\because R = A\frac{\sin\alpha}{\alpha}\right]$$

$$R' = A\frac{\sin\alpha}{\alpha}\frac{(\sin N\beta)}{(\sin\beta)}. \tag{12.16}$$

Thus, the resultant intensity at point P is given by

$$I = A^2\frac{\sin^2\alpha}{\alpha^2}\frac{(\sin^2 N\beta)}{(\sin^2 \beta)}. \tag{12.17}$$

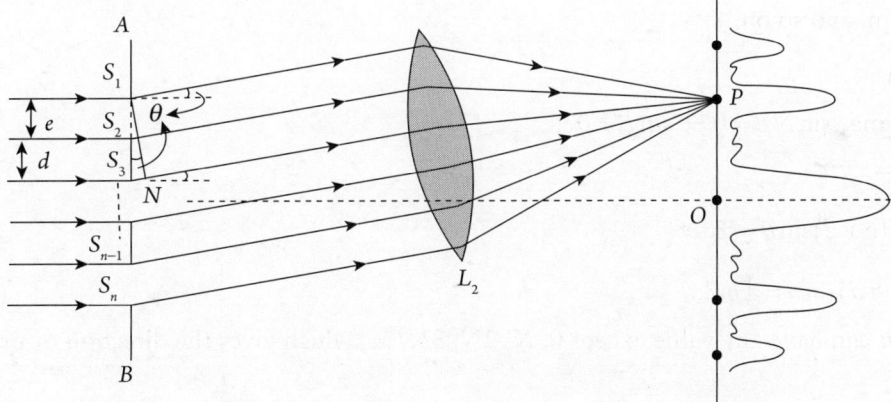

Fig. 12.7 Fraunhofer diffraction due to plane diffraction grating.

The factor $\left(A^2 \dfrac{\sin^2 \alpha}{\alpha^2}\right)$ gives the intensity distribution in the diffraction pattern due to a single slit, while the factor $\dfrac{(\sin^2 N\beta)}{(\sin^2 \beta)}$ gives the distribution of intensity due to interference caused by waves from N slits.

Conditions for Maxima and Minima
Principal Maxima

The intensity will be maximum when $\sin^2 \beta = 0$, that is, $\beta = \pm n\pi$, where $n = 0, 1, 2, 3, \ldots$.

Since at $n = 0$, $\beta = 0$, the condition $\dfrac{(\sin N\beta)}{(\sin \beta)} = \dfrac{0}{0}$, which is an indeterminate case. In order to find the value of this term, differentiating w.r.t. β, we get

$$\lim_{\beta \to \pm n\pi} \dfrac{\dfrac{d}{d\beta}(\sin N\beta)}{\dfrac{d}{d\beta}(\sin \beta)} = \lim_{\beta \to \pm n\pi} \dfrac{N \cos N\beta}{\cos \beta} = N.$$

The resultant intensity for principal maxima is given by

$$I = A^2 \dfrac{\sin^2 \alpha}{\alpha^2} N^2. \tag{12.18}$$

Now the condition for principal maxima is given by
$\sin \beta = 0 \Rightarrow \beta = \pm n\pi$

$\dfrac{\pi}{\lambda}(e+d)\sin\theta = \pm n\pi$

$$(e+d)\sin\theta = \pm n\lambda, \tag{12.19}$$

Where $n = 0, 1, 2, 3, \ldots$. At $n = 0$, the zero-order principal maximum is obtained at the center of the screen. Therefore, the other values of $n = 1, 2, 3, 4, \ldots$ correspond to the direction of the first-order principal maximum, second-order principal maximum, third-order principal maximum, and so on.

Minima
For minima, $\sin N\beta = 0 \Rightarrow \sin\beta \neq 0$

$N\beta = \pm m\pi$

$N\dfrac{\pi}{\lambda}(e+d)\sin\theta = \pm m\pi$

$$N(e+d)\sin\theta = \pm m\lambda, \tag{12.20}$$

Where m can have any value except $0, N, 2N, 3N, \ldots$, which gives the direction of principal maxima.

Secondary Maxima

It is clear from the minima criterion mentioned above that there are $(N-1)$ minima in between two primary maxima that follow one another. As a result, there are $(N-2)$ maxima between every two succeeding maxima. Secondary maxima are the (N–2) maxima. Differentiate equation (12.17) with respect to β and equate it to zero to determine the location of secondary maxima as

$$\frac{dI}{d\beta} = \frac{A^2 \sin^2 \alpha}{\alpha^2} \times 2\left[\frac{\sin N\beta}{\sin \beta}\right] \times \frac{N\cos N\beta \sin \beta - \sin N\beta \cos \beta}{\sin^2 \beta} = 0,$$

or $N \cos N\beta \sin \beta - \sin N\beta \cos \beta = 0,$

or $\tan N\beta = N \tan \beta.$

To find the intensity of secondary maxima from the triangle, as shown in Figure 12.8, we have

$$\sin N\beta = \frac{N \tan \beta}{\sqrt{1 + N^2 \tan^2 \beta}}.$$

Therefore, $\dfrac{\sin^2 N\beta}{\sin^2 \beta} = \dfrac{\dfrac{N^2 \tan^2 \beta}{1 + N^2 \tan^2 \beta}}{\sin^2 \beta}.$

$$\frac{\sin^2 N\beta}{\sin^2 \beta} = \frac{N^2 \tan^2 \beta}{(1 + N^2 \tan^2 \beta)\sin^2 \beta}$$

$$\frac{\sin^2 N\beta}{\sin^2 \beta} = \frac{N^2}{[1 + (N^2 - 1)\sin^2 \beta]}.$$

Putting this value of $\dfrac{\sin^2 N\beta}{\sin^2 \beta}$ in equation (4), we get the intensity of secondary maxima as

$$I = A^2 \frac{\sin^2 \alpha}{\alpha^2} \times \frac{N^2}{[1 + (N^2 - 1)\sin^2 \beta]}.$$

Thus, the above equation illustrates that the intensity of secondary maxima is proportional to $\dfrac{N^2}{[1 + (N^2 - 1)\sin^2 \beta]}$, whereas the intensity of principal maxima is proportional to N^2. Thus, the ratio of the intensity of these secondary maxima to the principal maxima is obtained by dividing the above equation by equation (12.18). We have

$$\frac{\text{Intensity of secondary maxima}}{\text{Intensity of principal maxima}} = \frac{I'}{I} = \frac{1}{[1 + (N^2 - 1)\sin^2 \beta]}.$$

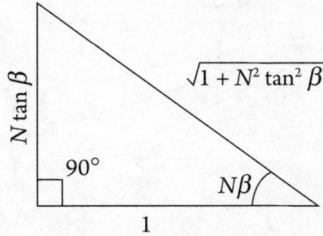

Fig. 12.8 Triangle for $\tan N\beta = N \tan \beta$.

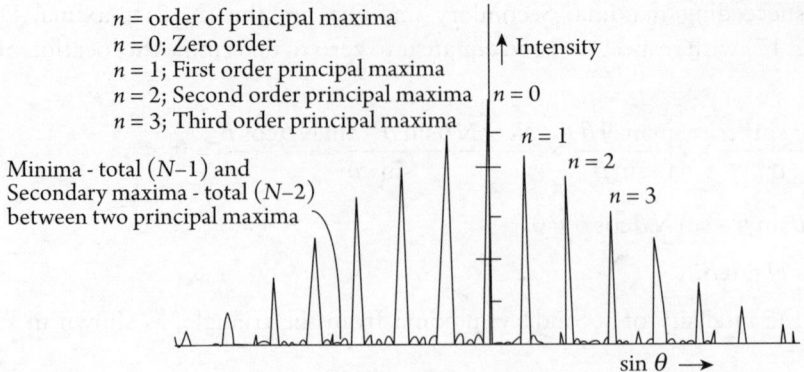

Fig. 12.9 Intensity distribution curve.

Hence, greater the value of N, the weaker are secondary maxima. In an actual grating, N is extremely large. Hence, secondary maxima are not visible in the grating spectrum.

The intensity distribution curve for plane diffraction grating is shown in Figure 12.9.

12.7 Formation of Spectra with Plane Diffraction Grating

The condition for principal maxima is given by

$(e+d)\sin\theta = \pm n\lambda.$

From the above equation, we can conclude that

1. Since $(e + d)$ and n are fixed, the angle θ is different for different wavelengths.
2. For fixed value of $(e + d)$, the angle θ is different for different values of n.
3. For $n = 0$, $\theta = 0$; for this, the zero-order principal maxima are obtained at the center of the screen, as shown in Figure 12.10.
4. The intensity of first-order principal maxima is greater than second-order principal maxima and so on.

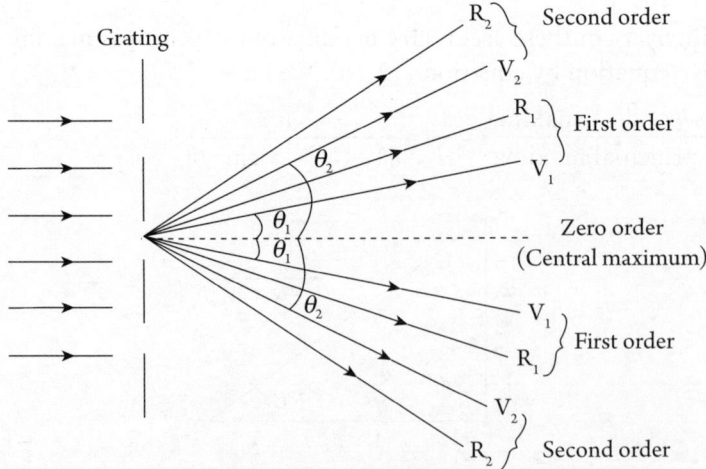

Fig. 12.10 Grating spectra.

Wave Optics

12.8 Absent Spectra for Diffraction Grating

The condition for principal maxima is given by

$$(e+d)\sin\theta = \pm n\lambda. \qquad (12.21)$$

The condition for interference maxima is given by

$$(e+d)\sin\theta = \pm n\lambda. \qquad (12.22)$$

The condition for diffraction minima coincides with interference maxima at the same angle θ. Therefore, the interference fringe will be absent.

On dividing equation (12.22) by equation (12.21), we get

$$\frac{e+d}{e} = \frac{n}{m}$$

$$n = \frac{(e+d)}{e}m, \qquad (12.23)$$

where $m = 1, 2, 3, \ldots$.

12.9 Maximum Number of Orders Possible in Grating Spectra

The condition for principal maxima is given by

$$(e+d)\sin\theta = \pm n\lambda,$$

$$n = \frac{(e+d)\sin\theta}{\lambda} \quad [\text{for max. } \theta = 90°],$$

$$n = \frac{(e+d)}{\lambda}.$$

If $(e+d) < 2\lambda$, then $n < \frac{2\lambda}{\lambda}$.

$n_{max} < 2$.

This means that for normal incidence, only first order is possible.

12.10 Determination of Wavelength of Light by Using Plane Diffraction Grating

The condition for principal maxima is given by

$$(e+d)\sin\theta = n\lambda,$$

$$\lambda = \frac{(e+d)\sin\theta}{n}.$$

The above expression can determine the wavelength of light used.

12.11 GRATING ELEMENT

The number of lines per inch written on the grating is known as grating element; it is denoted by $(e+d)$. If N is the number of lines per inch written on the grating, then the grating element is given by

$$e + d = \frac{2.54}{N}\text{cm}. \tag{12.24}$$

If N is the number of lines per inch written on the grating, then the grating element is given by $e + d = \frac{1}{N}\text{cm}.$ (12.25)

12.12 DISPERSIVE POWER OF PLANE TRANSMISSION GRATING

It is defined as the rate of change of the angle of diffraction θ with the rate of change of wavelength.

$$\text{Dipersive power} = \frac{d\theta}{d\lambda}.$$

The condition for principal maxima is given by

$$(e+d)\sin\theta = n\lambda. \tag{12.26}$$

On differentiating the above equation with respect to λ, we get

$$(e+d)\cos\theta \frac{d\theta}{d\lambda} = n$$

$$\frac{d\theta}{d\lambda} = \frac{n}{(e+d)\cos\theta} \tag{12.27}$$

$$\frac{d\theta}{d\lambda} = \frac{n}{(e+d)\sqrt{1-\sin^2\theta}}.$$

From, equation (12.26), $\sin\theta = \frac{n\lambda}{(e+d)}$. Therefore,

$$\frac{d\theta}{d\lambda} = \frac{n}{(e+d)\sqrt{1-\left(\frac{n\lambda}{e+d}\right)^2}}.$$

On solving, we get

$$\frac{d\theta}{d\lambda} = \frac{1}{\sqrt{\frac{(e+d)^2}{n^2} - \lambda^2}}.$$

This is an expression of dispersive power of grating; it is concluded from the above equation that:

1. Dispersive power is directly proportional to the order of spectrum.
2. Dispersive power is inversely proportional to grating element.
3. Dispersive power increases when θ increases.

12.13 Resolving Power of an Optical Instrument

The ability of an optical instrument to form separate images of two objects placed very close to each other is called the resolving power of the instrument.

The minimum separation between two objects that can be resolved by an optical instrument is called the limit of resolution of the optical instrument.

12.14 Rayleigh's Criterion for the Limit of Resolution

The two nearby point objects are said to be just resolved if the position of central maximum of one diffraction pattern coincide with the first minimum of the diffraction pattern of the other.

Rayleigh's criterion for the limit of resolution as just resolved is shown in Figure 12.11.

12.15 Resolving Power of Plane Diffraction Grating

The capacity of a grating to create distinct, extremely close-spaced diffraction principal maxima of a given wavelength is known as its "resolving power."

The ratio of the wavelength of spectral lines to the smallest wavelength difference between neighboring lines for which the spectral line can be just resolved at the wavelength λ.

Let us consider two spectral lines λ_1 and λ_2, which are very close to each other. The direction of the *n*th principal maxima for wavelength λ is given by

$$(e+d)\sin\theta_n = n\lambda_1 \tag{12.28}$$

$$N(e+d)\sin\theta_n = Nn\lambda_1, \tag{12.29}$$

where N is the number of lines on the grating.

The condition for first minima at an angle $(\theta_n + d\theta_n)$ is given by

$$N(e+d)\sin(\theta_n + d\theta_n) = m\lambda_1, \tag{12.30}$$

where $m \neq 0, N, 2N, 3N, \ldots$. The above equation is possible when $m = (Nn+1)$. Thus, equation (12.30) can be written as

$$N(e+d)\sin(\theta_n + d\theta_n) = (Nn+1)\lambda_1. \tag{12.31}$$

The direction for principal maxima at an angle $(\theta_n + d\theta_n)$ of wavelength λ_2 is given by

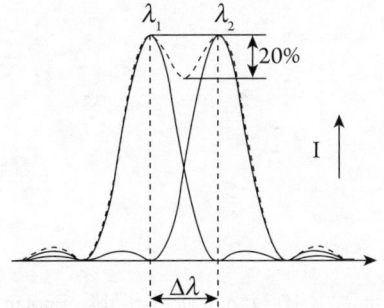

Fig. 12.11 Limit of resolution.

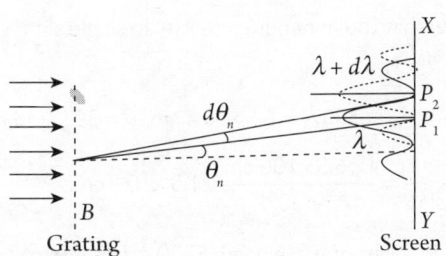

Fig. 12.12 Resolving power of plane diffraction grating.

$$(e+d)\sin(\theta_n + d\theta_n) = n\lambda_2$$

$$N(e+d)\sin(\theta_n + d\theta_n) = Nn\lambda_2. \tag{12.32}$$

According to the Rayleigh's criterion for limit of resolution, equations (12.31) and (12.32) must coincide with the just-resolved condition. Therefore,

$$(Nn+1)\lambda_1 = Nn\lambda_2.$$

On substituting $\lambda_1 = \lambda$ and $\lambda_2 = \lambda + d\lambda$, we get

$$(Nn+1)\lambda = Nn(\lambda + d\lambda)$$

$$\lambda = Nnd\lambda.$$

Thus, $R.P. = \dfrac{\lambda}{d\lambda} = Nn$

$$R.P. = \frac{\lambda}{d\lambda} = Nn = N\frac{(e+d)\sin\theta}{\lambda}. \tag{12.33}$$

This is an expression for resolving power of a plane diffraction grating. Further, the dispersive power is given by

$$\frac{d\theta}{d\lambda} = \frac{n}{(e+d)\cos\theta} \Rightarrow n = (e+d)\cos\theta\frac{d\theta}{d\lambda}.$$

Hence, the resolving power is

$$\frac{\lambda}{d\lambda} = Nn = N(e+d)\cos\theta\frac{d\theta}{d\lambda} = A\frac{d\theta}{d\lambda}.$$

This implies that

Resolving power (R.P.) = Aperture × Dispersive power (D.P.).

Finally, a distinction between R.P. and D.P. is the angular separation of two spectral lines, while R.P. is the closeness of two spectral lines that can be distinguished.

Solved Problems

Ex. 1: A light of wavelength 6500 Å falls normally on a slit of width a. The first minimum of spectrum is obtained for which value of "a" at an angle of 30°?

Solution:

We know that the minimum intensity in single slit is

$$a\sin\theta = \pm m\lambda.$$

Given: $\lambda = 6500\,\text{Å} = 6.5 \times 10^{-5}$ cm, $\theta = 30°$, and $m = 1$. Hence,

$$a = \frac{m\lambda}{\sin\theta} = \frac{1 \times 6.5 \times 10^{-5}\,\text{cm}}{\sin 30°} = 1.3 \times 10^{-4}\,\text{cm}.$$

Ex. 2: Light of wavelength 5500 Å falls normally on a slit of width 22×10^{-5} cm. Calculate the angular position of the first and second minima on either side of the central maxima.

Wave Optics

Solution:

The condition for minima is given by

$$e \sin\theta = \pm m\lambda.$$

Given $\lambda = 5500$ Å $= 5500 \times 10^{-10}$ m, $e = 22 \times 10^{-5}$ cm $= 22 \times 10^{-7}$ m. Then, for the first minima $m = 1$, we get

$$\sin\theta = \frac{m\lambda}{e} = \frac{1 \times 5500 \times 10^{-10}}{22 \times 10^{-7}} = 0.25$$

or, $\theta = \sin^{-1}(0.25) = 14.47°$.

For second minima $m = 2$, we get

$$\sin\theta = \frac{m\lambda}{e} = \frac{2 \times 5500 \times 10^{-10}}{22 \times 10^{-7}} = 0.5$$

or, $\theta = \sin^{-1}(0.5) = 30°$.

Ex. 3: Plane wave of wavelength 6×10^{-5} cm falls normally on a slit of width 0.2 mm. Calculate
  (a) Total angular width of central maxima.
  (b) Linear width of central maxima on a screen placed 2 m away.

Solution:

The condition for minima is given by

$$e \sin\theta = \pm m\lambda.$$

Given $\lambda = 6 \times 10^{-5}$ cm $= 6 \times 10^{-7}$ m, $e = 0.2$ mm $= 0.2 \times 10^{-3}$ m. Then, for the first minima $m = 1$, we get

$$\sin\theta = \frac{m\lambda}{e} = \frac{1 \times 6 \times 10^{-7}}{0.2 \times 10^{-3}} = 0.003$$

or, $\theta = \sin^{-1}(0.003) = 0.171°$.

a. The total angular width of the central maxima $= 2\theta = 2 \times 0.171° = 0.342°$.
b. The formula for the linear width of the central maxima on a screen is given by

$$2r = \frac{2\lambda f}{e} = \frac{2 \times 6 \times 10^{-7} \times 2}{0.2 \times 10^{-3}} = 120 \times 10^{-4} \text{m}.$$

Ex. 4: In a double-slit Fraunhofer diffraction pattern, the screen is placed 170 cm away from the slit. The width of the slit is 0.04 cm, and they are 0.08 mm apart. Calculate the wavelength of light if the fringe width is 0.25 cm. Calculate the missing orders also.

Solution:

The fringe spacing in the case of the interference pattern is given as

$$\omega = \frac{D\lambda}{2d} \Rightarrow \lambda = \frac{\omega \cdot 2d}{D}.$$

Given $2d = 0.4$ mm $= 0.04$ cm, $\omega = 0.25$ cm, $D = 170$ cm,

$$\lambda = \frac{0.25 \times 0.04}{170} = 5.88 \times 10^{-5} \text{cm} = 5880 \text{ Å}.$$

The condition for the missing order is given by

$$n = \frac{(e+d)}{e}m = \frac{0.08+0.4}{0.08}m = 6m.$$

where $m = 1, 2, 3, \ldots$. $n = 6, 12, 18, \ldots$

Hence, the 6th, 12th, 18th .... orders of inference fringes will be missing.

Ex. 5: Calculate the number of lines per cm in the grating if a diffraction grating used at normal incidence gives a line $\lambda_1 = 6000$ Å in a certain order, superimposed on another line $\lambda_2 = 4500$ Å of the next higher order, and the angle of diffraction is 30°.

Solution:

Since

$(e + d) \sin \theta = \pm n\lambda$

$(e + d) \sin \theta = n \times 6000.$

The next higher order is given by

$(e + d) \sin \theta = (n + 1) \times 4500.$

These orders are superimposed on each other, i.e.,

$n \times 6000 = (n + 1) \times 4500 \Rightarrow n = 3.$

Now,

$(e+d)\sin\theta = \pm n\lambda \Rightarrow \frac{1}{N}\sin\theta = \pm n\lambda$

$\frac{1}{N}\sin 30° = 3 \times 6000 \times 10^{-8} \Rightarrow N = 2778$ lines per cm.

Ex. 6: A diffraction grating used at normal incidence gives a green line (5400 Å) in a certain order of n superimposed on the violet line (4050 Å) of the next higher order. If the angle of diffraction is 30°. Calculate the value of order n. Also, find how many lines per cm are there in the grating?

Solution:

The direction of principal maxima for normal incidence is given as $(e + d) \sin \theta = n\lambda$. Let the nth order maxima of λ_1 coincide with the $(n + 1)$th order maximum of λ_2, then we have

$(e+d)\sin\theta = n\lambda_1 = (n+1)\lambda_2 \Rightarrow n\lambda_1 = (n+1)\lambda_2$

$n = \frac{\lambda_2}{\lambda_1 - \lambda_2} \Rightarrow (e+d) = \frac{\lambda_1\lambda_2}{(\lambda_1 - \lambda_2)\sin\theta}$

Given $\lambda_1 = 5400$ Å $= 5400 \times 10^{-8}$) cm, $\lambda_2 = 4050$ Å $= 4050 \times 10^{-8}$ cm.

$n = \frac{4050 \times 10^{-8}}{5400 \times 10^{-8} - 4050 \times 10^{-8}} = 3$

$(e+d) = \frac{5400 \times 10^{-8} \times 4050 \times 10^{-8}}{(5400 \times 10^{-8} - 4050 \times 10^{-8})\sin 30°}$

Therefore, the number of lines per cm $= \frac{1}{(e+d)} = \frac{(5400 \times 10^{-8} - 4050 \times 10^{-8})\sin 30°}{5400 \times 10^{-8} \times 4050 \times 10^{-8}} = 3086.$

Wave Optics

Ex. 7: How many orders will be visible if the wavelength of incident radiation is 5000 Å and the number of lines on the grating is 2620 to an inch?

Solution:

The direction of principal maxima for normal incidence is given as $(e + d) \sin \theta = n\lambda$. The maximum value of θ is 90°. Therefore, the number of orders visible with the grating is given by

$$(e+d) = n\lambda \Rightarrow n = \frac{(e+d)}{\lambda} = \frac{2.54}{2620 \times 5000 \times 10^{-8}} = 19.4$$

$n > 19$.

Therefore, the maximum number of orders visible in the spectrum is 19.

Ex. 8: A plane transmission grating has 40000 lines. Calculate its resolving power in the second order for a wavelength of 5000 Å.

Solution:

The resolving power of a grating is given by

$$R.P. = \left(\frac{\lambda}{d\lambda}\right) = Nn.$$

Given $n = 2$ and $N = 40000$,

$$R.P. = \left(\frac{\lambda}{d\lambda}\right) = 2 \times 40000 = 80000.$$

Ex. 9: A diffraction grating is just able to resolve two lines of wavelengths 5140.34 Å and 5140.85 Å in the first order. Will it resolve the lines 8037.2 Å and 8037.5 Å in the second order?

Solution:

The resolving power of a grating is given by

$$\left(\frac{\lambda}{d\lambda}\right) = Nn \Rightarrow N = \frac{1}{n}\left(\frac{\lambda}{d\lambda}\right).$$

In the given problem, $\lambda = \frac{5140.34 + 5140.85}{2} = 5140.595$ Å, $d\lambda = 5140.85 - 5140.34 = 0.51$ Å,

$$N = \frac{1}{1}\left(\frac{5140.595}{0.51}\right) = 10180.$$

Hence, the resolving power of the grating in the second order is $Nn = 10180 \times 2 = 20160$. The resolving power required to resolve the lines 8037.2 Å and 8037.5 Å in the second order is given by

$$R.P. = \left(\frac{\lambda}{d\lambda}\right)$$

$\lambda = \frac{8037.2 + 8037.5}{2} = 8037.35$ Å, $d\lambda = 8037.5 - 8037.2 = 0.30$ Å,

$$\left(\frac{\lambda}{d\lambda}\right) = \left(\frac{8037.35}{0.30}\right) = 26791.17.$$

Thus, the grating will not be able to resolve the lines 8037.2 Å and 8037.5 Å in the second order because the required resolving power of 26791.17 is greater than the actual resolving power of 20160.

Ex. 10: A plane transmission grating has 15000 lines per inch. Calculate the resolving power of the grating and the smallest wavelength difference that can be resolved with a light of wavelength 6000 Å in the second order.

Solution:

The resolving power of a grating is given by

$$\left(\frac{\lambda}{d\lambda}\right) = Nn.$$

Given $N = 15000$ and $n = 2$.

$$\left(\frac{\lambda}{d\lambda}\right) = 15000 \times 2 = 30000.$$

Given $\lambda = 6000$ Å; therefore, the smallest wavelength difference is

$$d\lambda = \frac{\lambda}{nN} = \frac{6000}{30000} = 0.2 \text{ Å}.$$

Previous Year Questions (University Examination): Short Questions

1. What do you mean by diffraction of light?
2. What is the difference between interference and diffraction?
3. Distinguish between Fresnel and Fraunhofer diffraction.
4. What do you understand by missing-order spectrum?
5. What do you understand by diffraction grating?
6. State the essential condition for diffraction of light.
7. A diffraction grating has 5000 lines per cm. What is its grating element?
8. Two objects are just resolved in the case of unaided eye. What is the minimum angular separation between the two objects?
9. Distinguish between single- and double-slit diffraction.
10. Define resolving power and limit of resolution.
11. What do you mean by dispersive power of a grating?
12. On what factors does the dispersive power of a grating depend?
13. What do you mean by chief characteristics of grating spectra?
14. What is the main difference between the dispersive power and resolving power of a grating?
15. What is the effect of wavelength on resolving power?
16. What is Rayleigh criterion of resolution?

Wave Optics

Previous Year Questions (University Examination): Long Questions

1. Why is the diffraction pattern generally not observed with an extended source of light?
2. Why does diffraction not occur when light passes through a window?
3. In the case of a single-slit diffraction, what happens when (a) wavelength and (b) slit width increase?
4. On the basis of superposition of waves, compare interference and diffraction phenomena.
5. Why is the diffraction of sound waves more evident in daily life than of the light waves?
6. In a single-slit diffraction experiment, the slit's width is doubled from its initial size. How does this impact the central diffraction band's size and intensity?
7. In a single-slit diffraction pattern, how is the width of the central bright maximum changed when (i) slit width is decreased, (ii) the distance between slit and screen is increased, and (iii) light of smaller wavelength is used.
8. Colored spectrum is seen when we look through a muslin cloth. Why?
9. What happens to the intensity as the order of secondary maxima increases?
10. What changes in the diffraction pattern of a single slit will be observed when the monochromatic source of light is replaced by a source of white light?
11. How would you differentiate between a prismatic and a grating spectrum?
12. Why is the prism spectrum more intense than the grating spectrum?
13. If the number of lines on your grating be doubled, what will happen to the grating element?
14. What will happen to the resolving power of a grating if the number of lines ruled on the grating increases?
15. Why do you not get a complete third-order spectrum in a grating?
16. What will be the possible orders available in the grating spectrum if the grating element lies between λ and 2λ?
17. What will be the effect on intensity of principal maxima of diffraction pattern when single slit is replaced by double slit?
18. What is meant by diffraction of light? Describe the feature of a single-slit Fraunhofer diffraction pattern.
19. Discuss the phenomena of Fraunhofer diffraction at a single slit, and show that the relative intensities of successive maxima are nearly
$$1 : \frac{4}{9\pi^2} : \frac{4}{25\pi^2} : \frac{4}{49\pi^2} : \dots$$
20. Explain the phenomenon of diffraction, and distinguish between Fresnel and Fraunhofer diffraction. Obtain intensities of the diffraction pattern in Fraunhofer diffraction due to a single slit.
21. In Fraunhofer diffraction at single slit, show that the intensity of first subsidiary maximum is about 4.5% of that of the principal maximum.
22. Derive an expression for the intensity distribution due to Fraunhofer diffraction at a single slit.

23. Find the ratio of the intensity of the secondary maximum that is adjacent to the central maximum relative to the central maximum for the single-slit Fraunhofer diffraction pattern.
24. Describe the feature of a double-slit Fraunhofer diffraction pattern? What are missing orders in double-slit Fraunhofer diffraction pattern?
25. Discuss Fraunhofer diffraction at a double slit. What is the effect of increasing the (i) slit width, (ii) slit separation, and (iii) wavelength?
26. Explain briefly how the Fraunhofer diffraction pattern is modified when a single slit is replaced by a double-slit arrangement. (Derivation is not required.)
27. Give the construction and theory of plane diffraction grating, and explain the formation of spectra by it. Additionally, explain what absent spectra in the grating are.
28. Prove that the number of lines per unit length has no effect on the angular width of the principal maximum in a plane diffraction grating, while the total number of lines on the grating does have an effect.
29. What is diffraction grating? What are the chief characteristics of grating spectra?
30. What do you understand by missing-order spectrum? Show that only first-order spectra are possible if the width of the grating element is less than twice the wavelength of light.
31. What do you understand by missing-order spectrum? Which particular spectra will be absent if the widths of transparencies and opacities of a grating are equal?
32. What is dispersive power of a plane diffraction grating? Derive the expression for it.
33. What do you understand by resolving power of an optical instrument? Explain Rayleigh criterion of resolution.
34. Define limit of resolution and resolving power. Derive an expression for the resolving power of a grating.
35. Define resolving power of a grating and obtain an expression for the resolving power of the grating.

Previous Year Questions (University Examination): Numerical Questions

1. Light of wavelength 5500 Å falls normally on a slit of width 22×10^{-5} cm. Calculate the angular position of the first two minima on either side of the central maximum.
2. In Fraunhofer diffraction due to a narrow slit, a screen is placed 2 m away from the lens to obtain the pattern. If the slit width is 0.2 mm and first minima lie 5 mm on either side of the central maximum, find the wavelength of light.
3. Light of wavelength 5000 Å is incident normally on a single slit. The central maximum spreads out at 30° on both sides of the direction of incident light. Calculate the slit width.
4. Light of wavelength 6000 Å falls normally on a straight slit of width 0.1 mm. Calculate the total angular width of the central maximum and also the linear width as observed on a screen placed 1 m away.
5. Plane wave of $\lambda = 6 \times 10^{-5}$ cm falls normally on a slit of width 0.2 mm. Calculate (i) the total angular width of the central maximum and (ii) the linear width of the central maximum on a screen placed 2 m away.

6. When monochromatic light is used to illuminate a single slit that is 0.14 mm wide, diffraction bands are seen on a screen 2 m distant. Calculate the wavelength of light if the second dark band's center is 1.6 cm from the bright band's center.

7. Light with the wavelengths 1 and 2 illuminates a single slit. One notices that the second diffraction minimum of 2 corresponds with the first minima found for 1 owing to Fraunhofer diffraction. What kind of relationship exists between 1 and 2?

8. Determine the angle at which the first dark band and the next brilliant band are generated in the Fraunhofer diffraction pattern of a 0.3-mm-wide slit illuminated by monochromatic light with a wavelength of 6000 nm.

9. A 50 cm focal length lens creates a Fraunhofer diffraction pattern with a single 0.3 mm wide slit. Determine the separations between the first dark band and the next bright band, which are located on either side of the center maxima. The light utilized has a wavelength of 5890 nm.

10. In a double-slit Fraunhofer diffraction pattern, the screen is placed 170 cm away from the slits. The width of the slits is 0.08 mm, and they are 0.4 mm apart. Calculate the wavelength of light if the fringe width is 0.25 cm. Also find the missing orders.

11. Deduce the missing orders for a double-slit Fraunhofer diffraction pattern if the slit widths are 0.16 mm and they are 0.8 mm apart.

12. Two parallel slits have widths 0.15 mm each, and the separation between them is 0.3 mm. They are illuminated normally by light of $\lambda = 6000$ Å, and the emergent light is focused by a convergent lens of focal length 100 cm. Deduce the positions of the first four interference maxima on one side in the focal plane of the lens.

13. How many lines per cm are there in a plane transmission grating which gives the first order of light of wavelength 6000 Å at angle of diffraction 30°?

14. White light (400 nm < λ < 700 nm) is incident on a grating. Show that, regardless of the the value of the grating spacing d, the second- and third-order spectra overlap.

15. A second-order spectral line is shown to deviate through 30° when a parallel beam of monochromatic light is permitted to impinge normally on a flat grating with 1250 lines per cm. Determine the spectral line's wavelength.

16. Find the angular separation of 5048 Å and 5016 Å wavelengths in second-order spectrum obtained by a plane diffraction grating having 15000 lines per inch.

17. The angle of diffraction for the second-order primary maxima for $\lambda = 5 \times 10^{-5}$ cm in a planar transmission grating is 30°. Determine how many lines are there on a centimeter of the grating surface.

18. A green line (5400 Å) in a specific order overlaid on the violet line (4050 Å) of the subsequent higher order is produced by a diffraction grating when employed at normal incidence. How many lines per centimeter of the grating are there if the diffraction angle is 30°?

19. Normally, 5000 Å light is incident on a diffraction grating. Find the difference between the first-order and third-order spectra's deviations. The grating surface has 6000 lines per centimeter.

20. If the incoming light's wavelength is 5000 Å and the grating's line density is 2620 in inches, how many orders will be visible?

21. A diffraction grating used at normal incidence gives a line $\lambda_1 = 6000$ Å in a certain order superimposed on another line $\lambda_2 = 4500$ Å of the next higher order. If the angle of diffraction is 30°, calculate the number of lines in a cm in the grating.

22. Monochromatic light from a He-Ne Laser ($\lambda = 6328$ Å) is incident normally on a diffraction grating of 6000 lines/cm. Find the angles at which one would observe the first-order maximum, second-order maximum, and so on.

23. How many orders will be observed by a grating having 4,000 lines per cm if it is illuminated by light of wavelength in the range 500 nm to 700 nm?

24. Show that only first-order spectra are possible if the width of the grating element is less than twice the wavelength of light.

25. A diffraction grating that has 4000 lines in a centimeter is used at normal incidence. Calculate the dispersive power of the grating in the third-order spectrum for wavelength 5000 Å.

26. A plane transmission grating has 40000 lines. Determine its resolving power in the second order ($n = 2$) for wavelength of 5000 Å.

27. Can D_1 and D_2 lines of sodium light be resolved for ($\lambda_{D1} = 5890$ Å and $\lambda_{D2} = 5896$ Å) in second order? The number of lines in the grating of 2 cm width = 4500.

28. A diffraction grating is just able to resolve two lines of wavelengths 5140.34 Å and 5140.85 Å in the first order. Will it resolve the lines 8037.2 Å and 8037.5 Å in the second order?

29. A plane transmission grating has 15000 lines per inch. Find the resolving power of the grating and the smallest wavelength difference that can be resolved with a light of wavelength 6000 Å in the second order.

30. What must be the minimum number of lines per cm in a half inch width grating to resolve the wavelength 5890 Å and 5896 Å in the first order?

31. Calculate the smallest width necessary for a grating with 1000 lines per cm to be able to resolve the sodium D-line components in the second order. Two sodium line components have wavelengths of 5890 Å and 5896 Å, respectively.

32. Find the smallest number of lines necessary in a plane diffraction grating to (i) resolve the sodium doublet (5890 Å and 5896 Å) in the first order and (ii) resolve it in the second order.

Multiple Choice Questions

1. Bending of light around sharp edges of an object is known as
   - (a) Scattering
   - (b) Diffraction
   - (c) Reflection
   - (d) Refraction

2. The total path diffraction depends only on the angle of diffraction in
   - (a) Fraunhofer diffraction
   - (b) Fresnel diffraction
   - (c) Both (a) and (b)
   - (d) None of the above

3. Which one of the following quantities remains constant during the diffraction process?
   - (a) Speed
   - (b) Wavelength
   - (c) Frequency
   - (d) All of the above

4. The distance between slit and screen is finite in
   - (a) Fresnel diffraction
   - (b) Fraunhofer diffraction
   - (c) Both (a) and (b)
   - (d) None of the above

5. In Fraunhofer diffraction, the distance between screen and slit is
   (a) Finite
   (b) Infinite
   (c) 1.25 cm
   (d) None of the above
6. In Fresnel diffraction, the wavefront incident on the slit is
   (a) Circular
   (b) Spherical
   (c) Cylindrical
   (d) Either cylindrical or spherical

Bibliography

1. Pecharsky, V. K., and Zavalij, P. Y. (2003). *Fundamentals of Diffraction*. Springer US.
2. Warren, B. E. (1990). *X-ray Diffraction*. Courier Corporation.
3. Mittemeijer, E. J., and Welzel, U., eds. (2013). *Modern Diffraction Methods*. John Wiley & Sons.
4. Schwartz, L. H. and Cohen, J. B. (2013). *Diffraction from Materials*. Springer Science & Business Media.
5. Mittemeijer, E. J. and Scardi, P., eds. (2013). *Diffraction Analysis of the Microstructure of Materials*, vol. 68. *Springer Science & Business Media*.

Keys
1. (b)   2. (a)   3. (d)   4. (a)   5. (b)   6. (d)

CHAPTER 13

Fiber Optics

13.1 INTRODUCTION

Communication through optical fiber is one of the remarkable discoveries of the twenty-first century that have brought a revolution in modern times. The transfer of information over long distances was earlier performed through copper wires and coaxial cables. The limitation of these devices, such as limited bandwidth, could not fulfill modern needs and hence were replaced by glass fiber. It was the effort of Alexander Graham Bell, who in 1880 used the light as a carrier of signal. Since the attenuation in the optical fibers was quite high, an attempt to minimize it was done to improve it, and today its features are so fantastic that optical communication through glass fiber with low loss has become a reality. A large number of advantages of optical fibers over the traditional wires and coaxial cables are not hidden now and have been accepted over the entire globe. Apart from their use in communication, optical fibers are widely used in other areas also. In a nutshell, we can therefore say that fiber optics is a backbone of communication infrastructure.

The optical fiber is a cylindrical waveguide system operating at optical frequency. It consists of a core at the center and a cladding outside the core. The core is generally a cylindrical dielectric glass, and cladding is the second dielectric cover usually of glass with a lower refractive index n_2, as shown in Figure 13.1.

For all cases of optical fiber, the refractive index of cladding is always less than the refractive index of the core; that is, $n_1 > n_2$.

- **The Core:** The core is the innermost part and ranges in diameter from 5 μm to 100 μm. The fibers are typically made of silica (SiO_2).
- **The Cladding:** Cladding with a slightly lower refractive index surrounds the core. 125 μm is the typical diameter.

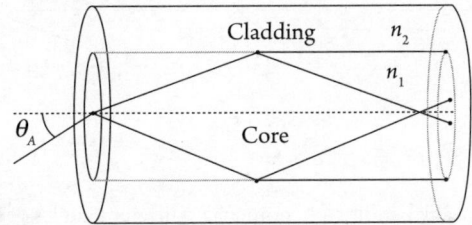

Fig. 13.1 Optical fiber with core–cladding interface.

Advantages of the Cladding

- It strengthens the fiber mechanically and guards against surface impurities that the fiber could come into contact with.
- The cladding can minimize the scattering loss of light brought on by dielectric discontinuities at the surface of the core.

13.2 Basic Principle of Optical Fiber

The purpose of optical fiber is to transport light from one place to another, as the conducting wire carries electrical current. Fiber optics is based on the principle of refraction, that is, total internal reflection. Light propagation from one medium to another is described by Snell's law.

The refractive index of a medium is defined as

$$n = \frac{C}{v},$$

where C is the velocity of light in vacuum, and v is the velocity of light in the medium.

When light passes from the denser to rarer medium, it bends away from the normal, as shown in Figure 13.2.

According to Snell's law,

$$n_1 \sin \varphi_1 = n_2 \sin \varphi_2, \tag{13.1}$$

where φ_1 and φ_2 are the angles of incidence and refraction, respectively.

If the angle of incidence φ_1 is increased, the refracted ray bends more and more away from the normal, and at a particular angle of incidence, the refracted ray passes perpendicular to the normal. This particular angle of incidence is known as critical angle (φ_c). When the angle of incidence is further increased, there is a total internal reflection. The value of critical angle of the two media is given by

$$n_1 \sin \varphi_c = n_2 \sin 90^0$$

$$\sin \varphi_c = \frac{n_2}{n_1}$$

$$\varphi_c = \sin^{-1}\left(\frac{n_2}{n_1}\right). \tag{13.2}$$

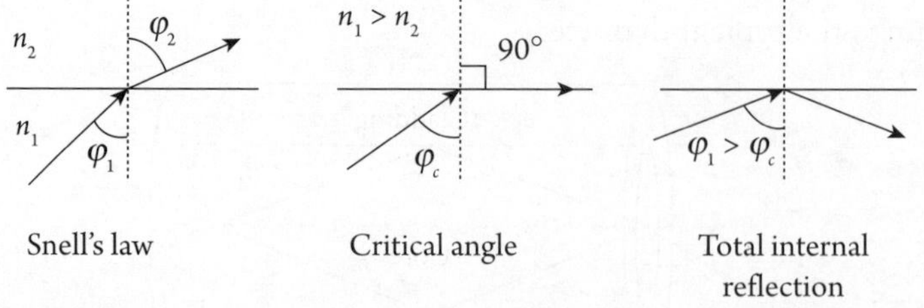

Fig. 13.2 Ray bending phenomena at different angles of incidence.

Fiber Optics

In optical fiber, for the ray entering the fiber, if the angle of incidence φ is greater than the critical angle φ_c, then there is total internal reflection at the interface. This ray will suffer total internal reflection at the lower interface also due to cylindrical symmetry of the fiber structure and therefore get guided through the core by repeated total internal reflection, as if the core of the fiber acts like a continuous layer of two parallel mirror. Thus, an optical fiber acts as a "light guide" and is also known as "optical wave guide."

13.3 Fiber Classification

The optical fibers can be classified mainly based on refractive index profile and modes of propagation.

Classification on Refractive Index Profile

Considering the refractive index profile of the core and cladding materials, fiber can be classified as
  i. Step index fiber (SIF)
  ii. Graded index fiber (GIF)
  iii. W index fiber (WIF)

Step Index Fiber (SIF)

Step index fiber (SIF) is an optical fiber having a core of constant refractive index n_1 and a cladding of slightly lower refractive index n_2. At the core-cladding contact, this type of fiber's refractive index profile abruptly changes, as shown in Figure 13.3. A mathematical of the refractive index profile is given by

$$n(r) = \begin{cases} n_1 & \text{for } r < a \text{ (core)} \\ n_2 & \text{for } r < a \text{ (cladding)} \end{cases}.$$

Graded Index Fiber (GIF)

Graded index fibers do not contain a core with a constant refractive index; instead, the core index $n(r)$ decreases with radial distance, from a maximum value of n_1 near the axis to a constant value n_2 beyond the core radius (a). The refractive index variation of GIF is shown in Figure 13.4.

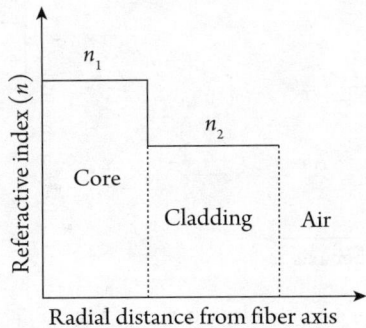

Fig. 13.3 Refractive index profile of step index fiber with radial distance from fiber axis.

$$n(r) = \begin{cases} n_1 = \left[1 - 2\Delta\left(\frac{r}{a}\right)^2\right]^{\frac{1}{2}} & \text{for } r < a \text{ (core)} \\ n_2 = [1 - 2\Delta]^{\frac{1}{2}} & \text{for } r < a \text{ (cladding)} \end{cases}.$$

When light travels through the core of a step index fiber, it does so in a Zig-zag pattern made up of straight line segments, but in a graded index fiber, the light rays are bent by refraction and curve inward toward the axis.

W-Index Fiber (WIF)

The width of the cladding with refractive index n_2 is made thick enough in this fiber to include all of the ends of the light energy field, which extend just past the core interface. A second, thicker layer of cladding with a refractive index of n_3 surrounds the first cladding with a refractive index of n_2, where $n_1 > n_3 > n_2$, as shown in Figure 13.5. This results in a W-shaped refractive index profile, which also strips the first cladding's leaky modes.

Modal Classifications

Considering the mode of transmission, fiber is of two basic types (Figure 13.5).
  i. Multimode
  ii. Single mode (SM)

In multimode fiber, the light can travel many different paths through the core of the fiber and can enter and leave the fiber at various angles. The number of modes of propagation in an optical fiber depends on the refractive index profile of core, cladding, its geometry, and the wavelength of light used. The core diameter of fiber can be reduced to achieve single mode of propagation with known values of refractive index and wavelength.

However, manufactures have concentrated mainly on three broad classes of fiber:
  a. Step index multimode fiber
  b. Graded index multimode fiber
  c. Step-index SM fiber

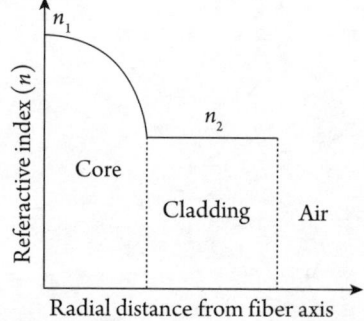

Fig. 13.4 Refractive index profile of graded index fiber with radial distance from fiber axis.

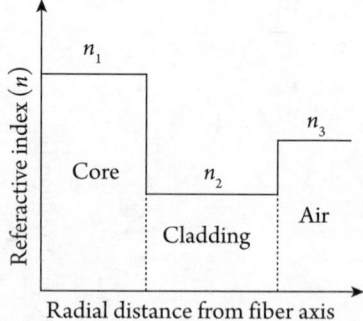

Fig. 13.5 Refractive index profile of W index fiber with radial distance from fiber axis.

Fiber Optics

13.4 ACCEPTANCE ANGLE, ACCEPTANCE CONE, AND NUMERICAL APERTURE OF A FIBER

All rays entering the fiber core will not continue to be propagated down the length. Only those rays that will have the angle of incidence greater than the critical angle at the core cladding interface are transmitted by total internal reflection. The light ray should impinge the core of the optical fiber within a maximum external incident angle $\theta = \theta_a$ to have the condition for propagation down the length. The path of a ray entering the fiber core at an angle $\theta = \theta_a$ for which the ray propagates grazing the core-cladding interface is shown in Figure 13.6. This angle θ_a is known as the angle of acceptance for fiber.

Angle θ_a can be derived using Snell's law. Let us consider the ray from air (refractive index n_0) is incident at an angle θ onto the perpendicular end face of the fiber for which the angle of refraction is θ'. Using Snell's law, we get

$$n_0 \sin\theta = n_1 \sin\theta' = n_1 \sin(90° - \theta'')$$

$$n_0 \sin\theta = n_1 \cos\theta'',$$

where θ'' is the angle of incidence on the fiber core-cladding interface. If this angle is less than critical angle, light ray will be refracted; otherwise, it will be totally reflected at the fiber wall if the angle of incidence is greater than the critical angle θ_c. Thus, light ray will propagate if it is entering on the core within the maximum incident angle θ_a. When $\theta'' = \theta_c$ and $\theta = \theta_a$, we get

$$n_0 \sin\theta_a = n_1 \cos\theta_c$$

or $n_0 \sin\theta_a = n_1 \left(1 - \sin^2\theta_c\right)^{\frac{1}{2}}.$ (13.3)

Since we know that

$$\sin\theta_c = \frac{n_2}{n_1}.$$

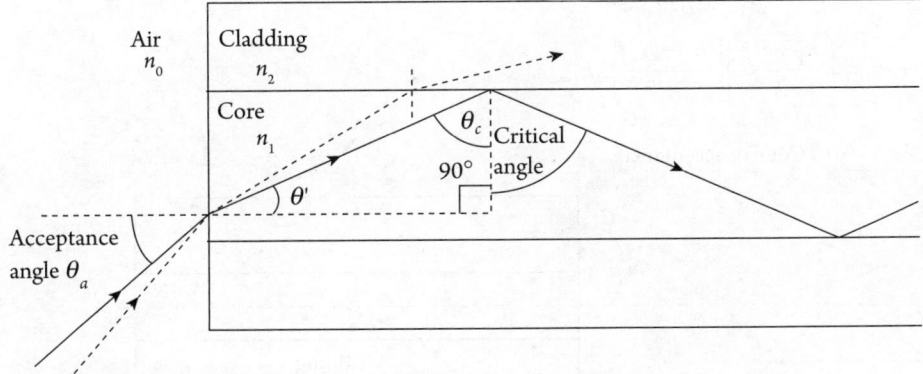

Fig. 13.6 Acceptance angle and light ray propagation in optical fiber.

From equation (13.3),

$$n_0 \sin\theta_a = n_1 \left(1 - \frac{n_2^2}{n_1^2}\right)^{\frac{1}{2}}$$

$$n_0 \sin\theta_a = \sqrt{n_1^2 - n_2^2}$$

$$\theta_a = \sin^{-1}\sqrt{n_1^2 - n_2^2} \quad [\because n_0 = 1 \text{ for air}].$$

Thus, the above expression is for acceptance angle θ_a. All the rays falling within the cone formed with acceptance angle θ_a as vertex angle would be transmitted down the fiber, and this cone is, as shown in Figure 13.7, known as acceptance cone for the fiber. The rays entering the fiber-core within an acceptance cone, specified by conical half angle θ_a are transmitted.

Acceptance angle is the maximum angle of incidence, specified by the conical half angle for the light rays to be transmitted through the optical fiber.

"Numerical aperture (NA) of a fiber measures its light gathering power." This is also known as the highest angle for acceptance of light into the core of the fiber.

$$\sin\theta_a = \sqrt{n_1^2 - n_2^2}.$$

In graded index fiber, numerical aperture, which is the function of radius, can be given as

$$NA(r) = \sqrt{n^2(r) - n_2^2},$$

where $NA(r)$ is a numerical aperture at a distance r from the center.

If we denote $n_1 - n_2 = \Delta n$ and $n_1 + n_2 = 2n$, we get

$$\sin\theta_a = \sqrt{2n\Delta n}.$$

The above expression is the numerical aperture of the fiber, while the acceptance angle is given by $\theta_a = \sin^{-1}\sqrt{2n\Delta n} = \sin^{-1}(NA)$.

We know that

$$\sin\theta_a = \sqrt{n_1^2 - n_2^2}.$$

Since $n_1^2 - n_2^2 = (n_1 + n_2)(n_1 - n_2)$,

$$n_1^2 - n_2^2 = \left(\frac{n_1 + n_2}{2}\right)\left(\frac{n_1 - n_2}{n_1}\right)2n_1,$$

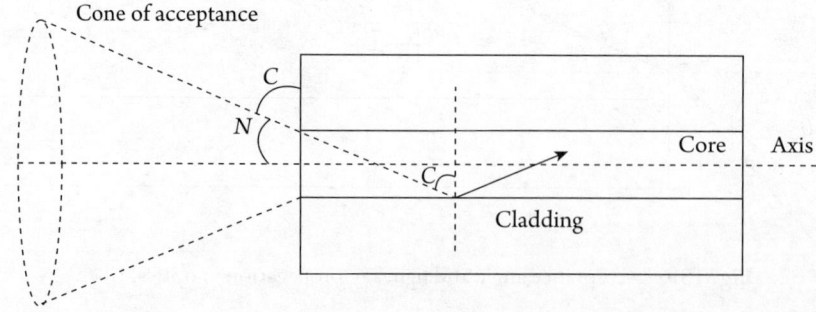

Fig. 13.7 Acceptance cone and numerical aperture of optical fiber.

Fiber Optics

$$n_1^2 - n_2^2 = \left(\frac{n_1 + n_2}{2}\right)\Delta \cdot 2n_1,$$

where $\Delta = \dfrac{n_1 - n_2}{n_1}$

Here, Δ is known as fractional change in refractive index or fractional refractive index. **Fractional refractive index is always positive because n_1 is always greater than n_2, which is a prime requirement for total internal reflection.** In order to guide light rays effectively through the fiber, $\Delta \ll 1$. Normally, its value is nearly 10^{-1}.

Assuming $\dfrac{n_1 + n_2}{2} = n_1$, the above equation can be written as

$$n_1^2 - n_2^2 = 2n_1^2\Delta.$$

Therefore, $\text{NA} = \sqrt{2n_1^2\Delta}$. \hfill (13.4)

The optical fiber's capacity to gather light is expressed as NA. It is a measurement of the amount of light that a fiber allows to pass in.

13.5 Modes in Optical Fiber

When light enters an optical fiber, total internal reflection causes waves with orientations higher than the critical angle to become stuck inside the fiber. However, not all of these waves go along the fiber. Typically, just a few ray directions are permitted to move. The fiber's modes are known for these permitted directions.

With the exception of the axial orientation, all the pathways are Zig-zag, as shown in Figure 13.8. Phase shift happens when a Zig-zag beam is repeatedly reflected off the fiber's walls. As a result, due to destructive interference, waves coursing along certain Zig-zag lines will be in phase and enhanced, while waves coursing along some other paths will be out of phase and diminished. Modes are the directions that light travels when the waves inside the fiber are in phase. The ratio d/λ, where d is the diameter of the core and λ is the wavelength of the wave being transmitted, determines how many modes a fiber can handle.

The higher-order modes indicate the wave propagates at angles close to the critical angle φ_c and lower-order modes indicate the wave propagates at angles much lower than the critical angle. The zero order ray travels along the axis and is known as the axial ray. The higher-order modes tend to send light energy into the cladding. This energy is lost ultimately.

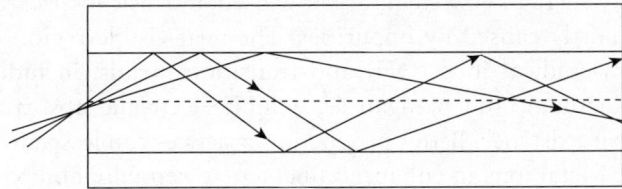

Fig. 13.8 Various propagation modes in optical fiber.

13.6 Modal Classification of Optical Fibers

The fibers are classified as follows:
- Single-mode fiber
- Multimode fiber

A **single-mode fiber** has a smaller core diameter and can support only one mode of propagation. On the other hand, a multimode fiber has a larger core diameter and supports a number of modes. Multimode fibers are further distinguished on the basis of index profile. Index profile is a plot of refractive index versus the distance. The index profile of a multimode fiber can be either a step index type or a graded index type. The index profile of a single-mode fiber is usually a step index type.

There are three major configurations and their index profiles.
- Multimode step-index fiber
- Multimode graded-index fiber
- Single-mode step-index fiber

13.7 Attenuation in Fibers

A fiber will gradually attenuate an optical signal as it travels through it. The optical output power from a fiber of length L divided by the optical input power is known as signal attenuation.

$$\alpha = \frac{10}{L} \log \frac{P_i}{P_0},$$

where P_i is the input power of optical signal launched at one end of the fiber, and P_0 is the output power of the optical signal emerging from the other end of the fiber. In case of an ideal fiber, $P_0 = P_i$ and the attenuation would be zero. The unit of measurement of attenuation is decibel/kilometer (dB/km).

13.8 Kinds of Attenuation

i. **Absorption by Material**

This includes absorption due to the light interacting with the molecular structure of the material, as well as loss because of material impurities. Even a highly pure glass absorbs light in specific wavelength regions. Strong electronic absorption occurs at UV wavelengths, while vibrational absorption occurs at IR wavelengths. Intrinsic absorption refers to these losses in absorption that are a fundamental characteristic of the glass itself. Where fiber systems are now used, intrinsic losses are negligible. Losses in fibers are mostly caused by impurities. The near-visible region of the spectrum is where hydroxyl radical ions (OH) and transition metals, including copper, nickel, chromium, vanadium, and manganese, exhibit electronic absorption. Heavy losses result from their existence. Better production practices can lessen losses brought on by contaminants. Metal ions in enhanced fibers are essentially nonexistent. Because OH ions cannot be properly reduced, they cause the most loss.

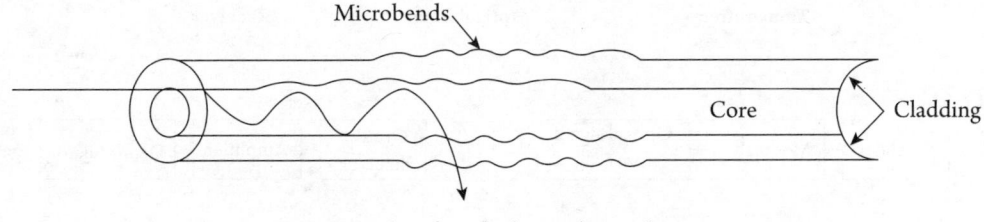

Fig. 13.9 Microbends in optical fiber.

ii. **Scattering**

Power loss occurs as a result of light scattering from an obstruction. Glass has regional differences in refractive index as a result of local variations in microscopic density. These variances, which are unavoidable and intrinsic to the manufacturing process, operate as barriers and disperse light in all directions. Rayleigh scattering is the term for this.

iii. **Waveguide and Microbend Losses**

The losses introduced during manufacturing or installation processes come in this category. Structural variations in the fiber, or fiber deformation, cause radiation of light away from the fiber. Microbend, very minute disturbances in core size, also causes loss of radiation of light, as shown in Figure 13.9.

13.9 MERITS OF OPTICAL FIBER

Optical fibers have many advantages that are not found in conducting wires. Some of them are listed below:

Silica (SiO_2), one of the most widely available and reasonably priced elements on earth, is used to make optical fibers. Compared to an analogous cable communication system, fiber optic transmission is less expensive overall. It is smaller, lighter, more flexible, and stronger, all at the same time. Use of it is secure. Unintentionally shorting out a wire communication link at high voltage lines might spark or ignite nearby flammable gases and cause significant damage. Fiber linkages prevent such mishaps since they are constructed from insulating materials.

It is resistant to RFI and EMI. The optical fiber entirely traps the light waves that are travelling down it; they cannot escape. Light cannot pair into the fiber from the sides either. Due to these characteristics, cross talk is less likely to occur when optical fiber is utilized. Transmission is therefore more private and secure.

Because of larger bandwidth, optical fibers have the ability to carry large amounts of information. **A telephone cable composed of 900 pairs of wire can handle 10000 calls, while 1 mm optical fiber can transmit 50000 calls. The optical fiber exhibits low loss per unit length.**

13.10 OPTICAL FIBER COMMUNICATION

The main components of a general optical fiber communication system are shown in Figure 13.10

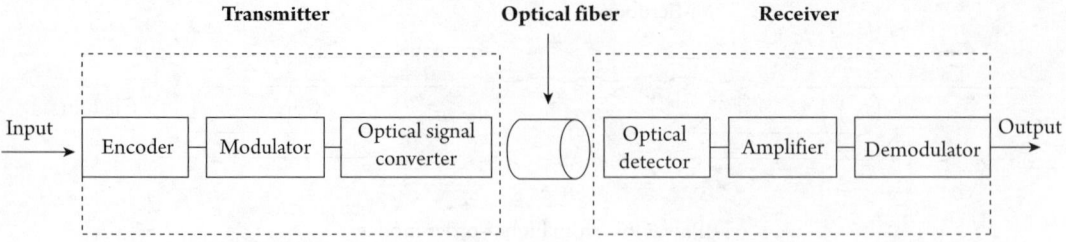

Fig. 13.10 Various components of an optical fiber communication system.

The transmission system has an information encoder or signal shaping circuit before the modulation or electronic driver stage that operates the optical source. Thus, the input signal is converted into optical signal. It is launched into an optical fiber from the launching end. The signal is received at the receiver end and is converted back into electrical signal by an optical detector. This electrical signal is then amplified and decoded or demodulated to obtain the information transmitted. There was rapid development in optical communication, especially after the availability of the laser and low-loss optical fibers. Theoretically, the optical system has a capacity of 10^5 times the microwave system, and this is one of the important reasons for its development. The block diagram of the transmitting system is given in Figure 13.10.

13.11 Applications of Optical Fiber

There are many advantages of using optical fibers over electrical cables. The optical fibers are widely used in the following systems:

1. In telecommunication networks, for example, telephone, cable TV, videophone, teleconferencing, and so on.
2. Undersea cable systems are used in international telecommunication networks.
3. They are used in digital transmission telecommunication networks.
4. In military applications, various factors are considered, such as size, weight, deployability, survivability in conventional and nuclear attack, security, and so on. In aircraft, ships, and tanks, optical fibers are used due to small size and weight. Optical fiber cables, being immune to electromagnetic interferences and noisy environment, are very useful in military mobiles.
5. Civil, consumer, and industrial applications include commercial television transmission (short-distance links between studio and broadcast center), closed-circuit television (CCTV) for security and traffic surveillance, networks for cable and common antenna television (CATV), links in automobiles between microcomputers for engine and transmission controls, power window and seat controls, control signals (for example, light), centralized locking, and door lamps for telemetry and control communications in industrial environments, and so on.
6. Optical sensor systems.
7. In local area networks.

13.12 Light-source Materials for Optical Fiber Communication

For optical fiber communication, lasers are used as carrier waves because they are highly monochromatic and allow a very large information bandwidth, while optical fibers are used as media for the transmission of optical waves. In early opto-electronic system used for fiber optics, it was traditional to use well-known GaAs–Al–GaAs systems for making lasers. These light sources are highly efficient, and good detectors can therefore be made using them. However, these sources operate in a range of ~0.9 μm, where attenuation is greater than that for larger wavelengths. Today, the modern system operates in the wavelength range of 1.3 and 1.55 μm minima. For such wavelengths, lasers can be made using InGaAs or InGaAsP grown on InP, and detectors can be made of the same materials. Such low-power coherent sources are particularly suitable for optical fiber communication.

13.13 Semiconductor Materials for Solar Cells

The solar cell is considered to be a major source for getting the energy utilizing the solar radiations. The first solar cell was developed by Chapin, Fuller, and Pearson in 1954 using a diffused p–n junction. Later, another solar cell using II–VI compound CdS was developed. The main concern is to get maximum conversion efficiency, low cost, and immunity from high-energy radiation when used in the upper atmosphere, such as in satellites.

In a conventional p–n junction solar cell having a single-band gap Eg, only those radiations contribute to the output whose energy exceeds the band gap Eg. Any material exhibiting a band gap between 1 and 2 eV can be considered as solar cell materials. The maximum efficiency is found to be 31% for a material having a band gap of 1.35 eV using III–V compound semiconductors. Many factors degrade the ideal efficiency and therefore practical obtainable efficiency, and thus practical obtainable efficiency is always lower. Recently, research has been undertaken to develop low-cost, flat panel solar cells. For terrestrial applications, and in order to maintain at least 10% conversion efficiency, thin-film CdS solar cells as well as amorphous Si solar cells are important.

A typical p–n junction solar cell is given below. It comprises a front ohmic contact stripe and fingers, a back ohmic contact that spans the whole back surface, an antireflection coating on the front surface, and a shallow p–n junction created at the surface (via diffusion).

The single silicon solar cell of surface area 2 cm^2 provides an open-circuit voltage of 0.5 to 0.6 V and a current (open circuit) of 30–60 mA; therefore, a need has been realized to contact such individual solar cells to make an array that could provide substantially high values of voltage and current. Such an array is shown in Figure 13.11, along with I–V characteristics and conversion efficiency values up to 60°C. Such an array delivers a power of about 10 and conversion efficiency of 11.5%. It has also been noticed that efficiency decreases linearly with temperature to about 200°C for silicon and 300°C for GaAs. For alethic applications, the high-energy particle radiation in other spaces produces defects in semiconductors that cause a reduction in output power. To improve the radiation tolerance, lithium has been incorporated, which can diffuse and combine the radiation-induced point defects and neutralize the defects.

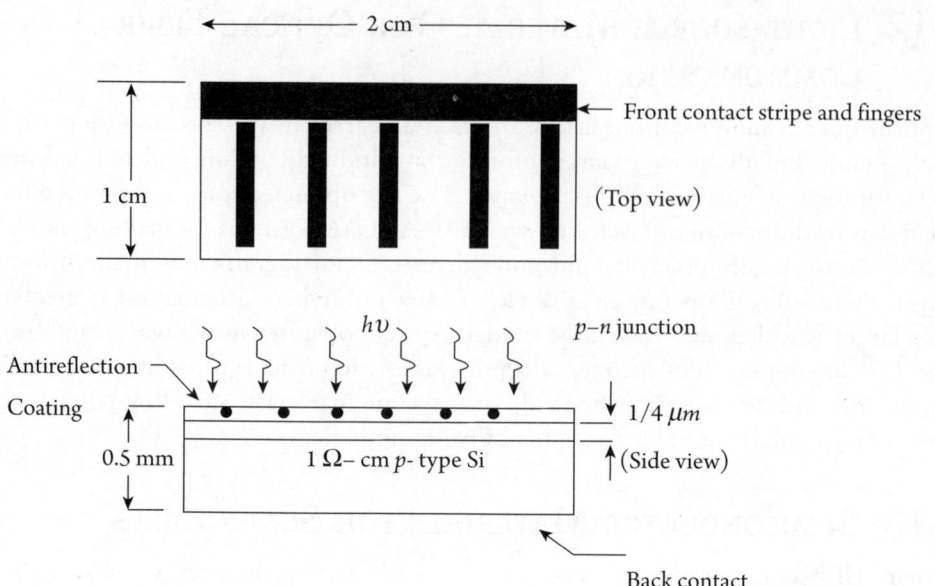

Fig. 13.11 Schematic representation of a silicon p–n junction solar cell.

Furthermore, the solar cell is converted with a cerium-dosed micro sheet to minimize the number of high-energy particles striking its surface.

Antireflection Coating

Antireflection coating increases short-circuit current and open-circuit voltage, which in turn boosts conversion efficiency. The addition of antireflection coating decreases total reflection to a few percent. The materials used for antireflection coating are explained in the section that follows.

Heterojunction, Interface, and Thin Film Solar Cells

The advantages of heterojunction solar cells over conventional p–n junction solar cells are listed below:
- Enhanced short wavelength spectral response
- Lower series resistance
- High radiation tolerance, if first semiconductor is thick in addition to being high in band gap.

A heterojunction solar cell is one in which a large band gap material has been added upon a p–n homojunction, as shown in Figure 13.11. It consists of p-type $Ga_{1-x}Al_x$ As/P-GaAs/n-GaAs. The wide gap semiconductor acts as a window, allowing photons of energy less than Eg_1. Those photons, which have energies in between Eg_1 and Eg_2, will create carriers in homojunction. If the absorption coefficient is high in the lower band gap semiconductor, the carriers are generated in the depletion region or close to it, so the collection efficiency is high. If the heteroface material is an indirect band gap material with $x > 0.4$ and is thin enough, many photons with $h\nu > Eg_1$ will pass through and be efficiently converted in the low-gap semiconductor. As x increases, the band gap Eg_1 also increases; therefore, the spectral response extends to higher photon energies.

One interesting heterojunction solar cell is the conducting glass–semiconductor heterojunction. The conducting glasses include oxide semiconductors, such as indium oxide with band gap 3.5 eV, tin oxide with band gap 3.5 eV, and indium tin oxide with band gap 3.7 eV. All these materials in thin film form have unique properties like good electrical conductivity and high transparency. They serve not only as a part of heterojunction but also as an antireflection coating.

Thin Film Solar Cells

Solar cells can also be prepared by depositing amorphous or polycrystalline layers of semiconductor materials, such as, CdS, Si, GaAs, and InP, on substrates such as glass, plastic ceramic metal, graphite, or metallurgical grade silicon. This can be done using any one technique such as evaporation plasma, vapor growth, or plating. Most light will be absorbed if the semiconductor thickness is greater than the inverse of the absorption coefficient, the diffusion length is greater than the film thickness, and the majority of photo-generated carriers can be collected if the diffusion length is greater than the film thickness.

The chief advantage of thin film solar cell is its promise of low cost with the use of relatively low-cost materials. The main disadvantages are low efficiency and long-term instability. The low efficiency is partly caused by the poor quality of semiconductor material grown on foreign substrates.

Another thin film solar cell is a heterojunction of $CuInSe_2/Cds$. The short circuit current and efficiency are dependent on grain size. The maximum short circuit current (~ 2.6 mA/cm^2) is obtained for 2.5 μm grain radius, and at this radius, efficiency is nearly 6.6%. Many ternary compounds are potential candidates for low-cost solar cell application, as discussed earlier.

Additionally, amorphous silicon is well-suited for use in solar cells. By RF light discharge decomposing silicon onto metal to ITO-coated glass substrates, layers between 1 and 3 meters thick are formed. Crystalline silicon has an indirect band gap of 1.1 eV, but hydrogenated a-si has an optical absorption with a direct band gap of 1.6 eV. This is a significant difference between the two types of silicon. On a thin sheet of hydrogenated amorphous silicon, p–n junction and Schottky-barrier solar cells have been created.

Solved Problems

Ex. 1: In an optical fiber, the core material has a refractive index of 1.6. The refractive index of clad material is 1.3. What is the value of the critical angle? Also, calculate the value of the angle of acceptance.

Solution:

Critical angle is given by

$$\sin \varphi_c = \frac{1.3}{1.6} = 0.8125$$

$$\therefore \varphi_c = 54.3°$$

Acceptance angle $\theta_0 = \sin^{-1}\left[\sqrt{n_1^2 - n_2^2}\right] = \sin^{-1}\left[\sqrt{1.6^2 - 1.3^2}\right]$
$= \sin^{-1}(0.87)$
$= 60.5°$

Angle of acceptance cone $= 2\theta_0$

Ans. $= 121°$.

Ex. 2: The numerical aperture of an optical fiber is 0.5 and the core refractive index is 1.54. Find the refractive index of the cladding.

Solution:

Cladding refractive index, $n_2 = \sqrt{n_1^2 - (NA)^2}$
$= \sqrt{1.54^2 - 0.5^2}$

Ans. $= 1.46$.

Ex. 3: What is the numerical aperture of an optical fiber cable with a clad index of 78 and a core index of 1.546?

Solution:

$NA = \sqrt{n_1^2 - n_2^2}$
$= \sqrt{1.546^2 - 1.378^2}$
$= \sqrt{0.491}$

Ans. $= 0.70$.

Ex. 4: A fiber cable has an acceptance angle of 30° and a core index of refraction 1.4. Calculate the refractive index of the cladding.

Solution:

$\sin \theta_0 = \sqrt{n_1^2 - n_2^2}$

$\therefore \sin^2 \theta_0 = n_1^2 - n_2^2$

$n_2^2 = n_1^2 - \sin^2 \theta_0$
$= (1.4)^2 - \sin^2 30°$
$= 1.96 - 0.25$

Ans. $n_2 = 1.308$.

Ex. 5: Optical power of 1 mW is launched into an optical fiber of length 100 m. If the power emerging from the other end is 0.3 mW, calculate the fiber attenuation.

Solution:

Attenuation, $\alpha = \dfrac{10}{L}\log\dfrac{P_i}{P_0} = \dfrac{10}{0.1 \text{ km}}\log\left(\dfrac{1\,mW}{0.3\,mW}\right)$

Ans. $= 52.3$ dB/Km.

Fiber Optics

Ex. 6: Calculate the fractional difference between the core and cladding refractive indices for a step index fiber having core and cladding refractive indices 1.55 and 1.50, respectively.

Solution:
We know that the fraction difference Δ is

$$\Delta = \frac{n_1 - n_2}{n_1},$$

where n_1 and n_2 are the refractive indices of core and cladding materials, respectively.

Here $n_1 = 1.55$ and $n_2 = 1.50$.

Hence, $\Delta = \dfrac{1.55 - 1.50}{1.55}$.

Ans. = 0.032.

Ex. 7: Calculate the numerical aperture, acceptance angle, and the critical angle of the fiber from the following data: μ_1 (core refractive index) = 1.50 and μ_2 (cladding refractive index) = 1.45.

Solution:
We know that

Numerical aperture, $NA = \mu_1 \sqrt{(2\Delta)}$,

where $\Delta = \dfrac{n_1 - n_2}{n_1}$

$\therefore \quad \Delta = \dfrac{1.50 - 1.45}{1.50}$

$= 0.033$.

So, $NA = 1.50 \sqrt{(2 \times 0.033)}$

$= 1.50 \times 0.257$

$= 0.385$.

Acceptance angle $\theta_0 = \sin^{-1}(NA) = \sin^{-1}(0.385) = 22.63°$.

According to Snell's law

$\sin \theta_c = \dfrac{\mu_2}{\mu_1}$ or $\theta_c = \sin^{-1}\left(\dfrac{n_1}{n_2}\right) = \sin^{-1}\left(\dfrac{1.45}{1.50}\right)$

or $\theta_c = \sin^{-1}(0.967)$

Ans. = 75.3°.

Ex. 8: If the fractional difference between the core and cladding refractive indices of a fiber is 0.0135 and numerical aperture is 0.2425, calculate the refractive indices of the core and cladding materials.

Solution:
We know that $NA = n_1 \sqrt{(2\Delta)}$

and $\Delta = \dfrac{n_1 - n_2}{n_1}$,

where n_1 and n_2 are the refractive indices of core and cladding materials, respectively.

Here, NA = 0.2425 and $\Delta = 0.0135$

$$\therefore n_1 = \frac{NA}{\sqrt{2\Delta}} = \frac{0.2425}{\sqrt{2 \times 0.0135}} = \frac{0.2425}{0.1643} = 1.476$$

and $\Delta = 0.0135 = \dfrac{n_1 - n_2}{n_1} = \dfrac{1.476 - n_2}{1.476}$

or $1.476 - n_2 = 0.0135 \times 1.476$

$n_2 = 1.476 - 0.02$

Ans. = 1.456.

Ex. 9: The main optical power launched into a fiber of length 6 km is 100 μW. The mean optical power at the fiber output is 5 μW. Calculate the signal attenuation per kilometer for the fiber.

Solution:

Signal attenuation = $10 \log_{10} \dfrac{P_i}{P_0} = 10 \log_{10} \dfrac{100 \times 10^{-6}}{5 \times 10^{-6}}$

$= 13.0$ dB.

The signal attenuation per kilometer = $\dfrac{13.0}{6} = 2.16$ dB/km. **Ans.** 2.16 dB/km.

Ex. 10: The sum of refractive indices of core and cladding is 2.89 and the difference is 0.03. Determine the NA for the optical fiber.

Solution:

NA = $(n_1^2 - n_2^2)^{1/2} = [(n_1 + n_2)(n_1 - n_2)]^{1/2}$

$= \sqrt{(2.89 \times 0.03)}$

Ans. = 0.29.

Previous Year Questions (University Examination)

1. Discuss the advantages and disadvantages of optical fibers over conventional communication transmission media.
2. With the help of a ray diagram, show how optical fibers can guide light waves? Explain what you understand by modes and numerical aperture.
3. Derive an expression for the NA of a SI fiber in terms of refractive index of the core and the relative refractive index difference between the core and the cladding.
4. Explain the difference between the step-index fiber and graded-index fiber.
5. What are the different types of attenuation losses in an optical fiber. Discuss the absorption losses.
6. Discuss the propagation of light wave through an optical fiber in detail. What do you mean by single-mode and multimode fiber? Explain clearly.
7. Write a note on optical fibers and their application.
8. What are the types of optical fiber? Mention the advantages of optical fiber communication.

Fiber Optics

9. What do you mean by "fiber optics"? Find expressions for acceptance angle and numerical aperture. Also, explain how a light signal propagates in a fiber.
10. Explain the terms acceptance angle and NA. What is meant by single-mode and multimode fiber?

Multiple Choice Questions

1. The fiber optics principle is based on
    (a) Reflection
    (b) Refraction
    (c) Total internal reflection
    (d) None of the above
2. The diameter of the core in single-mode fiber is
    (a) 70 μm
    (b) 110 μm
    (c) 10 μm
    (d) 10 m
3. Power loss in optical fiber is due to
    (a) Absorption of light
    (b) Scattering of light
    (c) Bending of optical fiber
    (d) All of the above
4. Numerical aperture of optical fiber is defined as
    (a) Core
    (b) Critical angle
    (c) Acceptance angle
    (d) Cladding
5. Which one of the following is a type of optical fiber?
    (a) Polarization preserving
    (b) Vapour axial deposition
    (c) Polished
    (d) None of the above
6. One of the optical fibers classified by the refractive index is
    (a) Step index fiber
    (b) Graded index fiber
    (c) W index fiber
    (d) All of the above
7. Optical fiber is also termed as
    (a) Coaxial cable
    (b) Optical waveguide
    (c) Twin pair cable
    (d) None of the above
8. Which parameter determines the maximum number of modes propagated through optical fiber?
    (a) K number
    (b) C number
    (c) V number
    (d) L number
9. One of the optical fibers classified based on material is
    (a) Ceramic
    (b) Copper
    (c) Glass
    (d) Aluminium
10. The source used in single-mode fiber is
    (a) LED
    (b) Sodium lamp
    (c) Mercury lamp
    (d) LASER

BIBLIOGRAPHY

1. Singal, T. L. (2017). *Optical Fiber Communications: Principles and Applications.* Cambridge University Press.
2. Ugale, S. P. and Mishra, V. (2012). *Fiber-Optic Communication: Systems and Components.* Wiley.
3. Kolimbiris, H. (2003). *Fiber Optics Communications: United States Edition.* Pearson.
4. Agrawal, G. P. (2015). *Fiber-Optic Communication Systems,* 3rd ed. (with CD ROM). Wiley.
5. Agarwal D. C. (2005). *Fiber Optic Communication.* S. Chand & Company.
6. Mynbaev, D. K. (2002). *Fiber-Optics Communications Technology,* 1st ed. Pearson Education.
7. Palais, J. C. (2008). *Fiber Optic Communications,* 5th ed. Pearson Education India.
8. Gerd, K. (2010). *Optical Fiber Communications,* 4th ed. McGraw-Hill Education.

Keys

1. (c)   2. (c)   3. (d)   4. (c)   5. (a)   6. (d)   7. (b)   8. (c)   9. (c)   10. (d)

CHAPTER 14

Physics Practicals

List of Experiments

1. Determine the energy band gap of N-type Germanium (Ge) semiconductor using the four-probe method.
2. Verify Stefan's fourth power law for black body radiation and determine the exponent of temperature.
3. Study of thermoelectricity: Determine thermopower of copper-constantan thermocouple.
4. Study the variation of magnetic field with distance along the axis of current-carrying coil and then estimate the radius of the coil.
5. Study of Carrey Foster bridge: Determine resistance per unit length of the bridge wire and of a given unknown resistance.
6. Determine specific charge (charge to mass ratio; e/m) for electron.
7. Study of tangent galvanometer: Determine reduction factor and horizontal component of the earth's magnetic field.
8. Determine the wavelength of sodium light using Newton's rings method.
9. Determine the concentration of sugar solution using half-shade polarimeter.
10. Determine the wavelength of spectral lines of mercury (for violet, green, yellow-1, and yellow-2) using plane transmission grating.
11. Determine the charge sensitivity and ballistic constant of a ballistic galvanometer.
12. Determine the wavelength of spectral lines of hydrogen and hence the value of Rydberg constant.
13. Draw the V-I characteristic of light-emitting diode (LED) and determine the value of Planck's constant.

Experiment 1: Band Gap of Semiconductor

Objective: To determine the energy band gap of n-type Ge-semiconductor using four-probe method.

Apparatus Used: The following apparatus are used in this experiment:
1. Probe arrangement (shown in Figure 14.1).
2. n-type Ge-crystal chip with nonconducting base.
3. Oven with temperature up to 200°C.
4. A constant current generator power supply.
5. Power source for the oven.
6. Thermometer.

Formula Used: Energy band gap is given by

$$E_g = 2 \times k \times 10^3 \times 2.3026 \times (\text{slope})$$

$$E_g = 0.396 \times (\text{slope}),$$

where slope $= \dfrac{AB}{BC} = \dfrac{\log_{10}\rho}{\dfrac{1}{T} \times 10^3}$, [Refer to Figure 14.2 for slope.]

where k is the Boltzmann constant, and its value is $8.6 \times 10^{-5} \dfrac{ev}{\text{deg}.K}$ and ρ is resistivity of n-type Ge-semiconductors.

Theory of the Experiment: The density of electrons in the conduction band is given by

$$n_e = 2 \times \left(\dfrac{2\pi m_e KT}{h^2}\right)^{3/2} \exp\left(\dfrac{E_f - E_c}{kT}\right).$$

Similarly, the density of holes in the valence band is given by

$$n_h = 2 \times \left[\dfrac{2\pi m_h KT}{h^2}\right]^{3/2} \exp\left(\dfrac{E_v - E_F}{kT}\right).$$

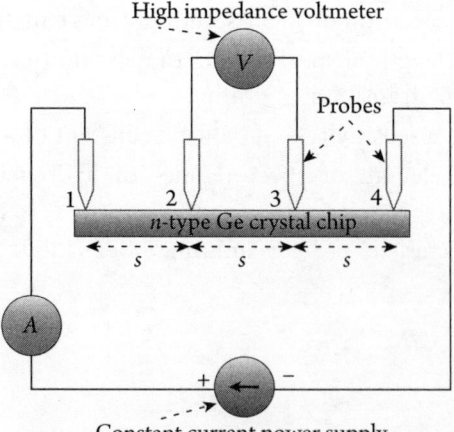

Fig. 14.1 Experimental setup of the four-probe method.

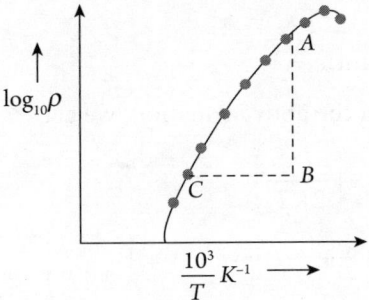

Fig. 14.2

In the case of intrinsic semiconductors, $n_e = n_h$.
Now, the Fermi level is given by

$$E_f = \frac{E_c + E_v}{2} + \frac{3}{2} KT \log\left(\frac{m_h}{m_e}\right),$$

where m_e is the mass of electron and m_h is mass of hole carrier.
When $m_h \cong m_e$, we find that

$$E_f = \frac{E_c + E_v}{2}.$$

Therefore, the Fermi level lies exactly halfway (middle) between the top of the valence band and the bottom of the conduction band. However, in reality, since $m_h > m_e$, the Fermi level is raised slightly as T increases.

In the case of N-type semiconductor, the concentration of electrons in conduction band is

$$n_e = (2N_d)^{1/2} \left(\frac{2\pi m_e KT}{h^2}\right)^{3/4} \exp\left(\frac{E_i - E_c}{2kT}\right),$$

where $(E_i - E_c)$ is called ionization energy of donors, N_d is the donor electrons, μ_n is the mobility of electrons, μ_p is the mobility of holes.

The conductivity is given by

$$\sigma = n_e . e(_n +_h)$$

$$\sigma = 2 \times \left(\frac{2\pi m_e KT}{h^2}\right)^{3/2} (m_h m_e)^{3/4} \exp\left(\frac{-E_g}{2KT}\right) e(\mu_n + \mu_h).$$

μ_p and μ_n have a temperature dependence and will largely cancel the $T^{3/2}$ temperature variation of term $\left(\frac{2\pi m_e KT}{h^2}\right)^{3/2} (m_h m_e)^{3/4}$; hence, its conductivity term σ is determined by the experimental term $e^{\left(\frac{-E_g}{2KT}\right)}$.

Therefore, $\sigma = \text{(constant)} e^{\left(\frac{-E_g}{2KT}\right)}$.

Resistivity, $\rho = \dfrac{1}{\sigma} = (\text{constant})e^{\left(\frac{E_g}{2KT}\right)}$.

On taking log both sides in the above equation, we get

$\rightarrow \log_e \rho = \log_e (\text{constant}) + \left(\dfrac{E_g}{2KT}\right)$

$\rightarrow 2.3026 \log_{10}\rho = 2.3026 \log_{10}(\text{constant}) + \left(\dfrac{E_g}{2KT}\right)$ \hfill (14.1)

$E_g = 2kT(2.3026 \log_{10}\rho - 2.3026 \log_{10}\text{constant})$.

From equation (14.1), if we plot graph between $\log_{10}\rho$ v/s $\dfrac{1}{T}\times 10^3$ as shown in Figure 14.2:

Slope = $\dfrac{AB}{BC} = \dfrac{\log_{10}\rho}{\dfrac{1}{T}\times 10^3}$.

Now, find out the slope value from the drawn graph on graph paper.
Then, band gap is given by

$E_g = 2\times k \times 2.3026 \times \dfrac{AB}{BC} \times 1000 \text{ eV}$

$E_g = 2\times 8.6 \times 10^{-5} \times 2.3026 \times \dfrac{AB}{BC} \times 1000 \text{ eV}$

$E_g = 0.396 \times \dfrac{AB}{BC} \text{ eV}$.

The resistivity is given as $\rho = R\dfrac{A}{l} = \dfrac{V}{I}\times \dfrac{A}{l} = Constant \times \dfrac{V}{I}$.

Procedure: Listed here are the steps that should be followed in this experiment:

1. Connect one pair green (+ve) and yellow (−ve) of probe to direct current source through milliammeter.
2. Other pair red and black to millivoltmeter.
3. Place four-probe arrangements in the oven.
4. Fix the thermometer in the given place in oven.
5. Adjust the current between 2 mA and 5 mA.
6. Measure the inner-probe voltage V with fixed interval of 5°C starting from 30°C at various temperatures up to 100°C.
7. Calculate the resistivity (ρ) and find out the various values of $\log_{10}\rho$ at different noted temperatures.
8. Calculate the values of $\dfrac{1}{T}\times 10^3$ at different noted temperatures by converting temperature to Kelvin.
9. Plot a graph between $\dfrac{1}{T}\times 10^3$ and $\log_{10}\rho$.

10. Find out the slope from the plotted graph.
11. Substitute the value of the slope in the formula to calculate the energy band gap.

Observation Table

Table 14.1 Observation table for resistivity

S.No.	Temperature Increase in °C	Voltage for given Current $I =$	Resistivity ρ = Constant $\left(\dfrac{V}{I}\right)$
1.			
2.			
3.			
4.			
5.			
6.			
7.			
8.			
9.			
10.			

Table 14.2

S.No.	Temperature in K	Resistivity (ρ)	$\dfrac{1}{T} \times 10^3$	$\log_{10}\rho$
1.				
2.				
3.				
4.				
5.				
6.				
7.				
8.				
9.				
10.				

Result:
From the graph, the energy band gap of n-type Ge-semiconductor is …eV.

$$\text{Percentage error} = \dfrac{E_{g(standard)} \sim E_{g(measured)}}{E_{g(standard)}} \times 100 = \underline{\qquad}.$$

Precautions and Sources of Error

1. Current through the sample should be constant.
2. Temperature should be taken carefully.
3. The semiconductor wafer is very brittle, so care must be taken when mounting on the platform below the four-probe arrangement.
4. Exert minimum pressure to get electrical contact of probe with the wafer.

Experiment 2: Stefan's Law of Black Body Radiation

Objective: Verification of Stefan's fourth power law for black body radiation and determination of the exponent of temperature.

Apparatus Used: The following apparatus are used in this experiment:

1. Voltmeter
2. Ammeter
3. Bulb
4. Connecting wire

Formula Used: In this experiment, the following equations are used for the verification of fourth power law:

1. For determination of $R(t)/R(t_d)$:

$$R(t) = R(t_D)\left[\frac{1 + at + bt^2}{1 + at_D + bt_D^2}\right], \tag{i}$$

   where t_D is the temperature at Draper point and $R(t_d)$ is the resistance of filament at Draper point.

2. For the verification of exponent of temperature:

$$\log_{10}P_R = \log_{10}C + \alpha\log_{10}T, \tag{ii}$$

   where α is the slope of graph plotted between $\log_{10}P_R$ and $\log_{10}T$.

Theory of the Experiment

A black body emits energy in the form of radiation at temperature T in K, and the power radiated is expressed as

$$P_R = \frac{d(E)}{dt} = A\sigma T^4 \text{ J/S}, \tag{14.2}$$

where A is the area of radiating surface (m^2), and σ is Stefan's constant ($J.S^{-1}m^{-2}deg^{-4}$). If the black body under consideration is surrounded by an atmosphere of temperature T_0, the net power loss $P = \frac{d(E)}{dt}$ is given by

$$P_R = A\sigma(T^4 - T_0^4). \tag{14.3}$$

If we assume the exponent of temperature to be "α" in general, instead of four as in equation (14.1) and $C = A\sigma$ is a constant, then equation (14.2) is reduced to

$$P_R = C(T^\sigma - T_0^\alpha). \tag{14.4}$$

As α is the value of exponent of temperature (≈ 4) for $T_0 \ll T$, we may neglect T_0^α as compared with T^σ and hence use approximation $P_R \approx CT^\alpha$; thus, the logarithmic of P_R on base 10 can be expressed as

$$\log_{10} P_R = \log_{10} C + \alpha \log_{10} T. \tag{14.5}$$

Thus, if the graph of logarithmic of the power radiated from a black body is plotted against the logarithm of its temperature, the slope of this graph gives the exponent of temperature. Therefore, we are required to measure the power radiated by a black body and its temperature to study Stefan's law of radiation, that is, to determine the value of α.

Measurement of Filament Temperature

We have the filament of a torch bulb as a black body. When a source of emf (battery) is applied across their filament, it emits radiation, and the temperature is raised relative to the surrounding temperature. The temperature of the filament of the bulb could be measured with an optical pyrometer, but we shall use an alternative method to measure the temperature of the filament, which is based on the fact that the resistance of the filament can be expressed approximately up to the second exponent of temperature as

$$R(t) = R_0\left[1 + at + bt^2\right], \tag{14.6}$$

where R_0 and R_t are the resistances of the filament at 0°C and t°C, respectively. The values of "a" and "b" are 5.1×10^{-3} deg^{-1} and 0.7×10^{-6} deg^{-2}, respectively, for the tungsten. Equation (14.6) can be used to measure the temperature "t" in °C if R_0 is known.

To avoid measurement of R_0, we prefer to choose another reference point that is known as the Draper point, which is 800 K (or 523°C). For judging the Draper point, we should keep the bulb in a dark background and increase the current slowly starting from very low values (nearly from zero). When the filament just becomes dull red, we achieve the Draper point. In an individual trial, one may go off by ±50 K in judging the Draper point, but the error in using R_0 is likely to be larger.

If R_D is the filament resistance at the Draper point, a graph between T and $\dfrac{R_t}{R_D}$ is obtained from the equation:

$$R_t = R_D\left[\frac{1+at+bt^2}{1+at_D+bt_D^2}\right] \Rightarrow \frac{R_t}{R_D} = \left[\frac{1+at+bt^2}{1+at_D+bt_D^2}\right]. \tag{14.7}$$

From this graph, one should read T from each observed resistance ratio. For tungsten, the value of "a" and "b" are given with equation (14.6). Use these to complete $\dfrac{R_t}{R_D}$ at $t = 600$, 800, 1000, 1200, 1400, and 1600°C. Now, plot a smooth curve between $\dfrac{R_t}{R_D}$ and T, known as Calibration graph, as shown in Figure 14.3.

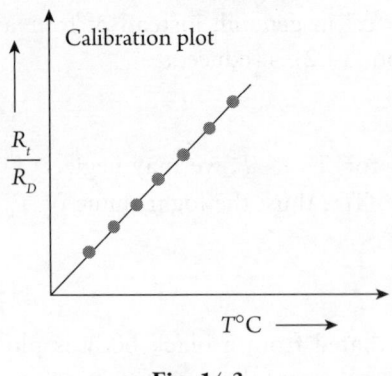

Fig. 14.3

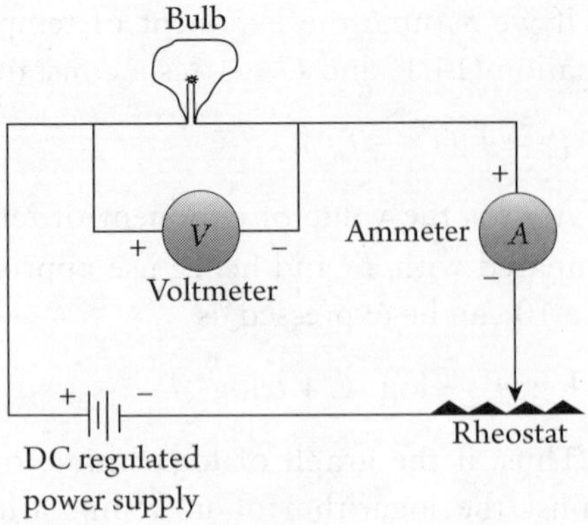

Fig. 14.4

Procedure: Listed here are the steps that should be followed in this experiment:

1. Connect the circuit, as shown in Figure 14.4.
2. Measure the voltage V corresponding to the Draper point carefully and repeat it several times to increase the accuracy of the measurement. Deduced mean of $\frac{V}{I}$ values. That is, R_D in judging the Draper point keeps the bulb in a dark background and increases the current slowly. When the filament just becomes dull red, you have the Draper point. In individual trials, one may go off by ±50 K in judging the Draper point. But the error in using R_0 is likely to be longer.
3. Measure the voltage V values corresponding to several I values, starting from below the Draper point to dazzling white glow. (Each time wait for about 2 minutes before taking a reading.) Each set gives $\frac{V}{I} = R_t$.
4. From R_t values calculate $\frac{R_t}{R_D}$, and draw the graph on a graph paper to find the related temperature in kelvin as $T = t + 273$.
5. Compute the products ($V \times I$), which gives input power P_R; take this as the radiated power.
6. Look for $\log_{10}T$ and $\log_{10}P_R$ from the observation Table 3 and plot a graph between them, as shown in Figure 14.5.

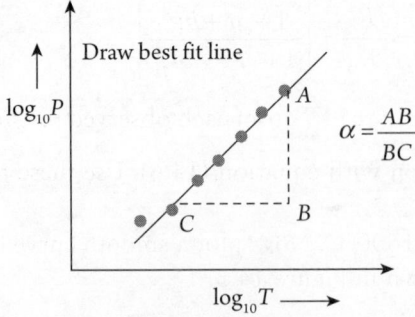

Fig. 14.5

Physics Practicals

7. This graph is a straight line only in the range of high temperature, where radiation occurs. The slope of this part gives α the power in Stefan's law.

Observation Table

1. Calibration table for the value of ratio $\dfrac{R_t}{R_D}$ by equation (14.7) at different temperatures.

S.No.	Values of t put in Equation (14.7) in °C	$\dfrac{R_t}{R_D}$
1.	600	
2.	800	
3.	1000	
4.	1200	
5.	1400	
6.	1600	

2. Observation table for the Draper point.

S.No.	V(Volts)	I(A)	$R_D(\Omega)$	Mean $R_D(\Omega)$
1.				
2.				
3.				

3. Observation table for power and R_t.

S.No.	V(Volts)	I(amp)	$R_t(\Omega)$	$\dfrac{R_t}{R_D}$	T from Calibration Graph (K)	$P = V \times I$	$\log_{10} P$	$\log_{10} T$
1.								
2.								
3.								
4.								
5.								
6.								

Result:

Measured slope (α_m) from the $\log_{10} T$ versus $\log_{10} P_R$ graph or the measured exponent of the temperature is ….

Percentage error $= \dfrac{\alpha_m - 4}{4} \times 100 = $ _____.

Precautions and Sources of Errors

1. Galvanometer should be sensitive.
2. During measurement of R_D the voltage and current should be minimum.
3. Heating device should be placed away from the hemisphere to avoid direct heating.
4. No kink should be present in the connecting wires.

Experiment 3: Determination of Thermopower

Objective: Study of thermoelectricity: determination of thermopower of copper-constantan thermocouple.

Apparatus Used: The following apparatus are used in this experiment:

1. Potentiometer
2. Resistance box (R.B.)
3. Copper and constantan wires
4. Three beakers
5. Electric heater
6. 2-volt lead accumulator
7. Thermometer
8. Galvanometer
9. Connecting wires

Formula Used: The potential difference across wire due to the current I is given by

$$\rho = iR, \tag{14.8}$$

where R is the resistance of the wire. For the experiment ρ is 55×10^{-3} V. Length of the wire is 10 m. Thus, the potential gradient along the wire is 55×10^{-6} V/cm.

The potential gradient is given by

$$E = \frac{V(\mu V)}{l},$$

where V is thermo-emf in μV and l is the length of the wire.

Theory of the Experiment

Thermoelectric Effect

There are three major kinds of **thermoelectric effects**. These are the **Seebeck effect**, the **Peltier effect**, and the **Thomson effect**.

In brief, all the above three thermoelectric effects have been described as follows:

The two dissimilar metals or alloys are taken as wire or strips and joined to form two junctions A and B. If these two junctions A and B are kept at different temperatures, emf is produced at the open ends, which can be detected by a micro voltmeter. This is called **Seebeck effect**.

Conversely, if a current is passed through this system, then these junctions A and B attain different temperatures—that is, one junction gets heated and the other junction is cooled. This is called the **Peltier effect**, which is used today in refrigeration.

If a conducting wire is melted to maintain a temperature gradient along its length, the two ends of the conductor show a potential difference. This is called **Thomson effect**.

In this experiment, we are concerned only with the **Seebeck effect**.

Measurement of the Thermo-emf

The order of magnitude of the thermo-emf lies in the microvolt region, and you need a device to measure emf in this range. Figure 14.6 shows the circuit diagram of the potentiometer. The potentiometer wire has a resistance of 55 Ω and is connected in series to a resistance of 1945 Ω and a 2-V battery, a lead accumulator. This 2-V is connected across 2000 Ω, and the current flowing in the circuit is 10^{-3} A. This current will produce the potential difference $\rho = iR$ across AB (where R is the resistance of the wire), that is, $\rho = 55 \times 10^{-3}$ V. Length AB is 10 m. Thus, potential gradient along the wire AB is 55×10^{-6} V/cm.

Figure 14.6 shows how the thermocouple is connected to the potentiometer circuit. The details are given here.

Take care that the hot junction in the case of the copper–constantan couple is connected to the positive end A of the potentiometer wire. The hot junction is dipped in water kept in a beaker and warmed by a heater, while the cold junction in water is maintained at room temperature. Before beginning this experiment, read the room temperature and record it. With these connections, you will get only one-sided deflection in the galvanometer "G."

Now switch on the heater till the water boils. Dip a thermometer and record this temperature. Now put the jockey on two ends A and B of the potentiometer; it must show opposite deflections in the galvanometer at 50°C when the sliding end is made to touch them. In case the opposite deflection in the galvanometer does not occur, check the connections of the circuit as shown in Figure 14.6. The mistake may be due to wrong connections of the battery, faulty connections of the thermocouple, a crack in one of the junctions, or the insufficient voltage supplied by the battery. Check this point if you can.

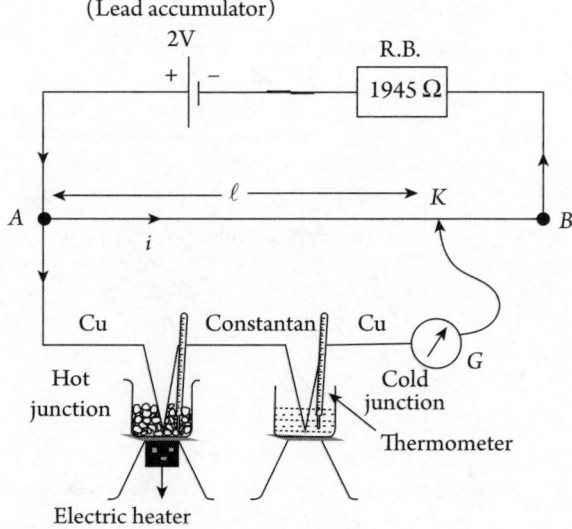

Fig. 14.6

Study of Temperature Dependence

After obtaining the two-sided deflection at A and B at temperature 50°C, find the null point K and record the AK length as L_1 in the table. Repeat it for increasing temperature by 5°C, starting from 50°C to 100°C. Now, switch off the heater and add a little cold water to the hot junction, lowering the temperature by 5°C. Again, get the null point and record L_2 length with decreasing temperature. Go on lowering the temperature and recording the new value of L_1 till the hot junction temperature falls to 50°C from 100°C. Mean length L is to be multiplied by the potential gradient ($E = 55 \times 10^{-6}$ V/cm) to find the value of thermo-emf in micro volts, which is plotted against the temperature difference, as shown in Figure 14.7. The slope on the X-axis gives the thermopower in microvolt /°C.

Observation Table

1. Room temperature T_c = _____ °C
2. Table for the determination of thermo-emf

S.No.	Hot junction Temperature T_H (°C)	$T_H - T_C$ (°C)	Length AK (cm)		$L = \dfrac{L_1 + L_2}{2}$	Thermo-emf (μV)
			L_1 (Increasing T)	L_2 (Decreasing T)		
1.						
2.						
3.						
4.						
5.						
6.						
7.						
8.						
9.						
10.						

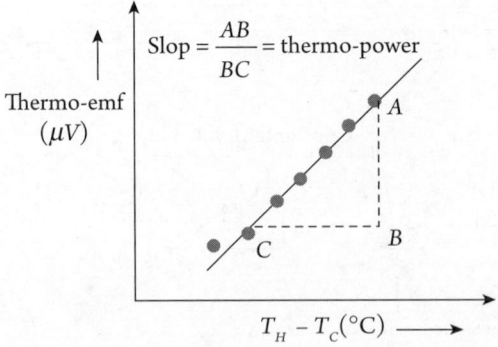

Fig. 14.7

Calculations: The thermo-emf of copper–constantan is calculated by the formula

$V = E \times l =$ _____ (μV).

The thermopower plotted by the graph is …(microvolts/°C)

Results:

1. The thermo-emf of copper–constantan is _____ (μV)
2. The thermopower plotted by the graph is _____ (microvolts/°C)

Precautions and Sources of Error

1. Galvanometer should be sensitive.
2. The heating device should be placed away from the hemisphere to avoid direct heating.
3. Ensure the steady temperature of the hemisphere with the help of two thermometers.
4. Kink-free connecting wires should be used for connections.

Experiment 4: Variation of Magnetic Field

Objective: To study the variation of the magnetic field with distance along the axis of the current-carrying coil and then to estimate the radius of the coil.

Apparatus Used: The following apparatus are used in this experiment:

1. Stewart and Gee's apparatus
2. Ammeter
3. Battery
4. Rheostat
5. Sprit level
6. Commutator
7. Plug key
8. Connection wire

Formula Used: The magnetic field on the axis of circular coil is given by

$$B = \frac{2\pi n r^2 i}{10\left(x^2 + \frac{y^2}{r^2}\right)^{\frac{3}{2}}},$$

where B is the magnetic field along the axis of a circular coil, and it is equal to the horizontal magnetic field ($H \tan\theta$), n is the number of turns in the coil, r is the radius of the coil, i is the current flowing in the coil (in ampere), and x is the distance of the point from the center of the coil.

Theory of the Experiment

The intensity of the magnetic field at a point P lying on the axis of a circular coil AB having n turns, at a distance x from the center of the coil O, is given by the formula

$$B = \frac{2\pi n r^2 i}{10(x^2 + r^2)^{\frac{3}{2}}},$$

where i is the current flowing through the coil. In case the current i is measured in amperes, then we have

$$B = \frac{2\pi n r^2 i}{10\left(x^2 + \frac{y^2}{r^2}\right)^{\frac{3}{2}}}.$$

The direction of the magnetic intensity P will be along OP, produced if the current flows through the coil in the anti-clock direction as observed from P. If the flow of current is clockwise, the field at P will be PO.

The magnetic intensity is maximum at the center O of the coil and is given by the relation

$$B = \frac{2\pi n i}{10 r}.$$

A graph showing the relation between the intensity of the magnetic field B and the distance x is given in Figure 14.8. The curve is first concave toward O, but the curvature becomes less and less. It changes sign at A and B and afterward becomes concave toward O. The points of inflection A and B where the curvature changes its sign lie at distance $\frac{r}{2}$ from the center, so the distance between A and B is equal to the radius of the coil.

If the field B is made perpendicular to the horizontal component of the earth's field H, the deflection magnetometer is then given by

$B = H \tan\theta$ or, $B \propto \tan\theta$

So, the graph between B and x will be similar to the graph between $\tan\theta$ and x, as shown in Figure 14.9.

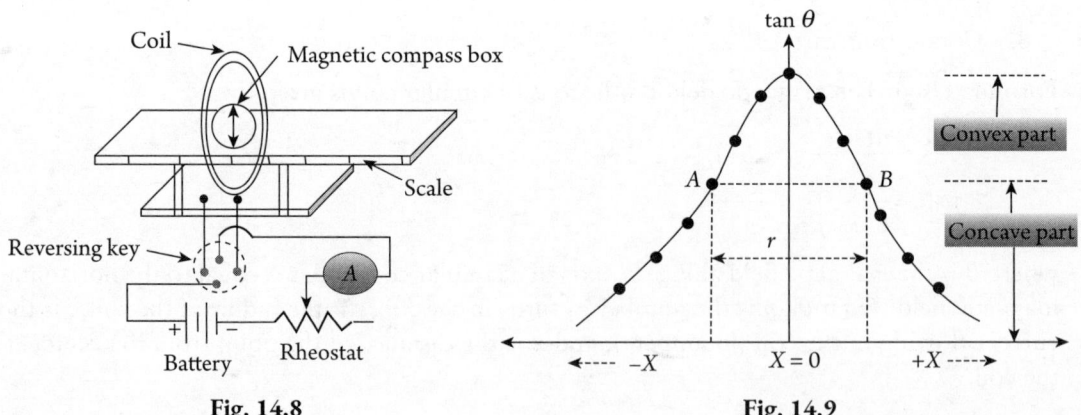

Fig. 14.8

Fig. 14.9

Physics Practicals

Procedure: Listed here are the steps that should be followed in this experiment:

1. Place the compass box on the sliding bench so that its magnetic needle is at the center of the coil. Level the base of the coil with the help of a spirit level (Figure 14.8).
2. Slide the instrument in the horizontal plane in such a way that the coil lies roughly in the magnetic meridian, and rotate the compass box till the pointer reads 0–0.
3. Using the commutator, flow the current in one direction and note down the deflection of the needle. Now reverse the direction of current and again note the deflection. If both the deflections are equal, then that means the coil is in magnetic meridian.
4. Adjust the current in the coil such that the deflection in the magnetic needle is of the order of 60°C–70°C. Note down the reading of both sides of the pointer. Reverse the direction of current, and again note down the reading of both ends of the pointer.
5. Move the compass box through 2 and 3 cm and note down the deflection. Again, move through 2 or 3 cm and continue to take reading till the compass box reaches the end of the bench.
6. Take the measurements for the other side of the bench.
7. Calculate θ according to the observation table and determine $\tan\theta$.
8. Plot the graph taking position (x) along the x-axis and $\tan\theta$ along the y-axis.

The distance between the two points of inflection on the curve will be the radius of the coil (Figure 14.9). The radius (r) of the coil can also be estimated by taking the mean value of the inner and outer radii of the coil.

Observation Table

1. Measurement of deflection (on the left side)

S.No.	Distance of Needle from the Center (x)	Current One Way θ_1	Current One Way θ_2	Current Reversed θ_3	Current Reversed θ_4	$\theta = \dfrac{\theta_1 + \theta_2 + \theta_3 + \theta_4}{4}$	$\tan\theta$
1.							
2.							
3.							
4.							
5.							
6.							
7.							

2. Measurement of deflection (on the right side)

S.No.	Distance of Needle from the Center (x)	Current One Way θ_1	Current One Way θ_2	Current Reversed θ_3	Current Reversed θ_4	$\theta = \dfrac{\theta_1 + \theta_2 + \theta_3 + \theta_4}{4}$	$\tan\theta$
1.							
2.							

(Continued)

S.No.	Distance of Needle from the Center (x)	Current One Way		Current Reversed		$\theta = \dfrac{\theta_1 + \theta_2 + \theta_3 + \theta_4}{4}$	$\tan\theta$
		θ_1	θ_2	θ_3	θ_4		
3.							
4.							
5.							
6.							
7.							

Calculations

1. The radius of the coil obtained from graph …cm.
2. The radius of the coil:
   i. Inner radius (r_1) = _____ cm
   ii. Outer radius (r_2) = _____ cm
   iii. Mean radius (r) = _____ cm
   Percentage error in measuring the radius of the coil is _____.

Results:

1. The radius of the coil is … cm.
2. The variation of magnetic field along the axis of the current-carrying coil is shown in the graph.

Precautions and Sources of Error

1. The current flowing in the coil should remain constant.
2. The deflection of the needle should be noted in the range of 30° to 75°.
3. There should not be any magnetic material near the apparatus.
4. The compass box should be adjusted in the horizontal plane by leveling it.
5. The coil should be exactly adjusted in the magnetic meridian.

Experiment 5: Carrey Foster Bridge

Objective: Study of Carrey Foster bridge: Determination of resistance per unit length of the bridge wire and a given unknown resistance.

Apparatus Used: The following apparatus are used in this experiment:
1. R.B.
2. Cylindrical resistance (unknown resistance)
3. Connecting wire

Formula Used: The resistance per unit length is given by

$$\sigma = \dfrac{R}{l_2 \sim l_1}$$

Here $l_2 \sim l_1$ is the difference between the lengths of the balanced position, R is the resistance of the R.B., and σ is the resistance per unit length.

The comparison between the two resistances is given by the formula

$$S + \sigma\left[l_2 \sim l_1\right]$$

Here $l_2 \sim l_1$ is the difference between the lengths of the balanced position, and S is the resistance of cylindrical resistance.

Theory of the Experiment

Resistance can be measured by comparison with other known resistors. Carrey Foster bridge is an electrical circuit based on Wheatstone principle. Wheatstone principle can be understood with the help of an electrical network, as shown in Figure 14.10. When the galvanometer shows no current, the resistors P, Q, R, and S satisfy the condition:

$$\frac{P}{Q} = \frac{R}{S} \tag{14.9}$$

A practical application of the Wheatstone principle is the meter bridge, shown in Figure 14.10, in which the branch ADC of Figure 14.10 is replaced by a meter-long resistance wire, with D as a variable contact point. If the null current in the galvanometer is observed for $AD = l$, equation (14.9) leads to

$$\frac{P}{Q} = \frac{l}{100-l}. \tag{14.10}$$

Equation (14.10) is correct under the following assumptions:

1. The wire is uniform with resistance per unit length "σ."
2. No end-errors exist in the resistance R and S due to connecting wires.

Take several measurements with P and Q by interchanging them, and calculate the mean of these measurements to reduce the error in measuring the ratio of P/Q. Once the ratio P/Q is measured, the unknown resistance S can be determined for known values of R using equation (14.9).

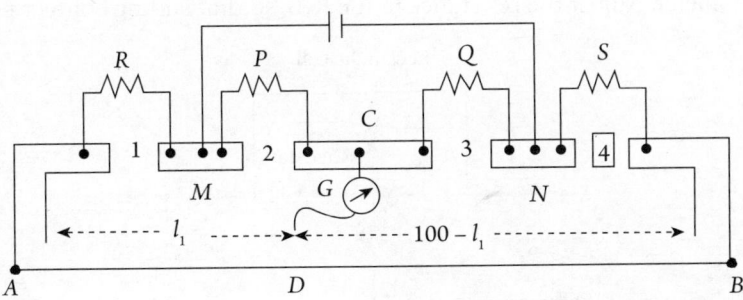

Fig. 14.10

Carrey Foster bridge

Figure 14.11 shows the bridge network. Resistance R' and S' in the outer gaps now get added to the wire lengths l and $(100 - l)$, respectively. If we assume the resistance α and β at the right end due to the connecting wire, respectively, then equation (14.9) now gives

$$\frac{P}{Q} = \frac{R_1 + l_1 + \alpha}{\left[R_1 + (100 - l_1)\sigma + \beta\right]}. \tag{14.11}$$

If R_1 and S_1 are interchanged in the gaps, and the new balance length comes as l_2, we get

$$\frac{P}{Q} = \frac{S_1 + l_2 + \alpha}{\left[R_1 + (100 - l_2)\sigma + \beta\right]}. \tag{14.12}$$

Comparing equations (14.11) and (14.12) and after simplification, we get

$$R_1 - S_1 = (l_2 - l_1)\sigma. \tag{14.13}$$

The remarkable feature is that the end errors α and β are eliminated. Also, P and Q do not appear in equation (14.13), and therefore a rheostat may be connected between A and C; the sliding contact D then makes the ratio $\frac{P}{Q}$ variable. Note also that we now get the differences of the resistance $(R_1 - S_1)$ rather than a ratio as in equation (14.10). This feature makes the Carrey Foster bridge superior and more precise compared to the meter bridge as far as precision in the determination of the resistance is concerned.

Procedure: Listed here are the steps that should be followed in this experiment:

1. Balancing the bridge: Complete the circuit diagram, as shown in Figure 14.11. Replace the resistance R_1 and S_1 with zero resistance by means of connecting thick conducting strips between the gaps. Fix the sliding contact point of the bridge wire and adjust the sliding point of the rheostat so that null current in the galvanometer is observed.

2. Estimating the resistance of the bridge wire: Connect an R.B. of the standard variable resistance in the left gap (in place of R_1) and zero resistant in the right gap (in place of S_1) (see Figure 14.11). Now, any value of the resistance in the R.B. will shift the sliding contact point C on the bridge wire to the left from the mid-point for null current in galvanometer. Adjust the resistance in the R.B. so that sliding contact point C is shifted

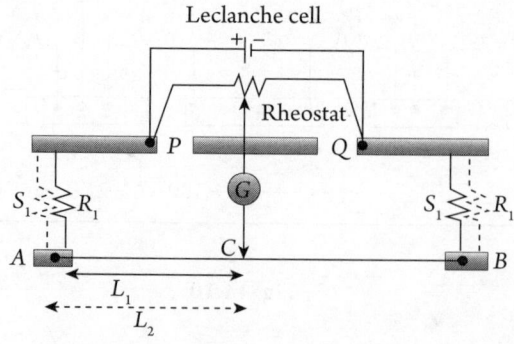

Fig. 14.11

extreme left about zero reading on the meter scale for null current in galvanometer. The value of the resistance in the R.B. is approximately the resistance of the bridge wire.

3. Determination of the σ: Select five or six values of resistance (R) between zero and the approximate resistance of the bridge wire with equal interval.

4. Select any one of the values of the resistance in the R.B. in the increasing order and adjust the sliding contact point C left to the mid-point of the bridge wire for null current in the galvanometer. Record the reading of the sliding contact point C on the meter scale as l_1 in the observation table for the corresponding resistance in the R.B. connected in the right gap and zero resistance in the left gap. Further, select the value of the resistances in increasing order in the R.B. again one by one and adjust the sliding contact point C right to the bridge wire for null current in the galvanometer. Record the reading of the sliding contact point C on the meter scale as l_2 in the observation table for the corresponding resistance in the R.B. Find out σ from plotting the graph between $(l_2 - l_1)$ and R. Also, find out the σ from each observation as well as average σ.

5. Determination of unknown resistance: Connect the unknown resistance in the left gap and a standard variable R.B. in the right gap. Set the sliding contact point C somewhere on the bridge wire and adjust the resistance in the R.B. for nearly null current in the galvanometer. Note this value of the resistance, say "S". Now adjust the sliding contact point C for null current in the galvanometer and record its reading on meter scale as l_1.

6. Interchange the unknown resistance and R.B. to the right gap and left gap, respectively, without disturbing the resistance value S'. Adjust the meter scale as l_2 again. Find out the unknown resistance as $[S_1 + (l_2 - l_1)\sigma]$. Repeat the process three to four times, and determine the average value of unknown resistance.

Observation Table

1. Least count of the meter scale = _____ cm.
2. Approximate value for the resistance of the bridge wire _____ Ω.
3. Table for the determination of σ

S. No.	Resistance Inserted in R.B. $R(\Omega)$	Balance Position of C (cm)		$l_1 \sim l_2$ (cm)	$\sigma = \dfrac{R}{l_2 \sim l_1}$	Average Σ
		l_1 (cm)	l_2 (cm)			
1.						
2.						
3.						
4.						
5.						

4. Table for the comparison of two equal resistances

S. No.	$S(\Omega)$	l_2 (cm)	l_1 (cm)	$l_1 \sim l_2$ (cm)	$S + \sigma[l_1 \sim l_2]$
1.					
2.					
3.					
4.					

Calculations

1. The resistivity is calculated by $\sigma = \dfrac{R}{l_2 \sim l_1}$, and its average value is _____ Ω-cm.

   The unknown resistance S calculated by $S + \sigma[l_2 \sim l_1]$ is _____ Ω.

Results

1. The resistivity of unknown wire is _____ Ω-cm.
2. The measured difference between two equal resistance is _____ Ω.

Precautions and Sources of Error

1. The unknown resistance should be soldered to thick copper strips for its connection to the bridge gap. The connecting wires and the copper strip should be thoroughly cleaned with sandpaper.
2. The connections should be tight, and the plugs of the R.B. should be twisted so that they are tight.
3. The battery key should be taken out when the reading is not being taken in order to avoid heating of the wires.

Experiment 6: e/m of Electron Using a Diode (Magnetron Valve)

Objective: Determination of specific charge (charge to mass ratio; e/m) for electron.

Apparatus Used: The following apparatus are used in this experiment:

1. Diode (magnetron valve)
2. Solenoid
3. Voltmeter
4. Ammeter

Formula Used: For the e/m ratio the relation is

$$e/m = \frac{8VL^2}{\mu_0 N^2 I_c^2 R^2},$$

where V is the known voltage given to the valve, L is the length of the solenoid, μ_0 is the permeability of the free space, N is the number of turns in the given solenoid, I_c is the critical current, and R is the radius of the solenoid.

Theory of the Experiment

Consider a magnetron valve. It is a diode that has a cathode and an anode in the form of a cylindrical plate. When a positive potential is applied to this anode, a field is established parallel to the axis of the cylindrical anode. After the thermionic emission from the cathode, the emitted electrons do not travel radially outward but follow a circular path. For a constant anode potential, the radius of the circular path can be changed by increasing the magnetic field.

At a critical magnetic field "B_c", it may so happen that the electrons just graze the cylindrical path and may not reach the anode. At this point, the anode current drops to zero. Obviously, this will happen when the diameter of the circular path is R_a, the radius of the cylindrical anode is shown in Figure 14.12. This causes variation in i_p.

Suppose (r, θ) are the polar coordinates of the electron. Let the radial velocity be r' (where $[r']$ represents $\dfrac{d}{dt}$) at time t after emission. So, if "B" is a magnetic field, it will experience a magnetic force B, which produces a transverse acceleration:

$$a_t = \left[\frac{d}{dt}(r\theta') + r'\theta'\right],$$

where the first term is contributed by the rate of change of transverse velocity (r, θ'), and the second term is due to the rate of change of "θ" of the radial velocity component. Hence, if m is taken to be the mass of the electron, then on integrating, we get

$$Be\, r' = m\frac{d}{d}(r\theta') + mr'\theta' = m[r\theta'' + 2r'\theta'],$$

or $Be\, r' = m\dfrac{d}{dt}(r^2\theta')$

$$r^2\theta' = \frac{Ber^2}{2m} + K.$$

If $t = 0$, $r\theta' = 0$ and $r' = 0$; when $r = R_f$, $K = -\left[\dfrac{BeR_f^2}{2m}\right]$.

$$\therefore r^2\theta' = \left[\frac{Be}{2m}\right](r^2 - R_f^2) = B_e r^2\left[1 - \frac{R_f^2}{r^2}\right]$$

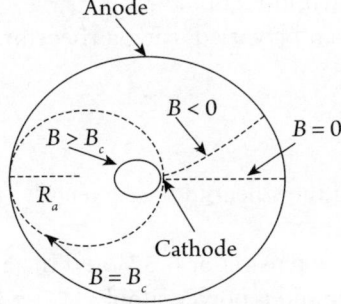

Fig. 14.12

or $\theta' = B_e \dfrac{\left[1 - \dfrac{R_f^2}{r^2}\right]}{2m}$. (14.14)

When $B = B_e$, $r' = 0$ at the moment of grazing $r = R_a$, and only the component $r\theta'$ of velocity remains. The velocity "v" of the electron is given by

$V = [(r')^2 - (r\theta')^2]$.

For $B = B_e$, $r\theta' = R_a\theta'$, $r' = 0$, so equation (14.14) reduces to

$R_a\theta' = B_c e R_a \dfrac{\left[1 - \dfrac{R_f^2}{r^2}\right]}{2m}$.

But if the potential applied to the anode is V, then the transverse component $R_a\theta'$ should be (assumed initial velocity of thermally emitted electrons to be zero)

$$M\left[\dfrac{(R_a\theta')^2}{2}\right] = V_e.$$ (14.15)

From equations (14.14) and (14.15),

$$mR_a^2 B_c^2 e^2 \dfrac{\left[1 - \dfrac{R_f^2}{r^2}\right]^2}{8\,m^2} = V_e$$

Taking $R_a \gg R_f$, we get

$$\dfrac{e}{m} = \dfrac{8v}{B_c^2 R_a^2} \quad \dfrac{\text{coulamb}}{\text{kg}}.$$ (14.16)

Experimental Setup

In an experimental arrangement, the magnetron valve is placed inside a long solenoid in such a way that its axis coincides with the axis of the cylindrical solenoid. The multilayer solenoid is made such that it is capable of carrying 10–15 A current, which can be supplied from a heavy duty current source, as shown in Figure 14.13.

The current in the solenoid I can be varied using a rheostat. The magnetic field produced in the solenoid is given by

$$B = \dfrac{\mu_0 NI}{L},$$ (14.17)

where N is the number of turns of the solenoid over a length "L", and μ_0 is the permittivity of free space.

The filament is heated using low tension or 6.3 V, and the anode can be applied a known voltage V using a potential divider and a power supply.

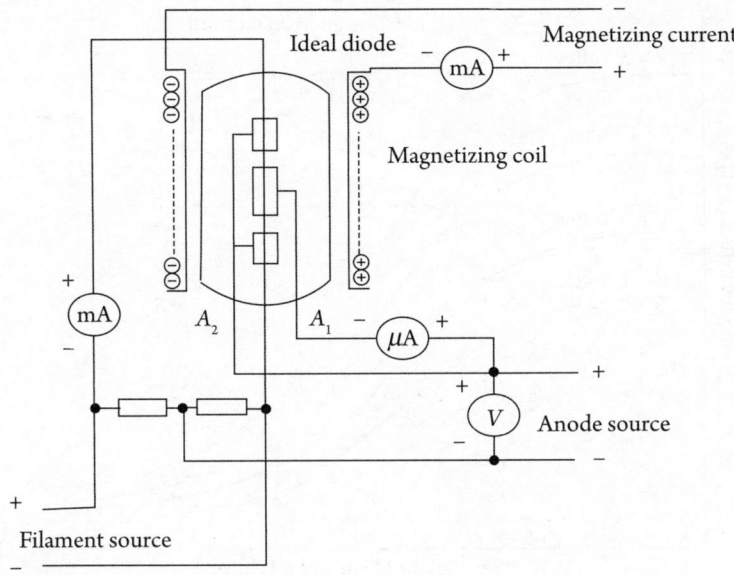

Fig. 14.13

Procedure: Listed here are the steps to be followed in this experiment:

1. The diode is placed inside the solenoid centrally so that the axis of the valve coincides with the axis of the solenoid. A suitable potential is applied to the anode valve after making circuit arrangements.
2. The maximum current is noted with zero current in the solenoid ($i = 0$). Now gradually the solenoid current i is increased, and the anode current is noted systematically till there is a sharp fall in the anode current.
3. At this stage, the solenoid current is I_c, and further increase is not required. A graph can be plotted of I_p (anode current) and I; the sharp fall point corresponds to I_c, as shown in Figure 14.14.
4. This procedure can be repeated for various values of V, now from equations (14.16) and (14.17). $e/m = \dfrac{8VL^2}{\mu_0 N^2 I_c^2 R^2}$.

The specific charge of electron can be calculated from the above equation.

Observation Table: The observations can be recorded as given here:

1. Number of turns in the given solenoid N = _____.
2. The length of the solenoid L = _____.
3. The radius of the solenoid R = _____.
4. The radius of the cylindrical anode R_a = _____.
5. Observation table for I_p

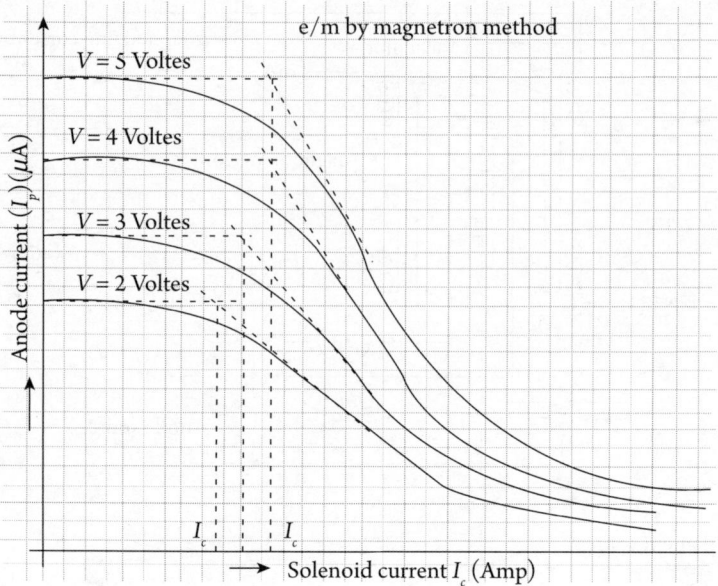

Fig. 14.14

S.No.	$V_1 = 30$ V		$V_2 = 35$ V		$V_3 = 40$ V	
	I_s	I_p	I_s	I_p	I_s	I_p
1.						
2.						
3.						
4.						
5.						
6.						
7.						
8.						

Calculations

1. From the graph, the critical current i_c is _____ amp.
2. The e/m of the electron is calculated by the formula

    $$e/m = \frac{8VL^2}{\mu_0 N^2 I_c^2 R^2} = \text{_____ coulomb/kg}$$

Results: The e/m of an electron using magnetron valve is _____ coulomb/kg.

Precautions and Sources of Error

1. The magnetron valve should be placed symmetrically inside the solenoid about its axis.

2. Voltage greater than that specified for the given magnetron valve should not be applied to its plate.
3. The current of solenoid should be varied steadily and should not be increased beyond the specified value for the given solenoid.
4. The low tension should be connected to the filament.

Experiment 7: Tangent Galvanometer

Objective: Study of tangent galvanometer: Determination of reduction factor and horizontal component of the earth's magnetic field.

Apparatus Used: The following apparatus are used in this experiment:

1. Tangent galvanometer
2. A coil
3. A resistor
4. A commutator
5. A key
6. A milliammeter
7. A spirit level and so on.

Formula Used: The reduction factor of the tangent galvanometer is given by

$i = K \tan\theta.$

where i is the current that passes through the coil, θ is the deflection produced in the needle, and the value of reduction factor K is given by

$K = \dfrac{H}{G}.$

Here, H is a horizontal component of the earth whose value is given by $B = H \tan\theta$, and G is called the galvanometer constant.

Theory of the Experiment

The tangent galvanometer consists essentially of a magnetic needle pivoted or suspended at the center of a circular coil having many turns of insulated copper wire and the plane of the coil being vertical. The magnetic needle is small, and the radius of the coil is large compared with the needle, so that the fields created by the current in the coil is uniform throughout the space in which the needle moves, and equal to the field at center of the coil. At right angle to the needle is attached a light aluminum pointer, which reads the deflection of the needle on the horizontal circular scale graduated in degrees; the needle, the pointer, and the scale are enclosed in a metal box with a glass cover to shield them from air draughts. Below the needle at the base of the box, a plane mirror is fixed, which helps avoid errors due to parallax in reading deflections. To level the instrument, the base is provided with a leveling screw.

The tangent galvanometer usually employed in a laboratory has 500 turns of copper wire wound on the circular frame, which terminates in four binding screws. Between the first and the second terminals, there are only two turns of wire that have a resistance of about 0.02 Ω. This is known as the ammeter coil and is used with strong current. Between the first and third terminals, there are 50 turns of wire that have a resistance of 1.1 Ω and are used with moderate currents. Between the first and fourth terminals, there are 500 turns of wire having a resistance of 226 Ω; this is called voltmeter coil and is used with weak currents.

Experiment

To use the instrument, the plane of the coil is set in the magnetic meridian. The needle and the coil should lie in the same vertical plane, with the pointer reading coinciding with the zero of the circular scale. Now, the current "i" is passed through the coil, and the deflection "θ" produced in the needle. Take the reading of the needle deflection corresponding to the current in the circuit on the scale. The magnetic field "B" at the center of the coil is perpendicular to the horizontal component of the earth's magnetic field "H." Hence, we have

$$B = H \tan\theta \qquad (14.18)$$

and $B = \left[\dfrac{\mu_0 n}{2r}\right] i = Gi,$ \qquad (14.19)

Where, G is called the galvanometer constant.
From equations (14.18) and (14.19)
$$Gi = H \tan\theta$$
$$i = \dfrac{H}{G} \tan\theta \qquad (14.20)$$

If $i = 1$, $G = B$, that is, the galvanometer constant is numerically equal to the strength of the magnetic field at the center of the coil due to the unit current.

In equation (14.20), K is called the reduction factor of the tangent galvanometer. This is a constant that, when multiplied by the tangent angle of deflection, gives the current in the galvanometer. Since its value is $\dfrac{2rH}{\mu_0 n}$, it depends on the radius of the coil and the number of turns in it and the value of the horizontal component of the earth's magnetic field. Hence, it is constant for a galvanometer at a particular place but varies for the same galvanometer from place to place.

When $\theta = 45°$, $\tan\theta = 1$, and hence $K = i$ is the reduction factor, which is numerically equal to the current required to produce a deflection of $45°$ in the galvanometer.

Procedure: Listed here are the steps that should be followed in this experiment:

1. The objective of the experiment is to study the tangent galvanometer and to determine the reduction factor of the two-turn coil in it. From the reduction factor, the value of H, the horizontal arrangement of the earth's magnetic field, is also to be computed. Also, Ohm's law is verified using the tangent galvanometer.

2. As indicated at the outset, in the plane of the coil, the tangent galvanometer is placed in the magnetic meridian and a current of i ampere is allowed to pass through it
   $i = K \tan\theta,$
   where K is the reduction factor and θ is the angle through which the needle is deflected away from the magnetic meridian.

Physics Practicals

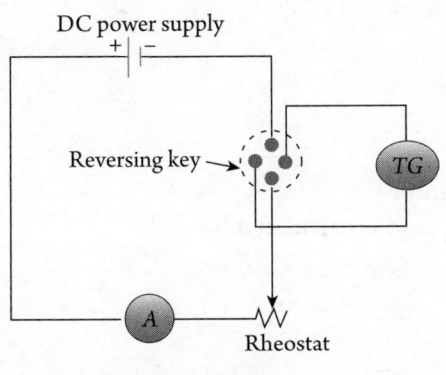

Fig. 14.15

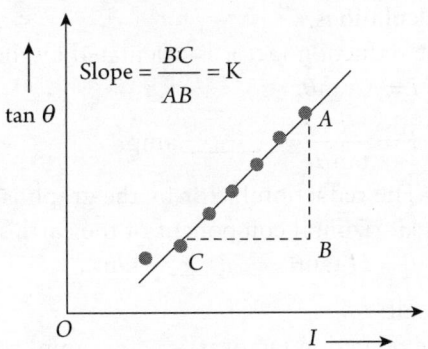

Fig. 14.16

3. Measure θ for various values of i such that θ lies between 25° and 65°. In this region, the reading of the deflection would be least liable to error.
4. While reading θ value avoids any error due to parallax.
5. Observations are to be repeated by changing the direction of current flow using a commutator, as shown in the circuit diagram in Figure 14.15.
6. Now, plot the graph between current (i) and average deflection. Slope of this plot will give value of K, as shown in Figure 14.16.

Observation Table

1. Least count of ammeter = _____.
2. Observation table for graph of i and $\tan\theta$

S.No.	Current (i)	Deflection in T.G θ(degree)					$\tan\theta$	K
		Clockwise Current		Anti-clockwise Current		Mean		
		θ_1	θ_2	θ_3	θ_4	$\theta = \dfrac{\theta_1 + \theta_2 + \theta_3 + \theta_4}{4}$		
1.								
2.								
3.								
4.								
5.								
6.								
7.								
8.								
9.								
10.								

Calculations
The reduction factor is calculated by the formula
$$i = K \tan\theta$$

$$K = \frac{i}{\tan\theta} = \underline{\qquad} \text{ amp.}$$

The reduction factor by the graph is _____ amp.

Horizontal component of the earth's magnetic field is calculated by
$$B = H \tan\theta = \underline{\qquad} \text{ wb/m}^2.$$

Result:
The reduction factor is _____ amp.

The value of the horizontal component of the earth's magnetic field at the laboratory is _____ wb/m².

Precautions and Sources of Error
1. The coil should be carefully adjusted in the magnetic meridian.
2. All the magnetic materials and current-carrying conductors should be at a considerable distance from the apparatus.
3. The current passed through the coil should be adjusted to produce a deflection of nearly 65°.
4. Current should be repeatedly checked, and for this purpose, an ammeter should be connected in series with the battery.
5. Parallax should be removed while taking readings of the pointer, and both ends of the pointer should be read.

Experiment 8: Newton's Rings

Objective: Determination of the wavelength of sodium light using Newton's rings method.

Apparatus used: A plano-convex lens, monochromatic light source, plane glass plate, microscope, and so on.

Formula used For the determination of wavelength (λ):

$$\lambda = \frac{D_{n+p}^2 - D_n^2}{4pR},$$

where D_n is the diameter of nth ring, D_{n+p} is the diameter of $(n + p)^{th}$ ring, and R is the radius of curvature of the plano-convex lens.

Theory of the Experiment
Introduction
When a plano-convex lens of large focal length is placed with its convex surface on a plane glass plate, a thin film of air is enclosed between the lower surface of the lens and the upper surface of the plate. When viewed under monochromatic light incident normally on the lens surface, circular interference fringes are observed as concentric circles with the point of

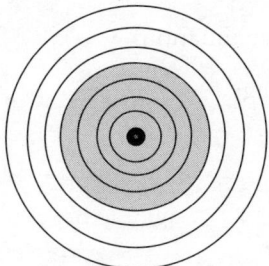

Fig. 14.17

contact as the center. These circular-ring-shaped fringes are popularly called Newton's rings and are shown in Figure 14.17.

Experiment

Consider Figure 14.18. In the diagram, ACB is the lens surface in contact with the glass plate. Let us suppose that thickness of the air film at a certain point P is $PQ = t$. This thickness remains constant along a circle of radius CP, having center C. From the theory of interference in the film, it is known that if "λ" is the wavelength of the incident light and "μ" is the refractive index of the film, then, at point P, the circular ring with radius CP will have bright or dark fringes, which can be deduced as

$$2\mu t \cos(r + \theta) = \frac{(2n+1)\lambda}{2}. \quad \text{(bright)}$$

$$2\mu t \cos(r + \theta) = \frac{2n\lambda}{2}. \quad \text{(dark)} \tag{14.21}$$

Here, r is the angle of refraction, and θ is the angle of the film. If $CO = r^2$ and the radius of curvature of the convex surface of the lens is R, then

$(CO)^2 = (CM) \cdot (MD)$
$r^2 = t(2R - t)$

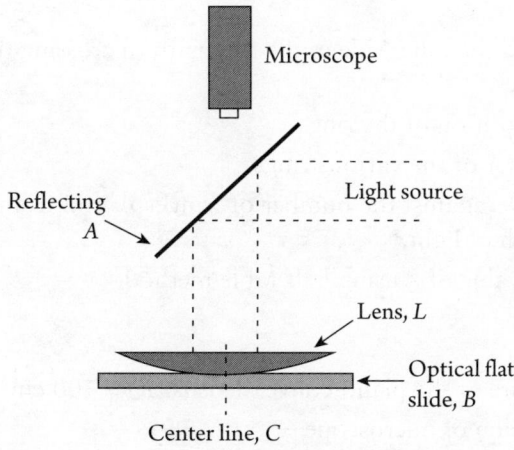

Fig. 14.18

$$t = \frac{r^2}{2R} \text{ (to a first order of approximation)}. \quad (14.22)$$

Substituting for t and simplifying:
For bright ring,
$$r_n^2 = \frac{n\lambda R}{[2\mu\cos(r+\theta)]}. \quad (14.23)$$

For dark ring, $r_n^2 = \dfrac{n\lambda R}{[\mu\cos(r+\theta)]}$ \quad (14.24)

If $\mu = 1$ (air film), $r = 0$ (normal incidence), and $\theta = 0$ (for large radius of curvature), equation (4.24) reduces to
$$r_n^2 = n\lambda R$$
Or $D_n^2 = 4n\lambda R,$ \quad (14.25)

where D_n is the diameter of the nth order dark ring. Obviously, the diameter of the $(n+p)^{th}$ order dark ring is

$$D_{n+p}^2 = 4(n+p)\lambda R. \quad (14.26)$$

From equations (14.25) and (14.26),
$$\lambda = \frac{D_{n+p}^2 - D_n^2}{4pR}.$$

This is an expression that provides a ready means of measuring "λ" of the monochromatic wave.

Procedure: Listed here are the steps that should be followed in this experiment:

1. Switch on the sodium light source and wait till it glows completely yellow.
2. Place the convex side of the lens over the glass plate.
3. Adjust the position of the glass-plate system to obtain the circular rings.
4. Now move the cross wire of the eyepiece on either side from center to up to the 24th fringe.
5. Now move toward the other side of the ring pattern crossing through the center in the steps of 3–3 fringes.
6. Write down the position of the rings.
7. Obtain the diameter of the various rings.
8. Plot a graph of D_n^2 against the number of fringes n, and from the slope of the graph, deduce wavelength of light.
9. Also, deduce the value of λ using half-table method.

Observations

1. Radius of curvature of the plano-convex lens is $(R) = 100$ cm.
2. Value of one division of microscope = _____.
3. No. of divisions on the Vernier scale of microscope = _____.
4. Least count of the microscope = _____.

Table: Determination of the diameter of Newton's rings:

S. No.	No. of Fringes (n)	Microscope Reading of the Rings		Ring Diameter $D_n = L - R$ (cm)	D_n^2 (cm)²	$D_{n+p}^2 - D_n^2$ (cm)²	λ (A°)
		Left-hand Side L (cm)	Right-hand Side R (cm)				
1.							
2.							
3.							
4.							
5.							
6.							
7.							
8.							

Calculations: The wavelength of sodium light is calculated by the formula:

$$\lambda = \frac{D_{n+p}^2 - D_n^2}{4pR}$$

$\lambda = $ _____ A°.

Percentage error = _____ %

Results: The wavelength of sodium light is _____ A°.

Precautions and Sources of Error

1. The lens and glass plate should be cleaned properly.
2. Lens of a large focal length should be used.
3. The point of intersection of cross wire should coincide tangentially with a particular ring.
4. The micrometer screw should always be screwed in a particular direction to avoid backlash error.
5. The amount of light for the source should be adjusted for maximum visibility of the rings and good contrast between dark and bright rings.

Experiment 9: Polarimeter

Objective: Determination of the concentration of sugar solution using half-shade polarimeter.

Apparatus Used: The following apparatus are used in this experiment:

1. Polarimeter tube
2. Sodium lamp
3. Sugar solution
4. Half-shade polarimeter

Formula Used: The specific rotation of the plane polarization of cane sugar solution in water is determined by

$$S = \frac{10\theta}{LC},$$

where θ is the angle of rotation of plane of vibration in degrees, L is the length of the polarimeter tube in centimeters, and C is the concentration of sugar solution in gm/cc.

Theory of the Experiment
Introduction

When plane-polarized light is made to pass through a certain substance, it is found that the plane of polarization of the emergent light is not the same as that of the incident light, and it is rotated through a certain angle. The property of crystal or substance is called "optical activity", and the phenomenon is called "optical rotation".

There are two types of optically active substances: those that produce clockwise rotation of the plane are known as *dextrorotatory* or right-handed substance and others that rotate the plane of polarization in the anticlockwise direction are called *laevorotatory* or left-handed substance.

If a monochromatic light from source S is made to pass through two Nicol prisms (their action is similar to Tourmaline crystals) PN and AN but in crossed positions, no emergent light can be received by the eye.

In this case, the polarization prism PN renders the beam polarized with its vibrations in the principal plane. In this analyzing prism, the positions are set with its principal plane perpendicular to the direction of these vibrations so that no light passes through AN. If a quartz crystal is cut with its refracting faces perpendicular to the optic axis and placed between PN and AN, with the two crystals still crossed, it can be observed that the light is no longer affected by the analyzing prism. In order to cut off the light, the analyzing prism is required to be rotated through some angle. So, the emergent light is still plane polarized, but its plane of polarization has been rotated.

It has been experimentally established that depending upon the arrangement of molecules, some quartz crystals are dextrorotatory, and some are laevorotatory.

The amount of rotation is found to depend on

1. Thickness and density of crystal.
2. Temperature and wavelength of light used

    (Jean-Baptiste Biot found that Rotation $\alpha \frac{1}{\lambda^2}$)
3. Concentration in case of solutions

Specific Rotation

The term "specific rotation" is used to bring the rotation of all optically active substances in a comparable form.

Specific rotation for a given temperature $t°C$ and for a light of given wavelength λ is defined as the rotation (in degrees) produced by a path of one decimeter length in a substance of unit density.

Physics Practicals

It θ is the rotation produced by decimeter length of a solution of density d/cc, then specific rotation S corresponding to some temperature t and light of wavelength λ is given by

$$S_\lambda = \frac{\theta}{l \times d} = \frac{\text{Rotation in degree}}{\text{Length in decimeter} \times \text{Concentration in gm/cc}}$$

Sugar is the most commonly used optically active substance, and the instrument used for measuring the optical rotation produced by the substance is called polarimeter. Basically, they consist of two Nicol prisms capable of rotation about the incident beam as axis. The optically active substance is placed between the two Nicol prisms. Generally, it is very difficult to locate the position of analyzing Nicol when no light is received, but with the used of Laurent's half-shade device, this difficulty is overcome.

Laurent's Polarimeter

It is an instrument used for finding optical rotation of certain solutions when used for finding the specific rotation of sugar; it is called a polarimeter. Also, if the specific rotation of sugar is known, then the concentration of the solution can be found.

Laurent's Half-Shade Device

It consists of two semicircular plates (C), one made of glass and another of quartz. Both glass and quartz are connected together, as shown in Figure 14.19.

The quartz plate is cut parallel to its axis, which is parallel to the line joining two plates YY'', termed as a half-wave plate. That is, its thickness is adjusted so that it introduces a path difference of $\lambda/2$ between the E-ray and the O-ray for sodium light. In other words, when light passes through the quartz plate, the O-ray gains a path difference of $\lambda/2$ compared to the path difference of the E-ray. The thickness of the glass plate is so adjusted that it transmits the same amount of light as the quartz plate.

Let the light after passing through the polarized P be incident normally on the half-shade plate and have vibrations along OP, but on passing through the glass half, the vibration will remain along OP, and on passing through the quartz half plate, these will split into E and O components. The vibrations of E component are along OY axis, and those of O component along OX axis.

On passing through the quartz plate, a phase difference of π and a path difference of $\lambda/2$ are introduced between the two vibrations. The O vibrations will advance in phase by π and will appear along OX instead of OX' on emergence. So, the resultant vibration after emerging from the quartz plate will be along OQ such that

$\angle POY = \angle YOQ$.

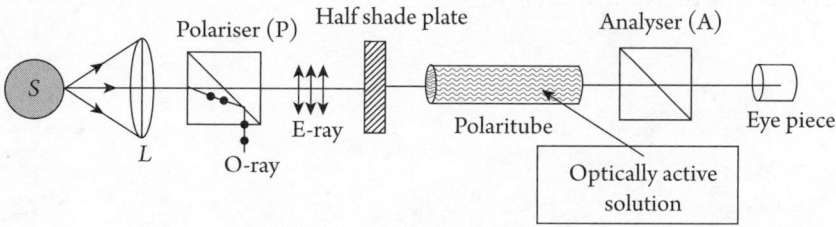

Fig. 14.19

In the analysis, Nicol is placed with its principal plane parallel to OP. The plane-polarized light through glass half will pass, and hence it will appear brighter than the quartz half from which light will be practically obstructed.

If the principal plane of Nicol is parallel (II) to OQ, the quartz half will appear brighter than the glass due to the above reason.

When the principal plane of analyzing Nicol is parallel (II) to OQ, the two halves will appear equally bright. It is because the two vibrations coming out of the two halves are equally inclined to its principal plane, and so the two components will have equal intensity.

Again, if the principal plane of the analyzer is perpendicular to the YOY' axis, the two halves are equally dark. The eye can easily detect the change when the two halves are equally dark, so the readings are taken for this position.

Procedure: Listed here are the steps that should be followed in this experiment:

1. First of all, clean the polarimeter tube and fill with clean water to minimize the air.
2. Place the polarimeter tube in an appropriate place.
3. Switch on the source of light and see through an eyepiece for the position of two halves of unequal intensity. Rotate the analyzer in the clockwise direction until the intensity of the two halves appears the same and note down the analyzer reading.
4. Rotate the screw in the anti-clockwise direction until the intensity of two halves appears the same and note down the reading.
5. Rotate the screw from mean position 180° in any direction and repeat steps (III) and (IV).
6. Fill the polarimeter tube with cane sugar solution and repeat the experiment for different concentrations of the sugar solution. Plot the graph between angle of rotation and concentration. Specific rotation can also be determined using this graph. Thus, the slope of the graph $= \dfrac{AB}{BD} = \dfrac{\theta}{C}$.

Observation Table

1. Room temperature = _____ °C.
2. Length of the solution tube = _____ dm.
3. Mass of the sugar dissolved = _____ gm.
4. Volume of the solution = _____ in c.c.
5. Least count of the circular scale = _____°.
6. Least count of the Vernier scale = _____°.

Reading on Analyzer for Equal Illumination in Degrees														Rotation in Degrees	
With Distilled Water							Strength of the Solution per 100 c.c.	With Sugar Solution							Angle (θ_1)	Angle (θ_2)
First Position			Second Position (180° apart)					First Position			Second Position (180° apart)				
Main Scale	Vernier	Total	Main Scale	Vernier	Total		Main Scale	Vernier	Total	Main Scale	Vernier	Total		

Calculations

The formula to calculate the specific rotation (S) is

$$S = \frac{10\theta}{LC} = \underline{\qquad}.$$

Results:

1. The specific rotation (S) of the cane sugar (for a given concentration per decimeter) solution at _____ °C is _____ °.
2. Estimate the same parameter α for the different concentrations of cane sugar.
3. Draw a graph between rotation and concentration of the solutions.
4. Error estimation as per the general instruction.

Precautions

1. There should be no air-bubble in the tube while filling it with solution or distilled water.
2. While taking one set of the observations, the polarizer should not be disturbed.
3. The cap of the tube should not be tightened beyond a limit as it may strain the glass. Strained glass may produce elliptically polarized light, which might interfere with the setting.
4. Two positions at ±90° may appear where the equal illumination remains for a long range. These readings should not be taken.
5. Switch off the lamp after completing the experiment.

Experiment 10: Grating Spectrum

Objective: Determination of wavelength of spectral lines of mercury (for violet, green, yellow-1, and yellow-2) using plane transmission grating.

Apparatus Used: The following apparatus are used in this experiment:

1. Spectrometer, as shown in Figure 14.20.
2. Plane transmission grating.
3. Mercury lamp.
4. Reading lens and so on.

Formula Used: The wavelength of spectral line is calculated using the formula

$e \sin\theta = n\lambda,$

where e is the grating element of grating and is given by $e = \dfrac{2.54}{Z}$ cm, Z is the line per inch on grating, n is the order of diffraction pattern, and λ is the wavelength of monochromatic light.

Theory of the Experiment

Plane Transmission Grating: Plane transmission grating is an arrangement equivalent in its action to a large number of parallel slits equal width separated equidistantly by opaque

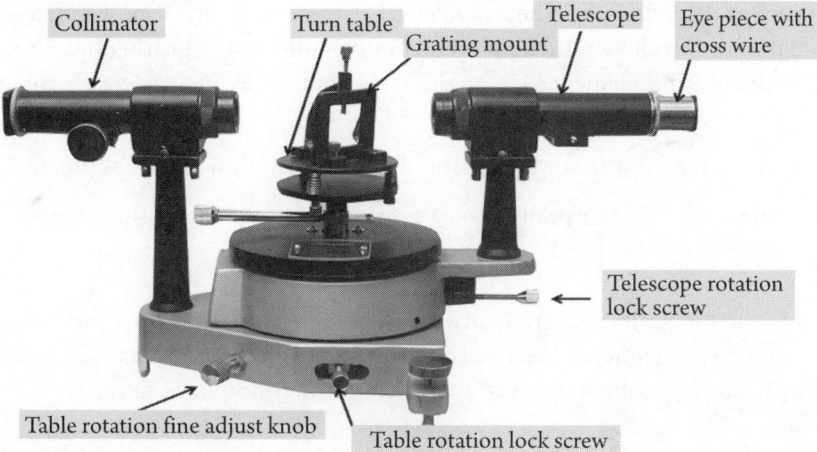

Fig. 14.20

surface of equal spacing. It is made up of transparent material like glass by ruling equidistant lines of equal width. The distance between two consecutive slits, that is, the sum of opaque and slit width, is known as grating element by "e", as shown in Figure 14.21.

Elementary Theory of Grating: Figure 14.21 represents the section of the grating and the plane wavefront incident normally on it. The plane wavefront, after scattering from the grating, is focused on the screen by a convex lens. According to Huygens' theory, each point in a slit acts as a secondary wavelet; therefore, the midpoint between consecutive slits S_1 and S_2 makes an angle θ with the normal to the slit. S_2K is perpendicular to S_1K; hence, the path difference between the two waves is

$$S_1K = S_1S_2 \sin\theta = e\sin\theta.$$

The same path difference will exist between the corresponding waves from any two points in consecutive slits separated by distance "e" grating element. Thus, if the path difference S_1K is equal to a whole number of wavelength "λ", then the waves interfere constructively producing maximum intensity in the direction given by angle θ. Hence, the condition for maxima is

$$e\sin\theta = n\lambda. \tag{14.27}$$

These maxima are known as principal maxima. For $n = 0$, we get zeroth order maxima. For $n = \pm1, \pm2, \pm3$, and so on, we get first, second, third, and so on order principal maxima lying symmetrically on either side of normal or the zeroth order.

Fig. 14.21

Determination of Wavelength: The wavelength of spectral line can be determined for a monochromatic or polychromatic light incident normally on the grating by using the formula $e\sin\theta = n\lambda$, if the grating element "e", angle of diffraction θ, and order of principal maxima "n" are known.

Procedure: Listed here are the steps that should be followed in this experiment:

1. Calculation of "e": Normally, every grating is provided by the number of lines per inch, say Z, ruled on it by manufacture. Thus, $e = \dfrac{2.54}{Z}$.

2. Finding the "n" and measurement of θ: The determination of θ and "n" is done with the help of spectrometer. We used the formula $e\sin\theta = n\lambda$, which is valid for normal incident. The experimental setup is shown in the following figure.

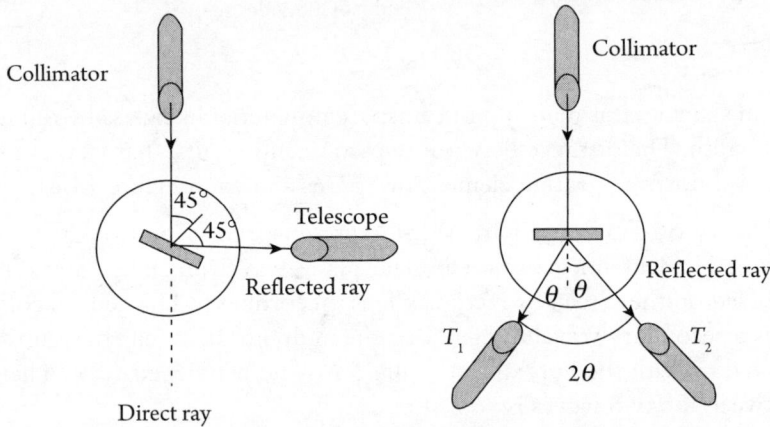

3. First, take spectrometer and adjust the level with the help of a leveling screw; next, focus the telescope using an eyepiece.
4. Adjust the collimating tube for parallel light. Now, focus the direct narrow slit position through the telescope.
5. Mount the grating at grating holder. Now, rotating the telescope in the left and right positions, we observe different color bands with different orders.
6. Set the cross wire on each left and right bands and note the readings of both orders.
7. Now put the values in the formula and find the result.

Least Count of Spectrometer

20 main scale division (M.S.D.) = 10°.

As shown in Figure 14.22, we have

1 M.S.D. = $\left(\dfrac{1}{2}\right)^\circ$ = 30'.

Total number of divisions on vernier scale: 30.

Least count = $\dfrac{\text{Value of 1 M.S.D.}}{\text{Total no. of divisions on vernier}} = \dfrac{30'}{30} = 1'$.

Physics Practicals

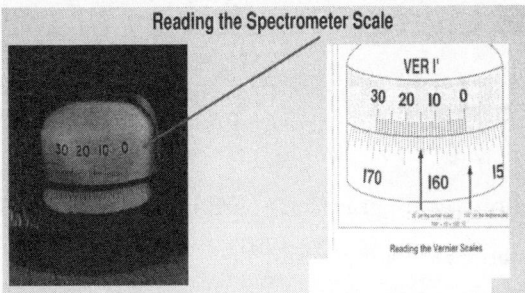

Fig. 14.22 Scale of measurement.

Observation Table

1. Number of lines per inch of grating = _____.
2. Least count of spectrometer = _____.
3. Table for determination of angle θ.

S.No.	No. of Order of Principal Maxima	Spectral Line	Left-side Reading		Right-side Reading		Difference	Angle
			Vernier (A)	Vernier (B)	Vernier (A′)	Vernier (B′)		

Calculations

The value of θ is calculated as follows:

$$\theta = \frac{(A \sim A') - (B \sim B')}{4} \text{ and } n\lambda = e\sin\theta.$$

Result:

1. The experimental value for wavelengths is = _____.
2. The standard value for corresponding lines is = _____.
3. % Error = _____.

Precautions and Sources of Errors:

1. The spectrometer must be properly aligned and focused.
2. The light source should be located about 1 cm from the collimator slit.
3. The telescope should be focused for an object located at infinity. Adjust the telescope focus knob so that the viewed image of the distant object is clearly in focus.

Experiment 11: Charge Sensitivity of a Ballistic Galvanometer

Objective: Determination of charge sensitivity and ballistic constant of a ballistic galvanometer.

Apparatus Used: The following apparatus are used in this experiment:

1. Ballistic galvanometer with lamp and scale arrangement
2. Two condensers of known capacity (generally of the order of 0.1 to 0.2 μF)
3. Morse key
4. Discharge key
5. A cell
6. Voltmeter
7. Ammeter
8. Connecting wires, and so on.

Formula Used: The logarithmic decrement is defined as

$$\lambda = \frac{2.303 \log\left(\frac{\theta_1}{\theta_{11}}\right)}{n-1},$$

where θ_1 is the first throw of the coil, θ_{11} is the eleventh throw of the coil, and n is the number of throw.

For the correction of deflection, we use the relation

$$\theta = \theta_1\left(1 + \frac{\lambda}{2}\right).$$

Theory of the Experiment

Damping and Damping Correction

In the case of an ordinary moving coil galvanometer, a constant current flows through the coil, soothe deflection is constant, and the pointer gives the constant reading. Ballistic galvanometers measure charge in the form of sudden discharge, and due to the impulse, a sudden kick is given to the coil. So, it is only the first throw that is effective in measuring the charge that flows through the coil.

After the first throw, the coil oscillates in the magnetic field with continuously decreasing amplitude. Due to electromagnetic induction in the coil, air resistance, and so on, there is a decrease in amplitude. Let θ be the actual deflection in the absence of damping and $\theta_1, \theta_2, \theta_3$ and so on be the successive observed throws to the right and left continuously, as shown in Figure 14.23.

For the different values of deflection, it is found that $\frac{\theta_1}{\theta_2} = \frac{\theta_2}{\theta_3} = \frac{\theta_3}{\theta_4} = d$, where d is called the decrement.

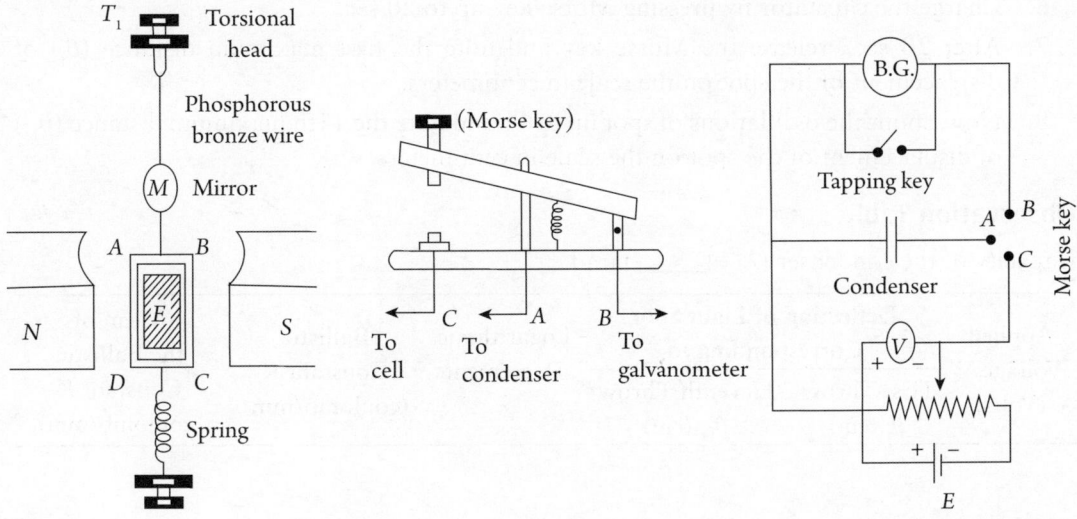

Fig. 14.23

Let $d = e^\lambda$ so that $\lambda \log_e d$, where λ is called the logarithmic decrement. Each complete vibration comprises two swings (that is, from extreme right to the left θ_1 and θ_2 and from extreme left to right θ_2 to θ_3).

$$\frac{\theta_1}{\theta_3} = \frac{\theta_1}{\theta_2} \times \frac{\theta_2}{\theta_3} = d^2 = e^{2\lambda} \text{ (for two swings)}.$$

Similarly for four swings,

$$\frac{\theta_1}{\theta_5} = d^2 = e^{4\lambda} \text{ and so on.}$$

Let θ be the true throw in the absence of damping, which is higher than the observed first throw θ_1. The motion of the coil from the mean position to the extreme right corresponds to a half swing:

$$\frac{\theta}{\theta_1} = \frac{d}{2} = e^{\lambda/2} = \left(1 + \frac{\lambda}{2}\right) \text{ approx.}$$

$$\theta = \theta_1 \left(1 + \frac{\lambda}{2}\right),$$

where θ_1 is the observed first throw and λ is the logarithmic decrement.

Procedure: Listed here are steps that should be followed in this experiment:

1. Connect the circuit, as shown in Figure 14.23.
2. First make a potential divider.
3. The two lower-end terminals of the given rheostat are connected to the 2 V battery.
4. Now 2 V is distributed along *AB*.
5. Adjust the spot on the scale at 0 cm.

6. Charge the capacitor by pressing Morse key up to 20 sec.
7. After 20 sec., release the Morse key and note the first maximum distance (θ_1) of displacement of the spot on the scale in centimeters.
8. Now, count the oscillations of spot up to 11 and note the 11th maximum distance (θ_{11}) of displacement of the spot on the scale in centimeters.

Observation Table

Capacity of the condenser (λ) = _____ farad.

Applied Voltage V (Volts)	Deflection of Light Spot Corresponding to		Logarithmic Decrement λ	Ballistic Constant K (coulomb/mm)	Mean of the Ballistic Constant K (coulomb/mm)
	First Throw θ_1 (cm)	Eleventh Throw θ_{11} (cm)			

Calculations

The value of λ and K are calculated using the following relations:

$$\lambda = \frac{2.303 \log\left(\dfrac{\theta_1}{\theta_{11}}\right)}{n-1}$$

$$K = \frac{CV}{d_1\left(1 + \dfrac{\lambda}{2}\right)}$$

Results:

1. The ballistic constant of the given ballistic galvanometer is = _____ coulomb/mm.
2. The charge sensitivity of the given ballistic galvanometer is = _____.
3. Percentage error = _____ %.

Precautions

1. The galvanometer should be leveled properly.
2. Coil of the galvanometer should oscillate freely.
3. A damping key/tapping key should be connected across the galvanometer.
4. The E.M.F. of the cell should be constant.
5. The condenser should be charged for a short time for every reading.

Experiment 12: Rydberg Constant

Objective: To determine the wavelength of spectral lines of hydrogen and hence the value of Rydberg Constant.

Apparatus Required

Spectrometer 6/7 inch with LC-30 Sec.,
Diffraction grating 15000 lines per inches (LPI),
Hydrogen lamp with power supply,
Spirit level and magnifying lens with LED.

Formula Used

The wavelength of the spectral line is given by

$$n\lambda = e \sin\theta,$$

where n is the order of the diffraction, e is grating element and θ is the angle of diffraction.

For the lines of the Balmer's series,

$$\frac{1}{\lambda} = R\left(\frac{1}{2^2} - \frac{1}{p^2}\right).$$

For red line $p = 3$

For blue line $p = 4$

For blue-violet line $p = 5$

For violet $p = 6$

Figure 14.24 shows hydrogen emission spectrum lines in the Balmer series lying in visible region. The red line at the right is H-alpha.

Theory of Experiment

The emission spectrum of atomic hydrogen is divided into a number of spectral series, with wavelengths given by the Rydberg formula. These observed spectral lines are due to the electron making transitions between two energy levels in an atom. Spectral emission occurs when an electron transitions, or jumps, from a higher energy state to a lower energy state.

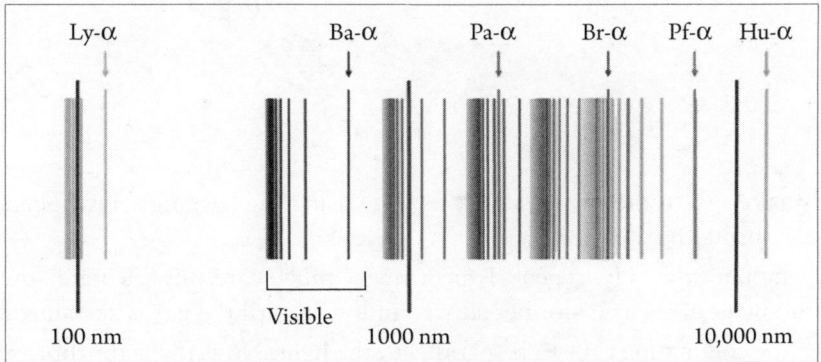

Fig. 14.24 The spectral series of hydrogen.

To distinguish the two states, the lower energy state is commonly designated as p', and the higher energy state is designated as p. The energy of an emitted photon corresponds to the energy difference between the two states. The energy differences between levels in the Bohr model, and hence the wavelengths of emitted/absorbed photons, is given by the Rydberg formula:

$$\frac{1}{\lambda} = RZ^2 \left(\frac{1}{p'^2} - \frac{1}{p^2} \right),$$

where Z is the atomic number, p is the upper energy level, p' is the lower energy level, and R is the Rydberg constant (1.09737×10^7 m^{-1}). This equation is valid for all hydrogen-like species, that is, atoms having only a single electron, and the particular case of hydrogen spectral lines are given by $Z = 1$.

Depending upon the value of p', the various observed spectral lines are named as follows:

Value of p'	Name of the Corresponding Series
1.	Lyman series
2.	Balmer series
3.	Paschen series (Bohr series)
4.	Brackett series
5.	Pfund series

Procedure
1. The experimental setup is shown in the figure shown below.

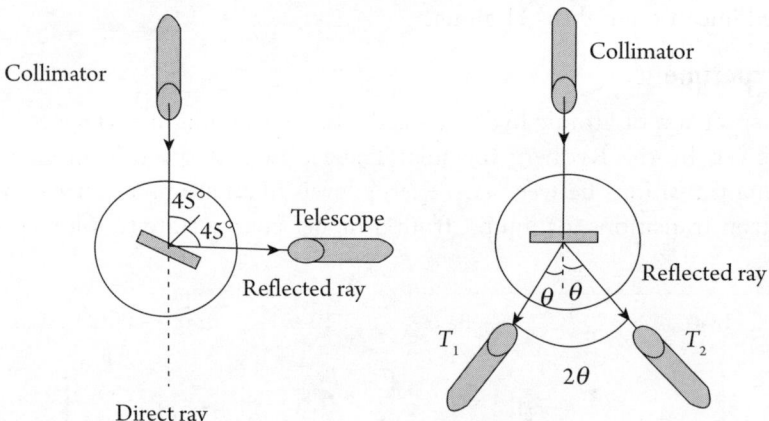

2. Firstly, take spectrometer and adjust its level using the leveling screw. Then, focus the telescope using the eye-piece.
3. Now, mount the Hydrogen Transformer tube carefully. Rotate the variable potentiometer of the transformer slowly and wait until the gas is sustained.
4. Adjust the collimating tube to be parallel to the light. Now, focus the direct narrow slit position through the telescope.

5. Mount the grating on the grating holder. Now, rotate the telescope left and right to observe the different colored bands in different orders.
6. Set the cross wires on the left and right bands, and note their readings for both orders.
7. Now, input the values into the formula and find the result.

Observation Table

1. Number of lines per inch of grating = _____.
2. Least count of spectrometer = _____.
3. Table for determination of angle θ

S.No.	No. of Order of Principal Maxima	Spectral Line	Left-side Reading		Right-side Reading		Difference	Angle
			Vernier (A)	Vernier (B)	Vernier (A')	Vernier (B')		

Calculation

The value of θ is calculated as follows:

$$\theta = \frac{(A \sim A') - (B \sim B')}{4}$$

and $n\lambda = e\sin\theta$.

For the lines of the Balmer's series:

$$\frac{1}{\lambda} = R\left(\frac{1}{2^2} - \frac{1}{p^2}\right).$$

Result:

1. The experimental value for Rydberg constant R = _____ 1/m.
2. Standard value for Rydberg constant $R = 1.097 \times 10^7$ m^{-1}.
3. % error = _____.

Precautions and Sources of Errors

1. The spectrometer must be properly aligned and focused.
2. The light source should be located about 1 cm from the collimator slit.
3. The telescope should be focused for an object located at infinity. Adjust the telescope focus knob so that the viewed image of the distant object is clearly in focus.
4. The hydrogen discharge lamp is powered by high voltage and the tube gets hot.
5. DO NOT touch the tube anywhere, especially near the ends where the electrical contacts are made.

Experiment 13: Planck's Constant

Objective: Draw the V-I characteristic for light emitting diode (LED) and determine the value of Planck's constant.

Apparatus Required: Planck's constant apparatus with LEDs of different color.

Formula Used: The following formula is used to determine Planck's constant:

$$h\nu = eV_0$$

$$h\left(\frac{c}{\lambda}\right) = eV_0 \quad \left[\because \nu = \frac{c}{\lambda}\right]$$

$$h = eV_0\left(\frac{\lambda}{c}\right), \tag{14.28}$$

where h is Planck's constant, e is the electronic charge, V_0 is the threshold voltage, λ is the wavelength of the LED, and c is the velocity of light.

Theory of Experiment

Planck's constant is one of the smallest constants used in physics, having a value of 6.6×10^{-34} J-s. The significance of this is that it reflects the extremely small scale at which quantum mechanical effects are observed. Planck's constant is one of the fundamental constants in modern physics. It relates the energy of a photon to its frequency. To determine this constant, we can use LEDs. LEDs are semiconductor diodes that involve direct band gap compound materials. These materials can be group III-IV or group II-VI of the periodic table. Diodes today come in a variety of colors. Each color is achieved by having a slightly different semiconductor material. We can select LEDs of different colors, including blue, green, red, yellow, and orange. The experiment is based on the fact that the energy of the photon relates to its frequency as:

$$E = h\nu,$$

where E is the energy of photon, h is Planck's constant, and ν is the frequency of the emitted photons. When the diode emits light, the voltage across the diode, V_0, is just enough to give energy to electrons to jump between two energy levels. Therefore,

$$V_0 e = h\nu,$$

where e is the electron charge and V_0 is the threshold voltage. The relation between the maximum wavelength, λ, and the turn on voltage, V_0, is

$$E = h\nu = hc/\lambda. \tag{14.29}$$

$$E = eV_0. \tag{14.30}$$

From equations (14.29) and (14.30), we get

$$hc/\lambda = eV_0,$$

or $h = eV_0\lambda/c$.

Here, λ is the wavelength of the LED and c is the velocity of light.

Procedure
1. The experimental setup is shown in the figure given below:

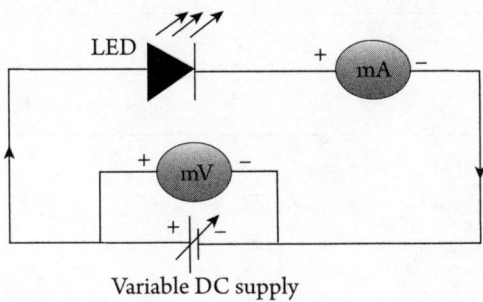

Fig. 14.25

2. Connect the circuit as shown in the diagram above.
3. Set the voltmeter at the range of 20 V and ammeter at 200 mA.
4. Connect the mains cord and switch "On" the power supply.
5. Now increase the DC voltage at the fixed interval of 0.1 V or 100 mV (from 2 V to 4 V).
6. Note the corresponding current by DC ammeter in the Observation Table.
7. Now take the current on the Y-axis and voltage on the X-axis and plot a graph between current and voltage, as shown in Figure 14.26.

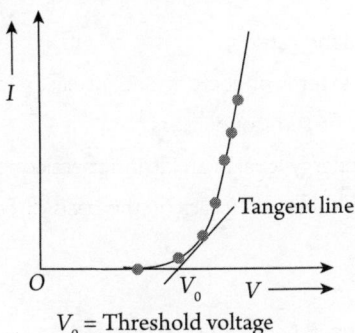

V_0 = Threshold voltage

Fig. 14.26

8. Note the reading of voltage at which the current flows suddenly through the LED; at this point, the graph suddenly changes direction. This point is known as the knee-voltage or threshold voltage.
9. Put this value in equation (14.28) and calculate the value of Planck's constant.

Observation Table

Determination of Planck's constant

S.No.	Blue LED		Green LED		Yellow LED	
	Voltage (V)	Current I (mA)	Voltage (V)	Current I (mA)	Voltage (V)	Current I (mA)

Calculation

$$h = \frac{eV_0\lambda}{c} = \underline{\qquad}.$$

Result:

1. The experimental value for Planck's constant = _____.
2. Standard value for Planck's constant $h = 6.63 \times 10^{-34}$ J-s.
3. % Error = _____.

Precautions and Sources of Errors

1. The connection must be made properly.
2. The potentiometer must be varied gradually and slowly to achieve the just-glow position of the LEDs.

Previous Year Questions (University Examination)

1. What do you mean by energy band gap?
2. Differentiate among conductors, semiconductors, and insulators on the basis of energy band gap.
3. Differentiate intrinsic and extrinsic semiconductors.
4. What is the position of Fermi energy level in an intrinsic semiconductor?
5. Differentiate P-type and N-type semiconductors on the basis of Fermi energy level?
6. What is p–n junction diode?
7. What is depletion layer?
8. What is an avalanche breakdown in p–n junction diode?
9. What is knee voltage or threshold volt in p–n junction diode?
10. What do you mean by four-probe method?
11. What is the formula used to determine energy band gap?
12. What is the energy band gap in n-type germanium and p-type silicon semiconductors?
13. Define semiconductors, and explain the role of impurities on the conductivity of semiconductor.
14. What is Stefan's law of black body radiation?

Physics Practicals

15. What is the Draper point in the verification of Stefan's law by the electrical method?
16. What do you mean by emissive power of black body?
17. What do you mean by calibration graph?
18. What is the difference between voltmeter and ammeter?
19. What is thermoelectricity and thermopower?
20. What is potentiometer? Explain its applications.
21. What is potential gradient? Write its formula.
22. What is balancing length?
23. Differentiate between galvanometer and ammeter?
24. What is seeback effect?
25. What is Peltier effect?
26. What is Thomson effect?
27. What is the difference between cell and battery, and also describe types of cells with examples?
28. Explain the variation of magnetic field along the axis of current-carrying coil.
29. What is the magnetic meridian and magnetic compass box?
30. What is formula of magnetic field at the center of current-carrying coil?
31. What is rheostat and reversing key (or four-way key) in the magnetic variation experiment?
32. What is the point of inflection?
33. What is the working principle of Carey Foster bridge?
34. What is the maximum value of the difference in the two resistances that can be measured by Carey Foster's bridge?
35. What do you mean by resistance and R.B.?
36. What do you mean by specific resistance?
37. What is tangent galvanometer?
38. What is the reduction factor of tangent galvanometer? Write the formula for reduction factor.
39. What is parallax error? How can it be minimized?
40. What are Newton's rings?
41. On which property of light are Newton's rings formed?
42. Why Newton's rings are circular?
43. Why is the central spot dark in Newton's rings ?
44. What is the least count of traveling microscope?
45. Write the formula used for the measurement of wavelength by Newton's rings method.
46. How are Newton's rings formed? Why does the spacing between rings decrease with increasing fringe order?
47. What are the applications of Newton's rings?
48. What do you mean by interference of light? Why does a sodium lamp glow with pink color in the beginning?
49. What is monochromatic light source? Write the wavelength of sodium light?

50. What is polarization of light?
51. What is specific rotation? Write the formula for it.
52. What is Nicol prism and the total internal reflection of light?
53. What are ordinary and extraordinary light?
54. What are polarizer and analyzer?
55. Write difference between the prism and grating?
56. What is plane transmission grating?
57. What is grating element? Write the formula for it.
58. What is the least count of spectrometer?
59. What is the formula used to determine the wavelength of spectral lines in diffraction grating experiment?
60. At Grating, L.P.I. is written over it. What does it represent? Also, define its value as used in the grating spectrometer experiment.
61. What variation would occur in grating spectra with increasing grating element?
62. What is polychromatic light? Give an example for it.
63. What is Ballistic galvanometer?
64. What do you mean by charge sensitivity of Ballistic galvanometer?
65. What is Ballistic constant? Write its formula with unit.
66. What is the capacitor? Why is it used in circuit?
67. What is damping key?
68. What is Morse key?
69. What do you mean by damped oscillations?
70. Explain hydrogen spectrum.
71. Write down the standard value of Rydberg's and Planck's constant with their units.
72. Write down the formula of percentage error?
73. What is LED? What materials are used to fabricate LED?
74. What is the formula used to determine Planck's constant in the experiment?
75. What do you mean by ohmic contact?

Multiple Choice Questions

1. The four-probe method is used to determine
   (a) High resistivity
   (b) Low resistivity
   (c) Moderate resistivity
   (d) All of the above
2. The energy band gap of any material is defined as
   (a) The energy difference between top of valence band and bottom of the conduction band
   (b) Energy in valence band only
   (c) Energy in conduction band only
   (d) None of these

3. The energy band gap of semiconductors is
   (a) 1 eV
   (b) 3 eV
   (c) 5 eV
   (d) All of the above
4. The energy band gap of silicon is
   (a) 1.1 eV
   (b) 3.6 eV
   (c) 5.1 eV
   (d) None of the above
5. The energy band gap of germanium is
   (a) 0.67 eV
   (b) 1.6 eV
   (c) 4.2 eV
   (d) 6.3 eV
6. The power radiated by a black body is proportional to
   (a) T
   (b) T^2
   (c) T^4
   (d) T^6
7. The Draper point is measured at
   (a) Maximum bulb glow
   (b) Minimum bulb glow
   (c) Average bulb glow
   (d) None of these
8. To determine the exponent of temperature for black body radiation, the graph is plotted between
   (a) $\log_{10} V$ vs $\log_{10} T$
   (b) $\log_{10} I$ vs $\log_{10} T$
   (c) $\log_{10} P$ vs $\log_{10} T$
   (d) $\log_{10} K$ vs $\log_{10} T$
9. The Draper point is a
   (a) Graph
   (b) Reference point
   (c) Calibration graph
   (d) None of the above
10. In Stefan's law experiment, the calibration graph is plotted between
    (a) R_t/R_0 vs $T°C$
    (b) I/R_0 vs $T°C$
    (c) V/R_0 vs $T°C$
    (d) None of the above
11. When two dissimilar metal wires are joined to form two junctions kept at different temperatures, an emf is produced. This effect is known as
    (a) Thomson effect
    (b) Peltier effect
    (c) Seebeck effect
    (d) Both (a) and (b)
12. When two dissimilar metal wires are joined to form two junctions and a current is passed through them, both junctions attain different temperatures. This effect is known as
    (a) Thomson effect
    (b) Peltier effect
    (c) Seebeck effect
    (d) None of the above
13. If a conducting wire is melted to maintain a temperature gradient along its length, the two ends of the conductor show a potential difference. This effect is known as
    (a) Thomson effect
    (b) Peltier effect
    (c) Seebeck effect
    (d) None of the above
14. The balancing length is obtained when
    (a) Galvanometer deflects in the left direction
    (b) Galvanometer deflects in the right direction
    (c) Galvanometer is at null point
    (d) None of the above

15. The point of inflexion in a graph is
    (a) Where the curvature changes its sign
    (b) Highest point in the curve
    (c) Lowest point in the curve
    (d) None of these
16. The magnetic meridian is defined as
    (a) Horizontal component of magnetic lines along the surface of the earth
    (b) Vertical component of magnetic lines along the surface of the earth
    (c) Inclined component of magnetic lines along the surface of the earth
    (d) None of the above
17. Carey Foster bridge is based on the principle of
    (a) Interference of light   (b) Diffraction of light
    (c) Wheatstone bridge   (d) None of these
18. Horizontal component of the earth's magnetic field is calculated by
    (a) $B = H \tan\theta$ (b) $H = B \tan\theta$
    (c) $I = V \tan\theta$ (d) None of the above
19. The central fringe in the Newton's ring experiment is
    (a) Bright   (b) Dark
    (c) First dark, then bright   (d) None of these
20. The specific rotation of sugar solution is measured by the formula
    (a) lc/θ (b) $10\theta/lc$
    (c) $\theta/10lc$ (d) None of these

Keys

1. (b) 2. (a) 3. (a) 4. (a) 5. (a) 6. (c) 7. (b) 8. (c) 9. (b) 10. (a)
11. (c) 12. (b) 13. (a) 14. (c) 15. (a) 16. (a) 17. (c) 18. (a) 19. (b) 20. (b)

Index

absent spectra, 332, 337
absolute zero temperature, 161–162
absorption
   by material, 358
   of radiation, 171–173
accelerated frames, 2–4
acceptance angle, 355–357
acceptance cone for fiber, 356
acoustic attenuation, 261–262
AC resistivity, 258–259
alkali-doped fullerenes, 264–265
amorphous semiconductors, 223
amorphous solids, 275–276
Ampere, Andre-Marie, 103
Ampere's circuital law
   displacement current, 111–112
   magnetic field intensity, 110
   steady state conditions, 111
Anderson localization, 223
angle of acceptance for fiber, 355
annihilation of matter, 22
antireflection coating, 362
applications of relativity, 39–40
arc discharge method, 239, 240
atomic packing factor (APF), 281
atomic radius, 281–284, 286–287
atomic spectroscopy, 63
atomic stability, 63
atoms per unit cell, 281–284, 286
attenuation
   absorption by material, 358
   in fibers, 358
   microbends, optical fiber, 359
   scattering, 359
   waveguide and microbend losses, 359
avalanche breakdown, 201
average speed of molecules, 149–150

ballistic galvanometer, 408–410
band gap, 370–374
band theory of solids
   different numbers of atoms, 209
   energy band gap, 209
   fermi energy, 210
   solids classification, 209–210
Bardeen, Cooper and Schrieffer (BCS) theory
   coherence length, 257
   Cooper pair formation, 256–257
   electron lattice electron interaction, 256
Bechgaard salts, 264–265
bending of light beam, 32
λ-BETS2X compounds, 265
β evaluation, 146
blackbody radiation, 63
black holes
   Cygnus X-1, 39
   detection, 39
   event horizon, 38
   Schwarzschild radius, 38
   singularity, 38
   white dwarfs and neutron stars, 39
body-centered cubic (BCC) structure
   atomic radius, 283
   atoms per unit cell, 283

CN, 283
PF, 283
representation of, 282
Bohr, Niels, 63
Boltzmann constant, 212
Boltzmann distribution rule, 174
Boltzmann partition function, 147
Bose–Einstein (BE) statistics, 143
  distribution law, 152–153
  energy distribution function, 153
  energy distribution law, 153–154
  fundamental principles, 152
buckminsterfullerene, 239
buckyballs
  applications of, 241
  arc discharge method, 239–240
  definition, 239
  properties of, 239
bullet trains, 268

canonical ensemble, 141–142
carbon nanotubes
  applications, 245
  CVD, 242–243
  description, 241
  HiPCO, 242–243
  honeycomb crystal lattice, graphene, 244
  MWNT/MWCNT, 242
  properties, 245
  structures of, 244
  SWNT/SWCNT, 242
Carrey Foster bridge, 384–388
carrier concentration, 222
Cartesian coordinate system, 2
centripetal acceleration, 4
chemical-vapor deposition (CVD), 242–243
chiral, 244
cladding, 351–352
Classical mechanics, 1
Clausius–Mossotti equation, 197–198
coherence length, 260
coherent sources, 297
compound semiconductors
  amorphous, 223
  Anderson localization, 223
  chemical elements, 222

dangling bond locations, 223
HIT solar cells, 225
hydrogenated amorphous silicon (a-Si:H), 224
lighted a-Se sections, 224
Staebler–Wronski effect, 224–225
thin film solar cells, 224
concept of curl, 107–109
concept of divergence, 106–107
concept of gradient, 105–106
concept of simultaneity, 16–17
conduction band, 160, 211–213
conductivity of semiconductors, 216–217
conservation of energy, 299–300
constancy of speed of light principle, 9
constructive interference, 297
coordination number (CN), 281–285, 287
critical current, 252–253
critical magnetic field
  type II superconductors, 252–253
  type I superconductors, 251–252
critical temperature (T_c), 250
crystal lattices, 276
crystalline solids, 275–276
crystallography
  amorphous solids, 275–276
  crystalline solids, 275–276
  definition, 275
crystal system structure
  BCC, 282–284
  FCC, 284
  NaCl crystal, 285–287
  SCC, 281–282
cubic crystal system, 278–279
current density, 252–253
Cygnus X-1, 39

damping correction, 408–410
Davisson and Germer experiment
  definition, 71
  diffracted beam plotted, 71–72
  Nickel single-target crystal, 71
  setup, 71–72
  wavelength calculations, 72–73
de-Broglie hypothesis
  definition, 64

free particle in kinetic energy, 65–66
matter waves, properties of, 66
velocity of, *see* velocity of de-Broglie waves
wavelength of, 64–65
defect breakdown, 201
degeneracy parameter, 162–163
density of states
  definition, 211
  electrons, 211
  holes, 211
dextrorotatory, 400
diameter of rings, 308–309
  bright rings, 309
  dark rings, 309–310
diamond structure, 286
dielectric breakdown, 200–201
dielectric constant, 191
  frequency dependence, 198–199
  vs. refractive index, 198
dielectric loss
  circuit diagram, 199
  definition, 199
  phasor diagram, 199–200
dielectric materials, 187
  constant, 187–188
  constant and electrical susceptibility, 191
  nonconducting materials, 187
  polarization, 189–190
dielectric strength, 200–201
diffraction grating
  absent spectra, 337
  definition, 332
  maximum number of orders, 337
  principal maxima, 334
  secondary maxima, 334–336
  spectra formation, 336
  theory, 333–334
  wavelength of light, 337
diffraction phenomenon, 325
  Fraunhofer, 325–330
  Fresnel, 325
diode (magnetron valve), 388–393
dipole moment per unit volume, 191
direct piezoelectric effect, 203
dispersive power, 338
doped fullerenes, 267

Doppler effect in light, 26–28
double-slit diffraction pattern, 330

$e^{-\alpha}$ determination, 146
E and H vectors
  perpendicular to each other, 122
  same phase, 122
eigen function, 81
Einstein's coefficient
  absorption of radiation, 173
  spontaneous emission of radiation, 173
  stimulated emission of radiation, 173
Einstein's mass–energy relation, 21–22
electrical susceptibility, 191
electric discharge process, 176–177
electric field, 104–105
electromagnetics (EM)
  data transmission, 103
  definition, 103
  electric and magnetic fields, 104–105
  fields, 103
  scalar and vector fields, 104
  waves, 104
electromagnetic waves, 222
  E and H vectors, 122
  transverse in nature, 120–121
  wave impedance, 122
electromagnetism, 112
  applications, 128–130
electronic polarization, 192–193
electrostriction effect, 203
energy band gap, 209, 370
energy distribution, 299–300
energy gap, 262–263
energy range E and E+dE, 140
energy–time and position–momentum uncertainty relations, 73–74
ensemble
  canonical, 141–142
  classification of, 141
  definition, 141
  grand canonical, 142
  micro-canonical, 141–142
entropy, 259–260
equation of continuity, 109–110
equivalence/relativity principle, 9

ether drag hypothesis, 4–5, 8
extrinsic semiconductors
   n-type, 215–216
   p-type, 216

Fabre salts, 264–265
face-centered cubic (FCC) structure, 284
Faraday, Michael, 103
   law, 124
fermi–dirac (FD) statistics, 143
   distribution function, 157–159, 211–212
   energy distribution law, 160–163
   fermi energy, 159–160
   fermi function, 159
   fermions, 157
fermi energy, 210
   level, 210
fermi function, 159
fermi level
   extrinsic semiconductors, 215–216
   intrinsic semiconductors, 214–215
fermi temperature, 162
   degeneracy parameter, 162–163
ferroelectric materials, 201–202
   in devices, 202
   hysteresis, definition, 202
filament temperature, 375
four-probe method, 370
fourteen Bravais lattices, see seven crystal systems
fractional refractive index, 357
frame of reference, 2
   inertial frames of reference, 2–3
   non-inertial frames of reference, 2–4
   space–time frame, 4
Fraunhofer diffraction
   absent spectra/missing spectra, 332
   central diffraction maximum, 329–330
   double (two) slits, 330–331
   due to single slit, 325–326
   maxima and minima, 327–329, 331–332
   path difference, 325–326
   plane diffraction grating, 332–336
free carrier density
   concentration of electrons, 212–213
   concentration of holes, 213–214

free electrons
   Fermi temperature, 162
   mean internal energy, 161–162
   in metal, 161
free space, Maxwell's equations, 119–120
   electromagnetic waves, 120–121
Fresnel diffraction, 325
fringe width, 306
Fulde–Ferrell–Larkin–Ovchinnikov (FFLO) phase, 267
fullerenes, 237–238

galaxies, 33
Galilean transformation equations, 9–10
Gauss theorem, 107
   of electrostatics, 124
   of magnetostatics, 124
generalized Ampere's law, 103
general relativity, 29
general theory of relativity, 1
graded index fiber (GIF), 354
grand canonical ensemble, 142
grating element, 338
grating spectrum, 336, 404–408
gravitational lensing, 32
gravitational mass, 30
gravitational red shift, 34–35
gravitational time dilation, 36
gravitational waves, 37
group velocity, 66–71

half-filled valence, 160
Hall effect
   applications of, 222
   conducting materials (metals), 218–220
   multiplier, 222
   in semiconductors, 220–221
Hall field, 218–219
Hall voltage, 218
hard superconductors, see type II superconductor
harmonic oscillator, 84–86
Heisenberg's uncertainty principle, 73–74
   binding energy, 76–77
   finite width, spectral lines, 78
   non-existence of electrons, 75

radius of Bohr's orbit, 75–76
zero-point energy, 77–78
helium-neon (He-Ne) lasers
   energy levels, 180–181
   setup, 180
heterojunction intrinsic thin-layer (HIT) solar cells, 225
heterojunction solar cells, 362–363
hexagonal crystal system, 280
high–loss dielectrics, 203
high-pressure carbon mono-oxide deposition (HiPCO), 242–243
high-temperature superconductors, 257–258
Hubble constant, 29
Hubble flow, 29

induced electric dipoles, 191
inertial frames of reference, 2–3
inertial mass, 30
insulators, 187
integral evaluation, 147
intensity distribution curve, 336
intercalated graphite superconductors, 267
interference of light, 297; see also thin films
internal fields
   constant, 197
   liquid and solids, 196–197
interplanar spacing in crystals, 290
intrinsic semiconductors, 214–215
inverse piezoelectric effect, 203
ionic polarization, 193
ionization energy of donors, 371
isotope effect, 263–264

k-BEDT-TTF2X charge-transfer complex, 265

laevorotatory, 400
lanthanum-barium cuprate (LBCO), 258
Laplacian operator, 79, 109
lasers; see also helium-neon (He-Ne) lasers; Ruby laser; three- and four-level laser systems
   absorption of radiation, 171–172
   applications, 181
   Einstein's A and B coefficients, 172–174
   spontaneous emission, 172
   stimulated emission, 172
   threshold conditions, 175
lattice constant, 287
lattice planes, 288; see also Miller indices
Laurent's half-shade device, 401–402
Laurent's polarimeter, 401
length contraction, 13–14
light amplification, 171
light cone, 25–26
light deflection, gravitational field, 30–31
light retardation, 36–37
Lorentz electric field, 197
Lorentz–Fitzgerald contraction hypothesis, 9
Lorentz–Lorentz equation, 198
Lorentz transformation equations, 10–13
loss factor, 202–203
low–loss dielectrics, 203
low-temperature superconductors, 258

Maglev trains, 268
magnetic fields, 40, 104–105
   variation, 381–384
magnetic flux density, 222
Maiman, T. H., 171
maser, 171
massless particles, 23
material fabrication
   bottom-up approach, 236–237
   top-down approach, 236–237
Maxwell–Boltzmann (MB) statistics, 142
   average speed, 149–150
   classical particles, 143
   distinguishable particles, 143
   distribution law, 144–146
   energy distribution function, 147–148
   energy distribution law, 148
   most probable speed, 149
   speed/velocity distribution law, 148–149
Maxwell's equations; see also free space, Maxwell's equations
   conducting media, 124–127
   in differential form, 112
   first equation, derivation of, 112–113, 115
   fourth equation, derivation of, 114–116
   nonconducting medium, 123–124
   second equation, derivation of, 113, 115–116

third equation, derivation of, 113–114, 116
Meissner effect (flux exclusion), 253–254
mercury isotopes, 263
merry-go-rounds, 4
meson decay, 15
Michelson–Morley experiment, 5–8
  negative results
    constancy of speed of light, 9
    ether drag hypothesis, 8
    Lorentz–Fitzgerald contraction hypothesis, 9
microbends, 359
micro-canonical ensemble, 141–142
microwave amplification, 171
Miller indices, 288–289
Minkowski space–time, 24–26
missing spectra, 332
mobility determination, 222
mobility of charge carriers, 219
modal classifications, 354
  multimode fibers, 358
  single-mode fiber, 358
Modified Ampere's circuital law, 124
monoclinic crystal system, 280
most probable speed, 149
moving inertial frame, 3
μ-space, 139–140
multimode fibers, 358
multimode graded-index fiber, 358
multimode step-index fiber, 358
multi-walled carbon nanotubes (MWNT/MWCNT), 242

nano
  definition, 233
  scale of measurement, 233–234
nanomaterials
  definition, 235
  fullerenes, 237–238
  nanoparticles, 239
  properties, 235
  types of, 235
nanoscale; *see also* nanomaterials
  material fabrication, 236–237
  property change, 235–236
nanoscience, 234

nanotechnology
  definition, 234
  history of, 234
neutrinos, 23
neutron stars, 39
Newton's ring method, 396–399
  air-wedge fringe configurations, 314
  definition, 307
  diameter of rings, 308–309
  experimental setup, 307
  lens and plate, 314
  pattern, 310
  plane glass plate, 314
  plano-convex lens, 306
  refractive index, 310
  small radius of curvature, 314
  in transmitted light, 311–313
  wavelength determination, sodium light, 310
non-inertial frames of reference, 2–4
nonpolar molecules, 188–189
non-primitive cells, 277
normalized wave function, 78
n-type semiconductors, 215–217
nuclear fission, 22
numerical aperture (NA), 356

observation tables, 373, 377, 380–381, 383–384, 387–388, 395–396, 402–404, 407, 413, 416
occupation index, 159
one-dimensional box
  eigen function, 81
  eigen value, 81
  potential function, 81
  wave function, 82–83
Onnes, H. K., 249–250
optical activity, 400
optical fiber; *see also* refractive index profile
  advantages, 359
  applications of, 360
  cladding, 351–352
  communication system, 359–360
  core, 351
  with core–cladding interface, 351
  light and optical wave guide, 353
  light-source materials, 361

modal classifications, 354
propagation modes, 357
ray bending phenomena, 352
refractive index of medium, 352
semiconductor materials, 361–362
optical pumping, 176
optical rotation, 400
organic field effect transistors (OFETs), 241
organic superconductors, 264–265
intercalated graphite superconductors, 267
phenanthrene type, 267–268
TTF, 267
orientation polarization, 194
orthorhombic crystal system, 279–280

packing factor (PF), 281–284, 287
pair production, 22
partition function, 147
Pauli's exclusion principle, 210
Peltier effect, 378
penetration depth, 127–128
perihelion shift of mercury, 36
phase space, 139
phase velocity, 66–71
phenanthrene-edge-type polycyclic aromatic hydrocarbon, 267
photoelectric effect, 63
photons, 33–34
piezo-electric effect, 201, 203
Planck's constant, 414–416
Planck's hypothesis, 63
Planck's radiation formula, 154–155
plane transmission grating, 404–405
plano-convex lens, 311–314
polarimeter, 399–404
polarizability, 191–192
electronic polarization, 192–193
ionic polarization, 193
orientation polarization, 194
space-charge polarization, 194–195
total polarization, 194–196
polar molecules, 188–189
population inversion, 174–175
poynting theorem, 117–118
physical significance of, 118–119
poynting vector, 117

primitive cells, 277
principle of equivalence, 29–30
p-type semiconductors, 216–217
pumping process
definition, 175
electric discharge, 176–177
optical, 176
stimulated emission, flow chart of, 176–177
pyro-electric material, 203

quantum confinement effect, 236
quantum mechanics, 84
applications, 85–86
quasars, 33

ratio of permittivity, 188
Rayleigh–Jeans law, 155
Rayleigh scattering, 359
Rayleigh's criterion, limit of resolution, 339
recessional red shift, 29
reciprocal lattice, 290
red shift of spectrum, 28–29
refractive index, 198, 310, 352
refractive index profile
GIF, 353–354
SIF, 353
WIF, 354
relativistic energy and momentum, 22–23
relaxation frequency, 199
relaxation time, 199
residual resistance, 250
resolving power, 339–340
rhombohedral crystal system, 279–280
root-mean-square speed, 150–151
Ruby laser, 171
construction, 178
energy levels of, 179
setup, 179
three-level solid-state laser, 178
Rydberg constant, 411–413

scalar field, 104, 109
Schrödinger's time-independent wave equation, 79–81
free particles, 80
Seebeck effect, 378

semiconductors
  applications, 226
  band gap, 370–374
  compound, see compound semiconductors
  Hall effect, 220–222
seven crystal systems
  Bravais lattices, 278–279
  CN, 281
  cubic crystal system, 278–279
  hexagonal crystal system, 280
  monoclinic crystal system, 280
  orthorhombic crystal system, 279–280
  rhombohedral crystal system, 279–280
  tetragonal crystal system, 279
  triclinic crystal system, 280
Silsbee's rule, 253
simple cubic structure (SCC), 281–282
simple cubic unit cell, 281
single-mode fiber, 358
single-mode step-index fiber, 358
single-slit diffraction pattern, 330
single-walled carbon nanotubes (SWNT/SWCNT), 242
skin depth, 127–128
soft superconductors, see type I superconductor
space-charge polarization, 194–195
space lattices/crystal lattices, 276
space–time frame, 4
  curvature of space, 31
  intervals (Minkowski), 24–26
  Lorentz transformation equations, 10–13
  two-dimensional coordinate system, 24
  unification of, mass and, 31–32
special theory of relativity, 1
specific heat, 261
specific rotation, 400–401
spontaneous emission of radiation, 172
Staebler–Wronski effect, 224–225
stationary inertial frame, 3
statistical mechanics; see also ensemble
  BE, 143
  density of microstates, 141
  examples, 139
  FD, 143
  MB, 142
  phase space, 139

  volume element, μ-space, 139–140
Stefan's Boltzmann law, 156–157
Stefan's law, black body radiation, 374–378
step index fiber (SIF), 353
stimulated emission of radiation, 172
Stokes' treatment, 300
superconducting quantum interference devices (SQUIDs), 268
superconductivity; see also AC resistivity; Bardeen, Cooper and Schrieffer (BCS) theory; doped fullerenes; high-temperature superconductors; superconductors
  acoustic attenuation, 261–262
  critical current and current density, 252–253
  critical magnetic field, see critical magnetic field
  definition, 249
  energy gap, 262–263
  entropy, 259–260
  history of, 249
  isotope effect, 263–264
  mechanical effects, 264
  specific heat, 260–261
  temperature dependence of resistivity, 250
  thermal conductivity, 261–262
  and transition temperature, 250–251
superconductors
  applications, 268
  characteristics, 264
  organic, 264–265
surface area-to-volume ratio, 235–236

tangent galvanometer, 393–396
temperature dependence, 380
tetragonal crystal system, 279
tetramethyltetraselenafulvalene (TMTSF), 265
tetramethyltetrathiafulvalene (TMTTF), 265
tetrathiafulvalene (TTF), 267
tetrathiapentalene (TTP), 267
theory of expanding universe, 28–29
theory of interference
  coherent sources, 297–298
  maximum and minimum intensities, 299
thermal breakdown, 201
thermal conductivity, 261–262

thermo-emf measurement, 379
thermopower determination, 378–381
thin films
    definition, 300
    solar cells, 363
    uniform thickness, 300–304
    wedge-shaped thin film, 304–305
Thomson effect, 378–379
three- and four-level laser systems, 177–178
three-dimensional box, 83
time dilation, 14–15
    experimental verification, 15–16
top-down approach, 236–237
total polarization, 194–196
transition temperature, 250–251
translational vectors, 276–277
transverse nature of EM wave, 121
triclinic crystal system, 280
twin paradox, special relativity, 16
two-dimensional (BEDT-TTF)2X, 265–267
type II superconductors, 252–253, 255–256
type I superconductors, 251–252, 254–255

unaccelerated frames, 2–3
unification of mass and space, 31–32
uniform acceleration, 30
uniform gravitation, 30
unit cells, types of, 278
unit cell with lattice parameter, 277

valence band, 211, 213–214
variation of mass with velocity, 18–21
vector field, 104, 109
velocity addition, 17–18
velocity of de-Broglie waves
    description, 67
    group and phase, 67–71
vortex state, 255

wave equation, 79–80
wave function, 78, 82–83
wave group, 66–67, 74
wave impedance, 122
wave packet, 64, 66
wave velocity, 66
wedge-shaped thin film, 304–305
Wheatstone principle, 385
white dwarfs, 35, 39
Wien's displacement law, 155–156
W-index fiber (WIF), 354
world line, 25–26

X-ray photons, 64

yattrium-barium cuprate (YBCO), 258

zero-point energy, 77–78
zero resistance, 249
Zig-zag, 244